studies in physical and theoretical chemistry 4

BIOMOLECULAR INFORMATION THEORY

studies in physical and theoretical chemistry

Other titles in this series

1 R. Ginell, **Association Theory**

2 T. Boublik, I. Nezbeda and K. Hlavatý, **Statistical Thermodynamics of Simple Liquids and Their Mixtures**

3 P. Hobza and R. Zahradník, **Weak Intermolecular Interactions in Chemistry and Biology**

studies in physical and theoretical chemistry 4

BIOMOLECULAR INFORMATION THEORY

Serafin Fraga, K.M.S. Saxena
and
Manuel Torres

Department of Chemistry, University of Alberta, Edmonton, Canada

ELSEVIER SCIENTIFIC PUBLISHING COMPANY
Amsterdam – Oxford – New York 1978

ELSEVIER SCIENTIFIC PUBLISHING COMPANY
335 Jan van Galenstraat
P.O. Box 211, 1000 AE Amsterdam, The Netherlands

Distributors for the United States and Canada:

ELSEVIER NORTH-HOLLAND INC.
52, Vanderbilt Avenue
New York, N.Y. 10017

ISBN 0-444-41736-2 (Vol. 4)
ISBN 0-444-41699-4 (Series)

Printed in The Netherlands

To
Ester, Nirmala, Joana

Pero ¿ es posible trabajar en algo serio y duradero, olvidando el enorme misterio del Universo y sin inquirirlo?

Miguel de Unamuno

PREFACE

Biological processes imply a mutual recognition of the systems involved, whereby the information stored in a system (or systems) is read by the other(s). Their comprehensive study must deal with both aspects of the problem: the information and its recognition.

Quantum-mechanical research has yielded, over the years, an astonishing wealth of data on biomolecules or their submolecular components, and the time has now come for a detailed inspection of the problems associated with the recognition processes. The necessary theoretical tools have been developed, and the availability of fast, large-capacity computers, coupled with efficient methods of data handling, ensure that the study of many interesting problems, postponed in the past, is now feasible.

Our efforts have been directed towards a unified presentation of the theory, a survey of the present, and a predictive look at the future. They will have succeeded if they encourage new work on this fascinating problem.

Serafin Fraga, K.M.S. Saxena, Manuel Torres

Edmonton, Alberta
Canada

ACKNOWLEDGMENTS

We acknowledge the kindness of the following organizations for the use of copyright material:

Academic Press Inc., Publishers (New York) and Academic Press Inc. (London) Limited; Aguilar, S.A. de Ediciones; Akademiai Kiado; The American Chemical Society; The American Institute of Physics; The Chemical Society (U.K.); Columbia University Press; International Union of Crystallography; John Wiley and Sons, Inc.; Marcel Dekker, Inc.; National Academy of Sciences (U.S.A.); North-Holland Publishing Company; Pergamon Press; Springer-Verlag; Taylor and Francis Ltd.

We acknowledge also the collaboration of Mr. E.N. Grams (University of Alberta Printing Services).

Table of Contents

INTRODUCTION

The field of quantum biology is now reaching a critical point in its development. In fact, one should say that it is coming of age.

Since 1945-46, when the first quantum-chemical treatments on the problem of chemical carcinogenesis were published [A. Pullman (1945), R. Daudel (1946)], many researchers have entered the field, a wealth of information has been collected, and no problem has escaped scrutiny.

Three periods may be distinguished in its development. Progress was slow and the difficulties many during the first stage. It was already 1951 when the expansion method was formulated for the self-consistent field theory, opening the way for the mass-production of accurate calculations. But still some years had to pass by before quantum biochemists would have access to first-generation computers. Unfortunately, they were small and slow and, although the goals were well defined and the theoretical tools available, many ambitious projects had to be postponed.

Waiting for those difficulties to disappear, when larger and faster computers would be available, would have meant many years of idleness. Instead, it was decided to proceed ahead, step by step, parallelling the advances in quantum chemistry. All the approximations were tried, and repeatedly the same systems were studied. Without doubt, many efforts were wasted, but new insights were gained and the ground prepared for the days ahead.

Large-capacity and extremely fast computers have now been with us for some years and efficient programs have been developed, so that accurate *a priori* calculations are now possible for many biomolecules, although not for all. In any case, molecules *per se* are not the primary goal; it is the study and understanding of biological processes, that constitutes the *raison d'etre* of this field.

Methods, developed for the study of individual molecules, would fail in such an endeavour, and a new approach must be adopted, in a realistic attack of the problem. On one hand, of course, it is imperative that calculations on biomolecules (at the molecular or submolecular level, as feasible) will be continued, so that a library of accurate and reliable data is built up. Such data banks will provide the input needed for the study of recognition processes, using suitable techniques.

It is with this research that the present work is concerned. The main goal is to review, justify, and expand the methods appropriate for a computational simulation of the operation (in all its aspects) of the genetic code, within the framework of the molecular information theory. Representative theoretical treatments have been chosen because of their accuracy or as illustrative examples of the research that should be pursued in the future.

The bibliography has been limited to those publications essential for the discussion, without any intention of achieving a comprehensive compilation. An existing overall examination of the field of quantum biology [E. Fraga and S. Fraga (1976)] constitutes a possible source of information on experimental data and theoretical results.

Biomolecular Information

'Molecular recognition is a very general
and important phenomenon in nature.'

D.F. Bradley

An information receptor recognizes information carriers, distinguishing the correct partner from the wrong carriers. The correct carrier is bound and the wrong ones are rejected.

Numerous examples of biopolymers acting as information receptors are known (enzyme-substrate complexes, codons-anticodons pairings, antibody-antigen associations, tastant recognition, drug binding,...) and well documented.

Polypeptides are the information receptors *par excellence*, being directly involved in the recognition processes that constitute the basis of life. Polynucleotides, on the other hand, provide the reservoir of genetic information (transmitted throughout the generations with the changes imposed by the environment) which is translated into the appropriate receptors.

An outline of the development and evolution of these information molecules provides the background for the examination of the characteristics of the accepted biological code and of its primary transmission and operation. An inspection of the problems in the study of the relationship between amino acid sequence and conformation and of illustrative results establishes the foundation for the theoretical formulation of the molecular information theory.

Chapter 1

The Code and Its Origins

Living organisms are capable of metabolism and self-reproduction and possess mutability [Oparin (1969)]. Such characteristics imply a considerable amount of self-instruction, that must proceed from some source of information.

Therefore, the study of life and of its genesis must deal, within the general framework of the evolution of matter, with the processes resulting in the development of such a high degree of information, characteristic of the living organisms. Several questions arise immediately in this connection: How far back must or can one go in the chain of evolution? What experimental information is available? Which are the main questions to be elucidated? From what points of view should they be studied?

A molecule will be considered as informational when, in fixed conditions, it is formed preferentially over other possibilities. That is, at every state of the evolution there is always some contribution of the environmental information and, therefore, it is not possible to trace back the statement that life started from random events.

The study of evolution at the present stage is relatively simple, in the sense that abundant experimental evidence is available. There is, however, no direct evidence of the early stages of evolution, so that the corresponding study will be, by necessity, of a more speculative character, being based on information for extraterrestrial systems at an early evolutionary stage or on observations from laboratory experiments that may simulate the prebiotic conditions.

Experimental observation, that has ultimately led to the adoption of a genetic code, has established that the deoxyribonucleic acids (DNA) constitute the source of self-information and that an interplay exists between nucleic acids and proteins.

Therefore, a study, that must explain the origin of a primitive code and the transmission and development of the information content to the level of the code presently accepted, must cover several distinct stages: formation of a prebiotic soup, formation of a self-reproducible system, and development of complex organisms. Both thermodynamic and kinetic arguments must be brought into the study of these stages of evolution.

PREBIOTIC STAGE

Environmental Conditions

The conditions on earth at the time of the primitive phase of evolution were probably very similar to those existing at present in other parts of the solar system. There is, therefore, an intrinsic interest in the study of the molecules found in the cosmic space, insofar as they may provide information on the origin of the amino acids and nucleotides.

Two main points must be considered regarding the synthesis of organic matter: the energy sources and the availability of the elements involved. On one hand, it is believed that, at the time under consideration, there were a variety of energy sources (electric discharges, radiation from the earth core, cosmic rays, and volcanic heat, among others, in addition to the solar radiation), resulting in conditions of extreme heat and radiation. On the other hand, the synthesis of organic matter requires, predominantly, the existence of H, C, N, and O, all of them present in the stars and, logically, in the primitive earth.

It has been shown [Oro (1972)] that, under extreme conditions of heat and radiation, molecules and radicals like CO, CN, CH, C_2, C_3, etc., would be stable. These molecules and radicals may be easily hydrogenated in a reducing atmosphere, yielding organic molecules.

Actually, astrophysicists have been able to detect many of these molecules, although the interstellar medium is composed basically of hydrogen. One must, therefore, conclude that their long life in the interstellar medium is due to the low frequency of collision and that in earth, for a number of reasons, higher concentrations must have been present.

Synthesis of amino acids

There are two possible paths for the study of the formation of amino acids in abiogenic conditions: the study of samples of extraterrestrial origin (meteorites and samples brought back from space missions) and laboratory studies in conditions simulating those believed to exist in the primitive earth.

In the search for amino acids in samples of extraterrestrial origin, methods had to be developed in order to deal with nanogram quantities. The method most widely used [Hare *et al.* (1970)] consists of the extraction with hot water, followed by hydrolysis and subsequent analysis in an ultrasensitive amino acid analyzer. [Other methods have also been used with similar results; see the work of Oro *et al.* (1970), Ponnamperuma *et al.* (1970), and Hamilton and Nagy (1972), and the review of Kvenvolden (1972)].

In these analyses, care must be taken in order to consider all the possible sources of contamination, such as bacteria (which will give different enantiomers) and fingerprints [Hamilton (1965)], as well as from the exhaust of the spaceship [Fox *et al.* (1976)]. That is, it is necessary to determine whether the amino acids, that may be found, are of biological or abiogenic origin; this distinction may be achieved on the basis of the values for the ratio $^{13}C/^{12}C$, by comparison to the known isotopic ratio in the earth biosphere [Craig (1953)], and of the percentage of *D*-enantiomers for optically active amino acids [Oro *et al.* (1971)].

Among the meteorites, the carbonaceous chondrites, considered to be low-temperature condensates of solar nebulae, are the most primitive samples known and contain the stable solar elements with exception of the volatile ones. It is estimated that 6% of the carbon content is in the form of organic matter of prebiotic origin [Anders *et al.* (1973)], not only as amino acids or its precursors but also as hydrocarbons [Gelpi and Oro (1970)] and fatty acids [Nagy and Bitz (1963)].

The results of the analysis of amino acids precursors in samples of lunar dust, brought back by different space missions, carried out by different groups [Nagy *et al.* (1970), Hare *et al.* (1970), Fox *et al.* (1970), Murphy *et al.* (1970)] are quite consistent. A set of six amino acids have constantly been found: glycine, alanine, glutamic and aspartic acids, serine, and threonine, in amounts of the order of 7-45 ng/g of lunar dust.

The similar amino acids content of lunar samples and meteorites and the fact that the ratios of amino acids content to the total carbon content are also comparable suggest a common evolution path and a common state of evolution in spite of the different cosmo-physical histories.

The synthesis of amino acids in laboratory simulations of primitive earth conditions came in the early stages of the study of molecular evolution. It was realized [Urey (1952)] that the primitive atmosphere must have been highly reducing, leading to the suggestion that electrical discharges could have converted the molecules in the atmosphere to excited species and to organic compounds. And, effectively, it was shown that the action of electric discharges on mixtures of water vapour, methane, ammonia, and hydrogen resulted in the formation of amino acids [Miller (1953)], the most abundant amino acids being glycine, α-alanine, β-alanine, and α-amino butyric acid. Similar experiments, but using different mixtures [Abelson (1957)], showed that the amino acid synthesis is possible only in a reducing atmosphere and that hydrogen cyanide represents an intermediate; in fact, it has also been shown that hydrogen cyanide is formed as well in the thermal decomposition of formamide [Harada (1967)] and that its dimers, trimers, and tetramers form the parent skeletons of glycine, alanine, and aspartic acid, which can be obtained by reduction and hydrolysis of those polymers.

On the other hand, thermolysis of a mixture of methane, ammonia, and water, with or without catalyst [Harada and Fox (1964), Oro (1963)] yielded the amino acids threonine, serine, proline, glycine, alanine, valine, leucine, isoleucine, tyrosine, and phenylalanine. And, after formaldehyde and ammonia were found in the galaxia, the possibility of amino acids formation by heating of their mixture was studied [Fox and Windsor (1970)]. The findings correlate satisfactorily with the values obtained from lunar dust.

The possibility of amino acid synthesis has been successfully attempted using every form of energy that could exist under the early earth conditions. Ultraviolet irradiation of mixtures of methane, ammonia, carbon monoxide, and water, with a hydrogen lamp that simulates the far-ultraviolet spectrum of the sun, yielded α-alanine, α-amino butyric acid, valine, and isoleucine. On the other hand, bombardment of a similar mixture with fast electrons has yielded glycine, alanine, and aspartic acid [Palm and Calvin (1962)] and cobalt-60 irradiation of aqueous solutions of hydrogen cyanide produced glycine, alanine, valine, serine, threonine, and aspartic and glutamic acids [Sweeney *et al.* (1976)].

All these experiments have shown the feasibility of the synthesis of amino acids or of their precursors under the primitive earth conditions. To date, however, it has not been possible to prove what the true events leading to the synthesis of amino acids were, and the probability is that, no matter what conditions existed, the synthesis occured along different paths.

The fact that the amino acid distribution obtained in laboratory tests agree with that found in samples of extraterrestrial origin confirms that the abiotic synthesis of amino acids constituted a very crucial step in the evolution of matter towards life. It is interesting to note that, whatever the method of synthesis employed, the amino acids glycine and alanine are always formed in predominant amounts, in agreement with the fact that glycine is the predominant amino acid found in lunar samples and meteorites.

The synthesis of amino acids is a consequence, in all probability, of the non-randomness and information content of the polyatomic molecules present in our galaxia. And the surprising fact, as mentioned above, that the same pattern of amino acids is formed under different conditions would seem to indicate the existence of a common cosmochemical path, the result of a kinetic rather than a thermodynamic control, since the molecules were in high-energy states and the collision rate in the interstellar space is very low.

Polymerization of amino acids

The information content of a molecule should be measured by its capability of selectively interacting with other molecules. And, consequently, it must be concluded that amino acids must have a high information content, being, as they are, very reactive molecules; as a matter of fact, they are rarely found as such in living organisms and in extraterrestrial samples they have been found in the form of amino acid precursors. The structure of the amino acids is such that they are very susceptible to environmental influences in their reactions. So, naturally, the second step in the evolution of matter must be associated with the reactions of amino acids, especially among themselves.

It has been shown [Fox (1969, 1976a), Fox and Harada (1958, 1960)] that amino acids may undergo a thermal polymerization in favourable conditions. The synthesis of polypeptides was achieved by heating of dry mixtures of amino acids contained in actual proteins, while in an alternate procedure a solution of amino acids was heated until the water was eliminated and the dry residue polymerized; polymers containing a set of 10-12 amino acids were obtained in this way. Two conditions are crucial in this polymerization: high concentration of amino acids and the presence of an acidic medium; both conditions can be met by using mixtures of pure amino acids and by the presence of amino acids such as aspartic or glutamic acids. It is not difficult to imagine the possibility that a similar situation existed in the primitive earth, with a volcanic activity believed to have been much greater than at present.

Polyamino acids may also have been formed by an alternate path [Mathews and Moser (1966, 1967)], which does not involve the polymerization of amino acids. Evidence has been found of the formation of polyamino acids in the base-catalyzed polymerization of hydrogen cyanide in anhydrous conditions and after subjecting mixtures of methane and ammonia to electric discharges.

This polymerization of amino acids is extremely important from two points of view. On one hand, of course, because it shows that protein-like polymers could have been formed in abiotic conditions. But what is of special significance is the fact that the polymers are not formed at random, which means a degree of selective capacity of the amino acids. The analysis of the polymers formed by heating of mixtures of amino acids, simulating prebiotic conditions, showed significant differences between the content of the amino acids in the original mixture and in the polymers. [For more details see the summary, presented by Fox (1975), of the evidence for the limited heterogeneity in thermal polyamino acids.]

The selective interaction between amino acids has also been shown on the basis of theoretical calculations [Carbo *et al.* (1978)]. The results of the calculations, presented in Table 1-1 and shown schematically in Fig. 1-1, indicate that the L-L or D-D complexes have a greater probability of survival than the mixed L-D systems. On the other hand, the considerable differences between the alanine-alanine and alanine-glycine interactions suggest a thermodynamic discrimination that might constitute the basis for a primitive biological code. [The authors would like to express their acknowledgement to Prof. Dr. R. Carbo for the communication of these results prior to their publication.] The systems considered are (*L*-ala)-(*L*-ala), (*L*-ala)-(*D*-ala), and (*L*-ala)-(gly), abbreviated as (LL), (LD), and (Lg). Figures 1-1a and 1-1b

Table 1-1. Values of the thermodynamic functions for various aggregates at different temperatures*

T(°K)	200	220	240	260	280	300
Interaction between *L*-alanine and *L*-alanine						
−<E>	3.53	3.34	3.13	2.91	2.69	2.48
S	4.04	4.93	5.84	6.73	7.54	8.26
−F	4.34	4.43	4.54	4.66	4.80	4.96
Interaction between *L*-alanine and *D*-alanine						
−<E>	3.34	3.15	2.94	2.73	2.53	2.34
S	4.57	5.48	6.37	7.20	7.96	8.61
−F	4.25	4.35	4.47	4.61	4.76	4.92
Interaction between *L*-alanine and glycine						
−<E>	6.90	6.87	6.83	6.79	6.74	6.87
S	2.38	2.54	2.70	2.86	3.04	3.24
−F	7.38	7.43	7.48	7.54	7.60	7.66

* As given by Carbo *et al.* (1978); < E > and F are given in kcal/mol and S in cal/°K mol.

present (from left to right) the interaction energies E_{LL}, E_{LD}, and E_{Lg}, and their differences E_{LD}-E_{LL}, E_{Lg}-E_{LL}, and E_{Lg}-E_{LD}, respectively, for different orientations of one molecule relative to the other, in steps of 60° (from 0° to 300°, given as coordinates). In the energy representations, the full circles denote negative interaction energies; for the energy difference representations, the full circles correspond to those cases where the potential energy for the first pair is negative and that for the second pair is positive, while the unshaded circles correspond to the opposite situation. Additional details on the actual relative positions, not needed for the discussion here, may be found in the original work.

The informational characteristics of the amino acids, manifested in their selective interactions, are greatly enhanced, as expected, in the polyamino acids where the structural capability of folding becomes important. Independently from the path followed when proceeding from the original polyatomic molecules to the amino acids and then to the polyamino acids, the process involves a change in information content: the functional information of the polyatomic molecules is complemented by the environmental information stored in the amino acids, resulting finally in the structural stereochemical information of the polyamino acids as evidenced by the enzime- and hormone-like activity of these proteinoids. [For more details see the summary given by Fox (1975).]

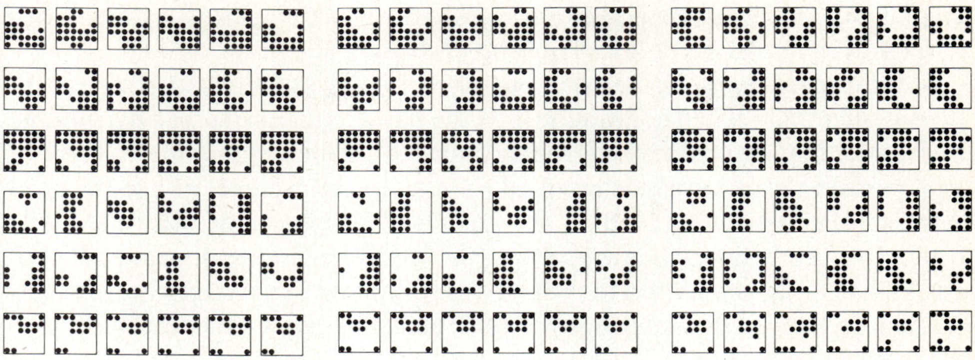

Figure 1-1a. Schematic representation of the interaction energies E_{LL}, E_{LD} and E_{Lg}. See the text for details [Carbo *et al.* (1978)].

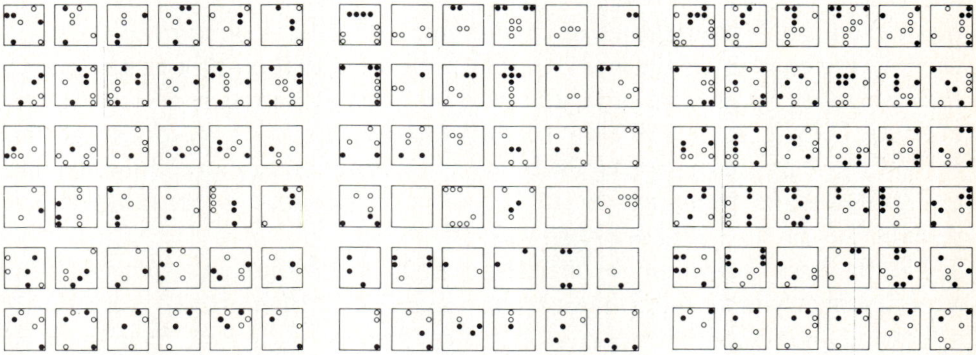

Figure 1-1b. Schematic representation of the differences E_{LD}-E_{LL} E_{Lg}-E_{LL}, and E_{Lg}-E_{LD}. See the text for details [Carbo *et al.* (1978)].

Formation of nucleotides

The formation of nucleotides in abiotic conditions must have been dependent on the presence of purine and pyrimidine bases as well as carbohydrates, leading to the formation of nucleosides which would have then interacted with phosphorus compounds to yield the nucleotides.

The presence of purines and pyrimidines, as well as of other nitrogen-containing compounds (such as amino derivatives of *s*-triazine and substituted guanidines), has been detected in water extracts of carbonaceous chondrites [Briggs (1961), Calvin (1961), Hayatsu (1964)].

Several mechanisms have been proposed for the synthesis of purines and pyrimidines and related compounds. At the basic level, purine and pyrimidine derivatives may be synthesized from mixtures of carbon monoxide, hydrogen, and ammonia, at a temperature of 600° C and in the presence of a nickel-iron catalyst [Hayatsu *et al.* (1968), Yang and Oro (1971)]; on the other hand, the formation of adenine and guanine in a Fischer-Tropsch type of process [Yang and Oro (1971)], is interpreted through the formation of hydrogen cyanide and cyanoacetylene. In fact, hydrogen cyanide seems to play a fundamental role in the synthesis of these systems. Mechanisms for the abiotic formation of adenine from hydrogen cyanide in aqueous ammonia have been suggested [Calvin (1961), Oro (1971)] and its synthesis [Oro and Kimball (1961), Ponnamperuma *et al.* (1963)], as well as that of guanine [Ponnamperuma (1965)], from hydrogen, ammonia, and other simple compounds, present in primitive earth conditions, has been reported. The possible formation of purines by oligomerization of dilute aqueous solutions of hydrogen cyanide has been proposed [Ferris (1968), Sanchez *et al.* (1967, 1968)]; the photochemical isomerization of diaminomaleonitrile to 4-aminoimidazol-5-carbonitrile constitutes the main step of this process, that must be carried out in the absence of oxygen, in order to avoid the photodestruction of the reaction product [Ferris *et al.* (1969)]. The search for suitable conditions for the formation of pyrimidines has not been as successful, although cyanoethylene, that might have been present in abiogenic conditions, is considered as a possible precursor [Sanchez *et al.* (1966)].

The presence of formaldehyde in the interstellar medium and its synthesis in experiments trying to reproduce abiotic conditions [Miller (1953)], makes it an appealing precursor of monosaccharides. The possibility of formation of carbohydrates by polymerization of formaldehyde in an alkaline medium has been confirmed [Mayer and Jaschke (1960)]. The experiments, that yielded a mixture of monosaccharides, are unrealistic because of the strong alkaline medium and the high formaldehyde concentrations needed; the strong objections [Horowitz and Miller (1962)], are mainly based on the fact that oceans were never highly alkaline in primitive earth conditions [Abelson (1966)]. A different approach to the polymerization of formaldehyde, more relevant to the prebiotic formation of monosaccharides, consists in refluxing formaldehyde solutions in the presence of naturally-occurring minerals, such as aluminum silicates, etc., yielding mixtures of monosaccharides [Gabel and Ponnamperuma (1967), Reid and Orgel (1967)].

The formation of nucleosides constitutes the next step in the process for the abiotic synthesis of nucleotides. However, while purines, pyrimidines, and monosaccharides can be obtained with relatively high yields in simulated prebiotic conditions, the methods developed so far for the synthesis of nucleosides have very poor yields. Adenosine and deoxyadenosine have been obtained by UV irradiation of adenine, ribose and phosphate and adenine, deoxyribose, and hydrogen cyanide, respectively [Ponnamperuma and Kirk (1963), Reid *et al.* (1967)].

The investigation of possible methods for the formation of nucleotides, from existing nucleosides, has been more successful. The two main reaction mechanisms proposed for the phosphorylation of nucleosides consist in the hypohydrous thermal reaction between inorganic phosphates and the nucleosides and an aqueous phosphorylation, involving intermediates formed by addition of inorganic orthophosphates to condensing agents. In the first method, nucleosides were heated with sodium dihydrophosphate in a sealed tube and it was observed that the yields of nucleotides decreased when monohydrophosphates were used [Rabinowitz *et al.* (1968), Ponnamperuma and Chang (1971)]; this difference has been attributed to the fact that dihydrophosphates can form polyphosphates under the conditions of the experiments, and that the polyphosphates are responsible for most of the phosphorylation, although the possibility of phosphorylation through orthophosphates cannot be excluded. The phosphorylation of adenosine in the second method [Schwartz and Ponnamperuma (1968)], which yields 2′-, 3′-, and 5′-phosphates, was accomplished by heating of an aqueous solution of adenosine and linear phosphate salts; the importance of this method is due to the fact that the phosphorylation is possible in a wide range of pH.

The main objection to the feasibility of these syntheses in abiotic conditions arises from the difficulty of incorporating phosphorus into the bio-evolutionary stream. It has been estimated that the concentration of phosphates in the primitive oceans would have been very low, especially taking into account the non-acidic pH estimated for the oceans [Rubey (1951)]. Two arguments have been given in order to explain a higher concentration of phosphates in the oceans: on one hand, it has been suggested that, in the very early stages, the sudden condensation of water could have determined a temporary acidity, resulting in larger concentrations of phosphates [Burger (1968)]; on the other hand, it could be that such an increase in concentration was due to calcium complexing [Schwartz (1971)].

SELF-ORGANIZATION

Life, as it exists today, is the result of a long process of evolution, which does not imply just the formation of the various components, discussed in the previous section, and their arrangement into known structures. In this connection it must be realized that the physical chemistry of living matter can only be understood by the application of the principles of the thermodynamics of open systems [Prigogine (1947), Glansdorff and Prigogine (1971)]. Although this subject is discussed in more detail in the following section, its implications, insofar as self-organization is concerned, must be examined here.

It is now well established that monomers could have been formed in a variety of conditions, that could have existed in the (abiotic) primitive earth. On the other hand, it has also been shown that, under certain abiotic conditions, these monomers can polymerize through a typical mechanism of linear chain propagation, although such a polymerization would result in very little information content being preserved. Alternatively, in open systems far from thermodynamical equilibrium, a mechanism *via* templates prevails [Babloyantz (1972)], which preserves the information content previously acquired.

It seems sensible, therefore, that phase-separated systems (which could function as open systems) appeared in the evolutionary stream at the earliest moment possible. That is, although the universe may be considered as a single open system, the phenomenon of the origin of life can only be explained on the

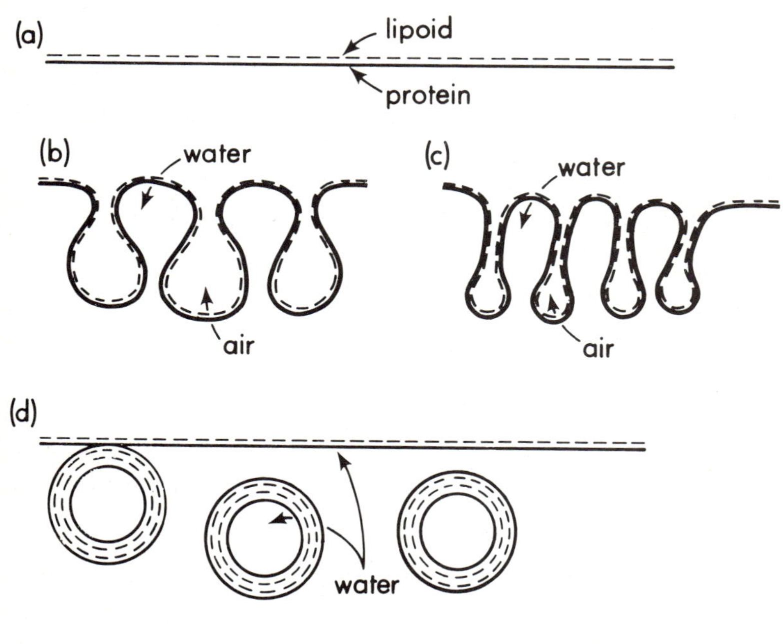

Figure 1-2. Schematic representation of the possible formation of protocellular lipoprotein vesicles [Goldacre (1958)]. [Reprinted with permission from 'Surface films, their collapse on compression, the shapes and sizes of cells and the origin of life', by R.J. Goldacre, in 'Surface Phenomena in Chemistry and Biology', edited by J.F. Danielli, K.G.A. Pankhurst, and A.C. Riddiford. Copyright by Pergamon Press, Ltd. (1958).]

basis of the appearance of individualized systems that will function as open systems.

The idea of the development of phase-separated systems in the early stages of evolution [Oparin (1924)] arose from that fact that the cell is a constant entity in all the living organisms and has led to the concept of protocell [Fox (1971)]. The formation of a cell implies the building of a membrane from hydrophobic materials; both lipids and proteins share this property and, as a matter of fact, lipids and proteins constitute the predominant components of present cell membranes.

Several models have been proposed in order to explain the physico-chemical properties of membranes [Shah (1972)] and how a cell could have been formed under primitive conditions. The early attempts [Goldacre (1958)] assumed that, in the early stages of the earth development, the sea contained large amounts of organic compounds, so that the action of the air at the interphase could have led to the formation of vesicles; water, and the organic compounds in it, could have been trapped within these vesicles, where any subsequent reactions would have been determined by the characteristics of the vesicular membranes (Fig. 1-2). Coacervates and protocells, discussed below, represent alternate models.

Coacervates

In a different approach, coacervate droplets have been suggested [Oparin (1965, 1966)] as a model of phase-separated systems. The significance of this proposal is due to the fact that studies of the activity of different substrates contained within these droplets have shown the existence of a behaviour similar to that of a cell; for example, when phosphorylase was present in droplets containing a solution of glucose-1-phosphate, starch was obtained.

A rather important fact observed in these experiments was the role played by the 'membrane', which allowed the diffusion of monomers and favoured the increase of the concentration of macromolecules within the droplets; in addition, it has been shown that the enzymatic activity is higher in the boundaries of the colloidal structure than in an equivalent dilute solution. Although these studies have been performed with contemporary substrates, their feasibility in abiotic conditions seems evident and, therefore, they provide a mechanism for the concentration (and mutual interaction) of organic polymers from a very dilute solution, which was most probably the case in the prebiotic soup [Bungenberg de Jong (1949)].

The relevance of phase-separated systems in the evolution of the information content is evident in the synthesis of polyadenine in coacervate droplets from ribonucleic acid and histone [Oparin (1966)].

It is interesting to note that the addition of polyphosphates to coacervate systems causes the division of the droplets and has, consequently, been proposed as a model for protobiotic division [Oparin *et al.* (1977), Orlovskii *et al.* (1977)].

Protocells

On the other hand, the formation of protenoid microspheres by contact of protenoids with cold water has also been observed [Fox and Dose (1972), Fox (1973, 1976b)]. The importance of this observation lies in the fact that the microspheres are formed from protenoid polymers which, as discussed in the previous section, could have been formed in the prebiotic soup under abiotic conditions and share some of the properties of present proteins.

These protenoid microspheres present properties that correspond to those of the present cells: they are capable of reproduction and their membranes have selective permeability, as observed by electron microscopy. The reproduction can take place in different ways: by formation of buds (which then separate and grow to parental size), binary fission, sporulation in the presence of calcium, partitive replication (with ejection of an endoparticle), and partition and recombination of two endoparticles.

As it would be expected, these protenoid microspheres present enzyme-like properties at the boundaries, in analogy to the coacervates but due, in this case, to the inherent enzyme-like characteristics of the protenoids. In addition, they have a high uniformity of size and tend to form ordered aggregates, with the possibility of communication through the junctions.

Using a protoribosome model, selective interaction has been observed [Fox and Dose (1972)] for the incorporation of phenylalanine into various aminoacyl adenylates, depending on the lysine content of the protenoid in the protenoid-polynucleotide particle. And, in addition, phenylalanine peptides have been obtained from phenylalanine and ATP (adenosine triphosphate) in the presence of protenoid-polynucleotide complexes.

In summary, taking into account the thermodynamical requirements for phase-separated systems in the evolution towards life and the experimental evidence, which supports the possibility of their formation at the early stages of evolution, one might logically conclude that protocells appeared at that time.

EVOLUTION

The formation of an organized system, recognized as a living organism, cannot be due to chance, as the probability of such an event would be vanishingly small. Therefore, the only alternative is a process which can lead to a gradual increase of complexity and orderliness. Such a process implies the existence of a code, capable of storing and transmitting information and of increasing it.

Consequently, a study of the genetic code must explain such characteristics as well as the process that has led, over the time, from the simple organization of a protocell to the presently accepted code. One is dealing, of course, with a historical study, where essentially only the final situation is known. Therefore, any attempt at an interpretation of the development of the genetic code may be valid, as long as it is logically (and causally) plausible. Such an interpretation might involve the following steps:

a) The protocells, discussed in the previous section, provide the appropriate environment for the formation of polynucleotides. A protocell may be considered as an 'information box', where the information carriers are confined but with walls permeable to the information monomeric units. The first step of the evolutionary process consists of the flow of units across the walls of the system and the formation of carriers, with the possibility of formation of mutants. It implies, therefore, both growth and selection, leading to a first generation of information carriers. Given the same environment, the different protocells would, statistically, develop the same information carriers.

b) Polypeptides could be formed in the prebiotic soup and, by extension, inside the protocells. Within the latter the existence of high concentrations of polynucleotides might constitute a decissive factor, insofar as to which polypeptides might be formed. One might envisage a situation where amino acids would be attached to polynucleotides and that recognition between polynucleotides would lead to the formation, first of all, of monopeptides and, in successive steps, of small polypeptides. Considering that the specificity of the interaction between polynucleotides is greater than the specificity of the interaction between a polynucleotide and an amino acid or a peptide, the sequence of amino acids in the resulting polypeptides would be determined by the polynucleotides present in the protocell.

c) The formation of polynucleotides and polypeptides, as considered above, would determine a growth of the protocell and its division on the basis of size. The resulting protocells would have the same polynucleotides as the original cell, and the process of growth and division could continue, given the same environment (from which the monomers flow into the protocells).

The three steps would simply lead to a proliferation of protocells, but a stationary situation would have been reached in the evolutionary process. One must consider, however, that additional interactions are possible.

d) The polynucleotides present in the protocells could recognize each other, yielding small deoxyribonucleic acids, which might in turn lead to longer DNA molecules. At this point, the interaction between proteins and DNA would determine the selectivity of the present code.

Three main concepts are invoked in the preceding outline and must be examined in detail: selection, complementarity, and specificity. A fundamental difficulty, in the second step, has been obviated by the consideration of a protoribosome [Fox and Dose (1972)] and therefore it will not be discussed here.

Selection

Only open systems [Prigogine (1947)], that is, systems exchanging both energy and matter with the environment, in a steady state far from equilibrium, may spontaneously evolve to organized forms. Such a system can fluctuate around multiple steady states, resulting in a decrease in entropy, and can be maintained as long as it is kept at a critical distance from equilibrium. On the other hand, the study of biological systems within the framework of irreversible thermodynamics has shown the importance of autocatalytic systems, such as those that reproduce *via* templates. An open system far from equilibrium may grow to more organized structures through autocatalysis ('dissipative structures'), in agreement with the physical laws [Glansdorff and Prigogine (1971)]. The relationship between dissipative structures and biological systems has already been investigated [Babloyantz and Nicolis (1972), Goldbeter and Nicolis (1972), Blumenthal *et al.* (1970)].

The formation of biological systems must be autocatalytic, since the rate of destruction is usually first order and the rate of formation must be at least of the same order and exceed the rate of decomposition. It must be realized, however, that an autocatalytic system in an equilibrium situation will not grow since both the forward and backward reactions will be catalyzed; in addition, an autocatalytic system which preserves its formation content would never evolve as it would generate only exact replicas of itself. Therefore, the system must possess a third characteristic that will allow it to increase its information content. The mutability, which constitutes such a condition, represents the property of a system to make erroneous copies, with a different information content, which will be able to replicate themselves.

Mathematical models for the prebiotic formation of biopolymers can be constructed [Babloyantz (1972)]. Given an open system, it may be assumed that there is a flow of monomers, *a* and *b* (e.g., nucleotides), which will be polymerized to A and B (polynucleotides). The formation of A would require the preliminary formation of V (assembled monomers) and V′ (stack of the same monomers). On the other hand, the system is assumed to function *via* a cross-catalytic cycle, so that A acts as a template for the formation of B (from *b*) and B acts as a template for the formation of A from V′. The system may then be described by the equations

$$\text{na} \leftrightarrow \text{V} \leftrightarrow \text{V}' \leftrightarrow \text{A}$$
$$\text{mb} \leftrightarrow \text{B}$$

Once A and B are formed, the system will include the autocatalytic cycle

$$A + mb \rightarrow A + B$$
$$B + V' \rightarrow A + B$$

with A and B being removed from the system by diffusion or by formation of pairs, AB. A treatment of the steady-state kinetic equations resulting from such a system shows that, in the neighbourhood of equilibrium, polymerization proceeds through chain propagation and changes slightly by addition of monomers; beyond a critical value of the monomer concentration, however, polymerization *via* templates takes over.

Within this context, evolution can be treated from a deterministic point of view in terms of the stability structures by consideration of the perturbation caused by the formation of erroneous copies. Given an open system of N substances, the evolution will be described by a set of differential equations, stable to random perturbations of those substances; if a new substance appears, e.g., due to mutation, the order of the differential system will increase and, consequently, the system will evolve.

For example, considering again the previous model, one can assume that, through mutation in the formation of A by B, a new species C is formed which can catalyze the synthesis of a new substance E, which in turn catalyzes both the formation of C and its own formation: that is, a new catalytic cycle has been formed. The new situation is then described by the equations of the previous system plus the new equations

$$B + V' \leftrightarrow B + C$$
$$le + C \leftrightarrow C + E$$
$$B + V' + E \leftrightarrow B + C + E$$
$$le + C + E \leftrightarrow 2E + C$$

as well as those corresponding to the diffusion of C and E. An analysis of the resulting system shows that once the amount of the new substance C, which is coupled to the previous cross-catalytic cycle, reaches certain value, the system switches to a new steady state dominated by C, due to the nonlinear coupling between C and E resulting in a selection pressure.

The problem of selection has also been studied from an informational point of view [see, e.g., the work of Eigen (1971), used as a basis for the following discussion].

The condition of a high information content requires that the carriers be polymers and the problem reduces to determining the conditions under which the biopolymers will self-reproduce with the possibility of mutations, that will increase the information content. [The carriers will be considered confined to the information box, as otherwise they would diffuse in the medium.]

At a given time during the growth of the system, monomeric units and carriers, of different lengths, will be present within the box. The total number of units, free and organized (forming part of the polymers), is given by

$$N = N' + N'' = \Sigma\Sigma\ p\ c_{ip} + \Sigma\ c'_m$$

where c'_m denotes the concentration of the m-th monomer, in free form, and c_{ip} is the concentration of the i-th polymer with *p* units; the polymers within the same class (i.e., with the same number of units) differ in the sequence of the units. The degree or organization of the system may be measured by

$$D = N'/N$$

and a constant overall organization may be said to exist when N and D are constant.

Growth of the system, in a physical sense, corresponds to an increase in the concentration of the carriers, but selective growth denotes the preferred increase in the concentration of only some of the carriers. A carrier with *p* units may be formed by addition of a specific unit to a carrier with (*p*-1) units or by duplication. Assuming statistically equivalent concentrations of all the units, the first possibility cannot lead to an increased concentration for any given carrier, so that selective growth will mainly be achieved by duplication.

The overall rate of formation of an information carrier, I_{ip} (i.e., the i-th carrier with *p* units), may be expressed, in a first approximation, as

$$\text{rate}(i,p) = (F_{ip} - R_{ip})\ c_{ip}$$

Restricting the discussion to carriers of class *p*, the index *p* may be dropped for simplicity, so that one can rewrite the above expression as

$$\text{rate}(i) = (F_i - R_i)\ c_i$$

where the two terms give the rates of duplication and removal, respectively. Taking into account the possibility of errors in the duplication, the corresponding term may be written as

$$F_i = k\ A_i\ Q_i$$

where k is a first-order rate constant, A_i denotes an amplification factor, and Q_i represents a quality factor that expresses the fraction of correct copies formed. On the other hand, the removal term is given by

$$R_i = k\ D_i + d_i$$

where D_i denotes the individual decomposition term and d_i is a dilution term.

The above expression for the rate is incomplete, as it does not take into account the formation of the carrier under consideration by errors in the duplication of other carriers of the same class. The complete expression, modified by addition of the corresponding term, is then

$$\text{rate}(i) = (F_i - R_i)\, c_i + \Sigma'\, f_{ij}\, c_j$$

where the summation over *j* extends to all the carriers, except the one considered, of the same class *p*.

Substitution of the above definitions leads, finally, to

$$\text{rate}(i) = k\,(A_i\, Q_i - D_i)\, c_i + \Sigma'\, f_{ij}\, c_j - d_i\, c_i$$

which should describe the behavior of the system regarding metabolism, self-reproduction, and mutability, as well as some selective behavior.

Disregarding the contribution of the dilution term (which can be positive or negative), one can see that the system will grow if

$$A_i\, Q_i > D_i$$

but will not evolve, for which it should fluctuate around a steady state, for which $\Sigma\, \text{rate}(i) = 0$. A steady-state situation imposes the condition of constant overall organization as well as an additional constraint on the total dilution: taking into account (as the errors in the duplication must result in carriers of the same class) that

$$\Sigma\, k\, A_i\, (1 - Q_i)\, c_i = \Sigma\Sigma'\, f_{ij}\, c_j$$

one can write

$$\Sigma\, k\, A_i\, c_i = \Sigma\,(k\, A_i\, Q_i\, c_i + \Sigma'\, f_{ij}\, c_j)$$

so that, when the steady-state situation has been reached, one has

$$\Sigma\, \text{rate}(i) = k\, \Sigma\, (A_i - D_i)\, c_i - \Sigma\, d_i = 0$$

or

$$d = k\, \Sigma\, (A_i - D_i)\, c_i$$

where d represents the total dilution.

Denoting by

$$P_i = A_i - D_i$$
$$X = (\Sigma\, P_i\, c_i)/\Sigma\, c_i = (\Sigma\,(A_i - D_i)\, c_i)/\Sigma\, c_i$$
$$S_i = A_i\, Q_i - D_i$$

the excess production, mean productivity, and selective value, one may write

$$d = k\, \Sigma\, P_i\, c_i = k\, X\, \Sigma\, c_i$$

or, equivalently,

$$d_i = k\, X\, c_i$$

so that the rate equation becomes

$$\text{rate}(i) = k\,(S_i - X)\, c_i + \Sigma'\, f_{ij}\, c_j$$

If the second term is negligible, growth will occur whenever the selective value for a given information carrier, l, is greater than the mean productivity. Taking into account the steady state condition,

$$\Sigma\ k\ (S_i - X)\ c_i = 0$$

one can see that, as the system grows,

$$X \rightarrow S_l$$

On the other hand, if the second term is not negligible, one can write

$$\Sigma\ (k\ (S_i - X)\ c_i + \Sigma'\ f_{ij}\ (c_j/c_i)\ c_i) = \Sigma\ k\ (S_i + (1/k)\ \Sigma'\ f_{ij}\ (c_j/c_i) - X)\ c_i = 0$$

so that

$$S_l \rightarrow S_l + (1/k)\ \Sigma'\ f_{lj}\ (c_j/c_l)$$

For a small second term,

$$X \rightarrow S_l + (1/k^2)\ \Sigma\ (f_{lj}\ f_{jl})/(S_l - S_j)$$

which represents a selective equilibrium, valid for a master carrier, l, if $Q_l \simeq 1$. If an error copy, l′, is produced with a selective value, $S_l' > S_l$, the original metastable equilibrium will collapse to a new equilibrium and the system will evolve to a new optimum.

The above formulation, which holds when large numbers of carriers are involved, suffers from two important limitations, as the formation of a mutant is a non-deterministic process and the growth is subject to fluctuations. The rates of formation and removal are independent of each other and therefore they are not self-regulatory; for a self-regulatory fluctuation, the larger the deviation in one direction, the larger it is the probability for its reversal and the probability of distribution is symmetrical around the point of equilibrium and time-independent [Munster (1959)].

A stochastic treatment may be developed for stationary linear models [Bartholomay (1958a,b, 1959)]. The problem corresponding to the equation

$$\text{rate}(i) = (F_i - R_i)\ c_i$$

[with F_i and R_i constant] is represented by the stationary Markov chain (c_{it}, $0 \leqslant t < \infty$), where the time *t* is a continuous parameter and c_{it} is a random population variable, which refers to a discrete sequence of states. The change from the state *i*, at t = 0, to the state *j*, at a time *t*, is described by a transition probability $p_{ij\,t}$, from which expectation values as well as mean variances for the population of any state at time *t* may be determined.

It can be shown that, as long as the rates of formation and removal are equal and the selective values are identical for all the species, the population of the system will die out, independently of the existence of reproduction errors. Only when the rates of formation and removal are different, there is possibility of reproduction errors, and the system is maintained in a steady state, such a system will be stable and have a reproducible behavior. The system would select the strain with a maximum $S_l > 0$, but if such species exists in only a few copies and there are fluctuations, there may still be a possibility of decay and of uncertainty in the selection.

In the case of an equilibrated distribution, with $S_l = X$, the species with a selective advantage, $S_l' > S_l$, for which n copies are present at $t = 0$, has a finite probability (but not certainty) of being selected. The probability of this species dying out is given by

$$P_{no}(t) = (1 - (F_l - R_l)/(F_l - R_l \exp((R_l - F_l)\, t)))^{2n}$$

which, at $t \rightarrow \infty$, transforms into

$$\lim P_{no}(t) = (R_l/F_l)^{2n}$$

As a consequence, if a single mutant l′ is formed with a selective advantage, $S_l' > S_l$, in a previously equilibrated population $[S_l = X]$, the single copy has a probability, R_l'/F_l', of dying out and a probability, $1\text{-}R_l'/F_l'$, of survival; with $F_l' > R_l'$, the probability of extinction will decrease with increasing n. Therefore, small selective advantages have a rather low probability of survival and the treatment for linear models does not predict a point of no return.

In any case, neither the original rate equation for uncontrolled growth nor its final form, for the superposition of selective strains, can be considered as a true linear differential equation. The amplification, quality, and decomposition factors impose restrictions on the selection and self-organization of the system, especially insofar as the chain length is concerned. In addition, it is improbable that all the monomeric units would be present in similar concentrations and, if one exists in a much larger concentration, its inclusion would be so highly favoured that very uniform sequences, without any coding capacity, would result.

Perhaps such a simple code played a role in the early stages of evolution, but it would have been at a disadvantage with respect to a code using ‘complementary instruction’.

Complementarity

The replication process, considered above, implies a very special mode of recognition of the monomeric units: each monomer recognizes itself, although there may be a possibility of recognition of others, yielding erroneous copies.

There is, however, another possible scheme of recognition. At the simplest level, it implies a dual recognition of pairs of monomers; that is, if monomer *a* recognizes *b*, then *b* recognizes *a*. As in the preceding scheme, the possibility of recognition errors may exist.

Within the framework of such a scheme, involving complementary recognition, the replication involves an alternation between what have been denoted as positive ($+i$) and ($-i$) copies. If the information carriers are single-strand polymers, the complete replication involves a double step: transcription of the original copy into its negative and transcription of the negative copy into the replica of the original. In the case of double-strand polymers, the replication takes place in a single step, yielding two replicas of the original.

The rate equations for complementary instruction are

$$\text{rate}(+i) = k\,(A_{+i}\,Q_{+i}\,c_{-i} - D_{+i}\,c_{+i})$$
$$\text{rate}(-i) = k\,(A_{-i}\,Q_{-i}\,c_{+i} - D_{-i}\,c_{-i})$$

and the general behavior of such a system is analogous to the one described above, substituting c_i and S_i with

$$c_i = \begin{pmatrix} c_{+i} \\ c_{-i} \end{pmatrix}$$

and

$$\mathbf{S}_i = \begin{pmatrix} -D_{+i} & A_{+i}\,Q_{+i} \\ A_{-i}\,Q_{-i} & -D_{-i} \end{pmatrix}$$

respectively.

The importance of complementary recognition arises from the fact that, even in the case of excess concentration of a monomeric unit, two units must always be incorporated in the system. Mixed sequences will be formed as a consequence of duplication errors and the code will increase its information content if one of the errors has a selective advantage.

The nucleotides of adenine and thymine and guanine and cytosine, respectively, may form (complementary) pairs, but although recognition is a fast process, the stability of such pairs (involving hydrogen-bonding) is very low in polar media, with the result that the system has a small capability of information content. On the other hand, polynucleotides may form a double-stranded helical structure, with increased stability and information content capability.

Mutability vs. Recognition

Nucleic acids have then the possibility of complementary instruction and, consequently, of forming the basis for a code. Without catalytic help, however, only short chains can be of evolutionary significance due to the lack of high recognition abilities.

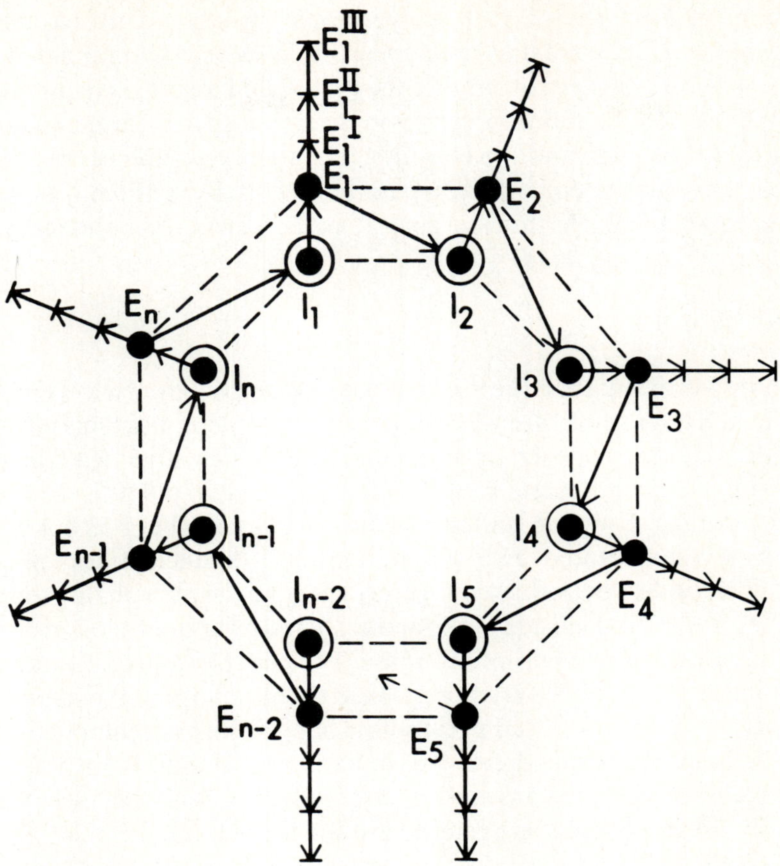

Figure 1-3. The self-instructive catalytic hypercycle involving the information carriers, I_i, and the polypeptide chains, E_i [Eigen (1971)].

Proteins, on the other hand, possess a high functional specificity and recognition capability, due to their spatial structures, but cannot lead to selective advantages because of their difficulty in producing mutants.

Therefore, it would seem that a coupling between nucleic acids and proteins would provide the logical path for the evolution from the protocell, containing both types of polymers, to the present situation.

A model, proposed for such a coupling [Eigen (1971)], is based on a closed hypercycle (Fig. 1-3). The set of polynucleotides (formed by complementary strands) reproduce themselves (that is, the strands reproduce each other) with the catalytic help of the previous polypeptide chain and, in turn, provide information for the synthesis of the following peptide in the cycle, which, being closed, is autocatalytic.

A system making use of such a hypercycle possesses the characteristics needed for evolution, insofar as the hypercycle is autocatalytic and non-linear, with competition for selection among the cycles, and responds to mutations of any polynucleotide. In addition, selection is very sharp and can account for the universality of the code and the chirality. The theory combines the stochastic character of evolution, corresponding to the essentially random generation of error copies, with the deterministic aspect, as the system will necessarily evolve once the error has appeared.

THE GENETIC CODE

It is now accepted that the nucleic acids, deoxyribonucleic (DNA) and ribonucleic (RNA) acids, are the carriers of the genetic information.

According to a proposed model [see, e.g., Crick and Watson (1954)] confirmed by X-ray diffraction studies [Langridge *et al.* (1960)], the DNA are formed by double stranded, uniform helices, with a height of 33.6 A per turn. Each strand is constituted by a polynucleotide, the nucleotides (phosphate esters of nucleosides) being linked together through the phosphate groups. The nucleosides are *N*-*β*-glycosides of 2-deoxy*D*-ribofuranose (deoxyribose), the aglycon being a puric or a pyrimidinic base. The phosphate group is attached to the positions 3′ and 5′ of deoxyribose, linked to the nitrogens at positions 9 and 1, respectively, of the puric and pyrimidinic bases. The molecular planes of the nucleotidic bases are perpendicular to the axis of the double helix; the distance between the planes of two consecutive bases is 3.36 A, with ten nucleotides per turn. The consecutive bases are rotated around the helix by 36° relative to each other.

A number of deoxyribonucleotides are known to occur in DNA, but the fundamental nucleotides involved are those of adenine, guanine, cytosine, and thymine. The total number of adenine-nucleotides in a DNA is equal to the number of thymine-nucleotides and the number of guanine-nucleotides is equal to the number of cytosine-nucleotides. The two strands in DNA are linked by hydrogen-bonds between the bases: adenine is hydrogen-bonded to thymine and guanine to cytosine. The selection by nature of these four units to code the genetic message merits some comments, taking into account that a binary system would have been sufficient. An answer to this question may be given on the basis of the information content [Seybold (1976)]. Assuming that the storage of the information requires a 'word' of *n* units, the possible arrangements of N 'letters' is given by

$$D = N^n$$

and the corresponding information content by

$$I = \log_2 D = n \log_2 N$$

If the (energetic) cost, C, of producing the message is taken as proportional to the product n N, the optimum value of N may be obtained by maximization of the functional

$$I'(N) = N^n + \lambda (n N - C)$$

The result is N = e (= 2.71828 . . .): that is, a 3-letter alphabet should be used, with words as long as necessary. However, the requirement of complementarity, discussed in the preceding section, dictates that the alphabet must have an even number of letters, so that the choices reduce to either 2 or 4: the latter may have been adopted because of the shorter words needed, with corresponding smaller energy requirements and chances of error during its replication.

Molecular Characteristics

The theoretical study of DNA molecules, faced with a number of considerable practical difficulties, has been simplified to that of their component units, most of the efforts having been directed towards the treatment of the individual bases, although some work has been carried out on the hydrogen-bonded pairs of bases, stacked bases, and nucleosides. Some considerations have enhanced the practical usefulness of the corresponding results (which, otherwise, would not have other interest than that associated with the description of a given molecule).

There have been two schools of thought regarding the feasibility of quantum-mechanical calculations for large molecules. On one hand, it was emphasized that such calculations would require enormous amounts of computer time using computers with an extremely large capacity of storage [Christoffersen (1972)]. On the other hand, a more optimistic view holds that *a priori* computations will become routine practice, as a result of a very efficient numerical analysis and of the advances in computer technology [Clementi (1972)]; in fact, it has been pointed out that it will even be possible to evaluate, in an economical way, rather accurate functions.

Both points of view may be accepted as correct, depending on what is understood by a 'large molecule'. Within the context of this work it is appropriate to distinguish between 'molecule' and 'polymolecule': the bases, pairs of bases, nucleosides, and nucleotides may be considered as molecules, while polynucleotides constitute examples of polymolecules. In such a case, the above statements may be rephrased in the following way: *a priori* calculations for molecules are feasible, and will become even more so in the future (up to a certain limit), but *ab initio* computations on polymolecules are not, at present, realistically possible. This statement should be further qualified by distinguishing between small and large polymolecules; even granting the feasibility of the calculations for small polymolecules, one is left to wonder about the computing facilities that would be required for some large polymolecules; it is in this connection that other methods, as described below, may be found to be appropriate.

With the above considerations in mind (that is, thinking of calculations on small polymolecules), it is then pertinent to mention that the use of linear combinations of the molecular orbitals of the component units, at the self-consistent field level, represents the most appropriate method for the study of polynucleotides [Flurry and Breen (1970)]: it offers, on one hand, the advantage of being appropriate for both homo- and hetero-polymers and, in addition, it yields results in satisfactory agreement with experimental data. A complete treatment for large polynucleotides will be time-consuming and expensive, but a short-cut may be introduced by consideration of only a few of the molecular orbitals of each monomer; the underlying assumption is, of course, that the lower molecular orbitals of the units are not affected (to a given extent) by the formation of the polymer. Calculations of this type should be undertaken (if practically feasible) whenever a complete information (orbital energies, etc.) is required.

A simpler approach, however, will suffice for conformational and stability studies. Such a treatment is based on the assumption that the various constituents retain, to a large extent, their identity. That is, within the framework of the approximations discussed in the following chapters, a study of the polynucleotides (including double-strand polynucleotides) might be undertaken on the basis of the results for their constituents: in a simplistic approach, such constituents may be taken as the bases, the two sugars (ribose and deoxyribose), and the phosphoryl residue, but it would seem more appropriate to base such calculations on the results for the pairs of bases and the sugar-phosphate-sugar backbone [see, e.g., the work of Matsuoka *et al.* (1978), Tosi *et al.* (1978a,b), and Vasilescu *et al.* (1977)].

Results should be available for all the systems in order to completely test the validity of the assumption. Until all the information has been collected, it will be necessary to justify the approximation on the basis of the results obtained for uridine (Fig. 1-4), uracil, and ribose in *a priori* calculations [Pullman and Berthod (1977)]. Comparison of the values for the energies of the occupied orbitals (Table 1-2) and for the total charges on the atoms (Figs. 1-5 and 1-6) clearly supports the preceding argument, although it must be emphasized that such a constancy is not observed for every property; in particular, greater changes are found for the electrostatic molecular potential, which explains the greater proton affinity of uridine (due to changes in the vicinity of the carbonyl group) and the low proton affinity of the oxygen O(1′) in uridine.

An additional confirmation is provided by the observation [Ladik *et al.* (1977)] that the energy levels of base pairs may be predicted 'rather accurately' from those of the separate bases, but the situation is not as satisfactory as for the nucleotides, as seen in Fig. 1-7. In any case, this difficulty is irrelevant, since it would be recommended to use the results for the hydrogen-bonded pairs of bases [see, e.g., the work of Clementi (1971) for results of an *a priori* calculation on guanine-cytosine and relevant references].

Table 1-2. Energies (hartrees) of the occupied molecular orbitals in uridine and its constituents*

Uridine	*Uracil*	*Ribose*	*Uridine*	*Uracil*	*Ribose*
20.599		20.553	0.843	0.866	
20.567		20.539	0.829	0.831	
20.563		20.544	0.811		0.795
20.558		20.537	0.797		0.773
20.548	20.562		0.770		0.740
20.544	20.552		0.765	0.777	
15.737	15.741		0.755		0.731
15.701	15.708		0.734	0.740	
11.622	11.630		0.718	0.720	
11.568	11.580				0.714
11.520		11.442	0.704		0.683
11.498	11.525		0.696		0.668
11.489		11.464	0.678	0.688	
11.487		11.459	0.674		0.649
11.477		11.454	0.670		0.631
11.475		11.456	0.655		0.623
11.411	11.429		0.636	0.643	
1.499		1.450	0.624	0.631	
1.477	1.487		0.618	0.628	
1.433	1.442		0.602		0.580
1.430		1.405	0.582	0.590	
1.425		1.396	0.572		0.545
1.411		1.390	0.563		0.541
1.349	1.354		0.540		0.518
1.287	1.290		0.528		0.498
1.164		1.130	0.515		0.456
1.155	1.159		0.510		0.486
1.120		1.107	0.500	0.509	
1.031		1.002	0.490		0.471
0.992	0.997		0.470	0.480	
0.947	0.956		0.451	0.456	
0.914		0.891	0.405	0.423	
0.901		0.877			

* As given by Pullman and Berthod (1977). All the energies are negative. One orbital of the ribose has no correspondent in uridine due to the fusion of the two rings in the nucleoside.

Regarding the availability of data, it should be mentioned that *a priori* results have been obtained for the remaining bases (adenine, guanine, cytosine, and thymine) and the phosphoryl moiety. The effective charges (Appendix A7) for the remaining puric and pyrimidinic bases are presented in Figs. 1-8

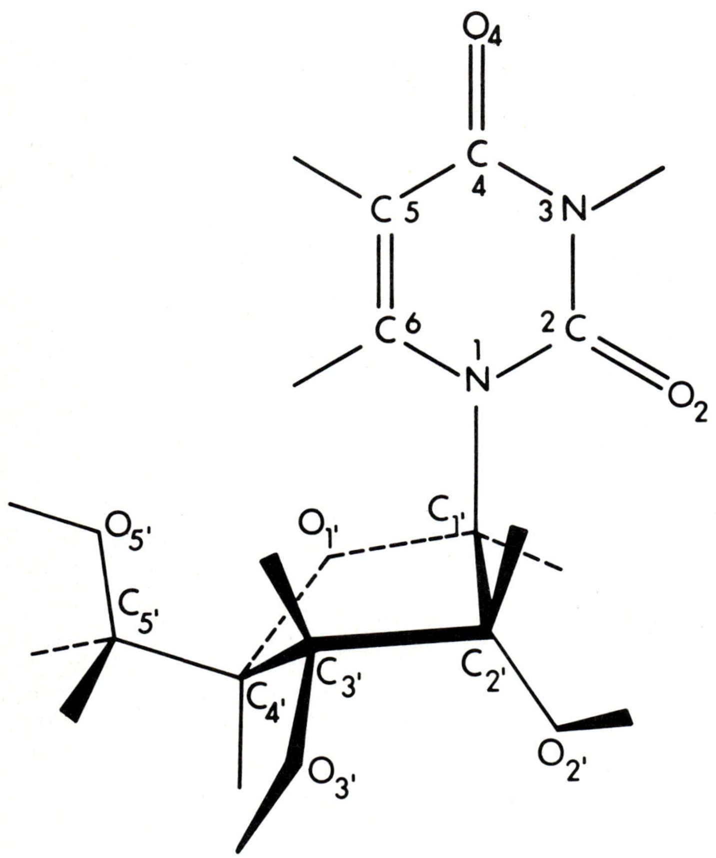

Figure 1-4. Schematic representation of the conformation and numbering system of uridine [Pullman and Berthod (1977)].

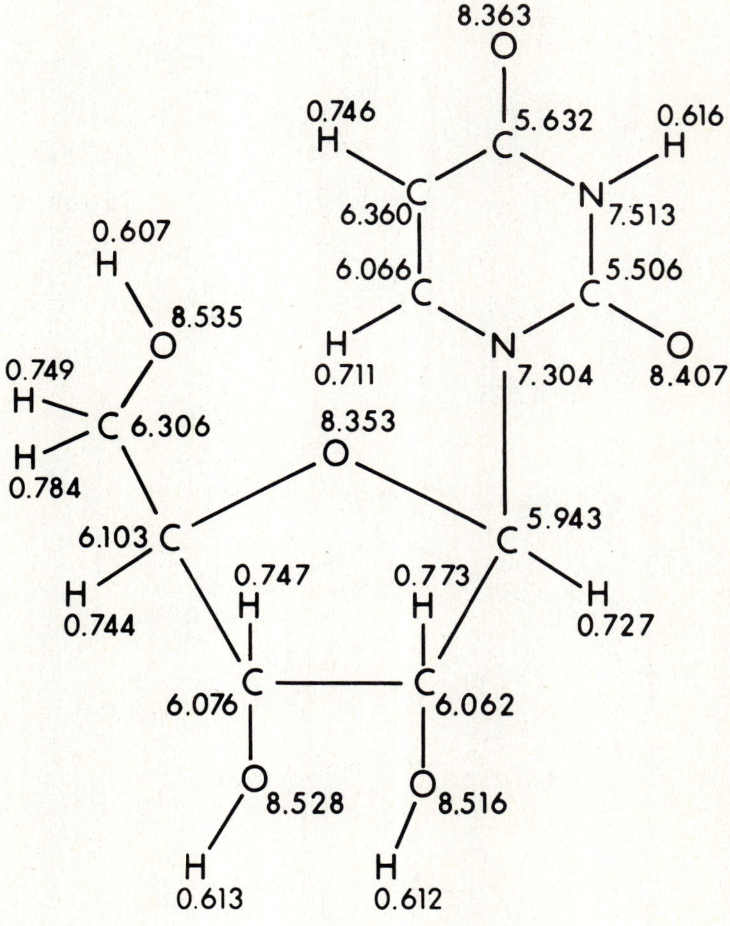

Figure 1-5. Total charges in uridine [Pullman and Berthod (1977)].

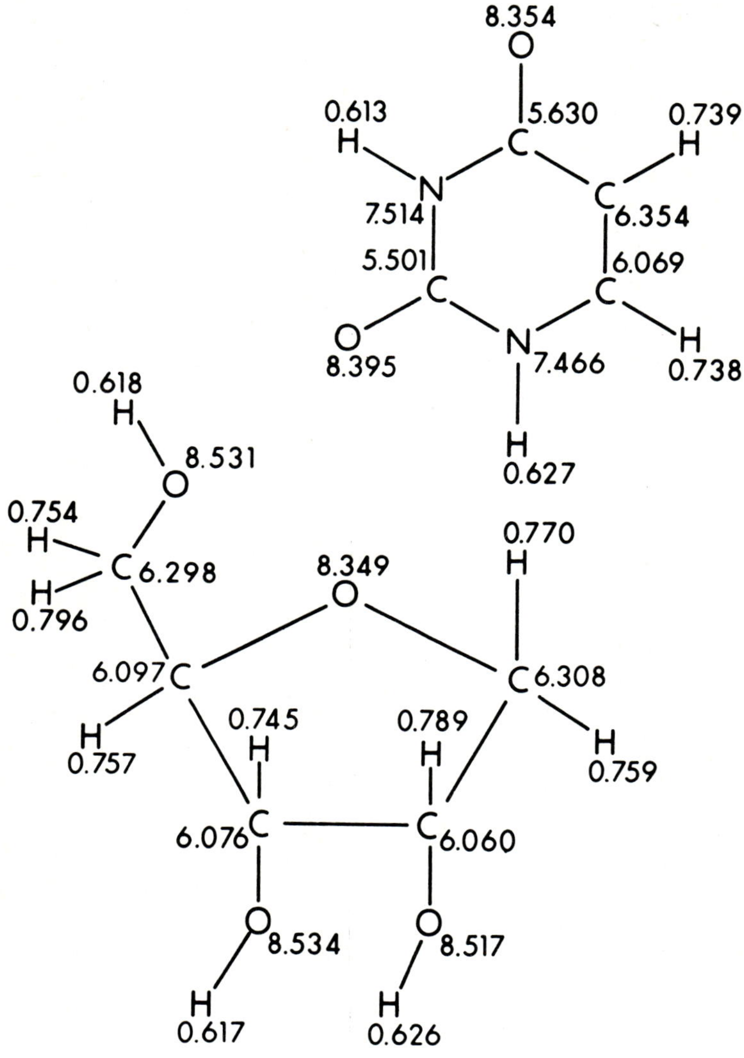

Figure 1-6. Total charges in ribose and uracil [Pullman and Berthod (1977)].

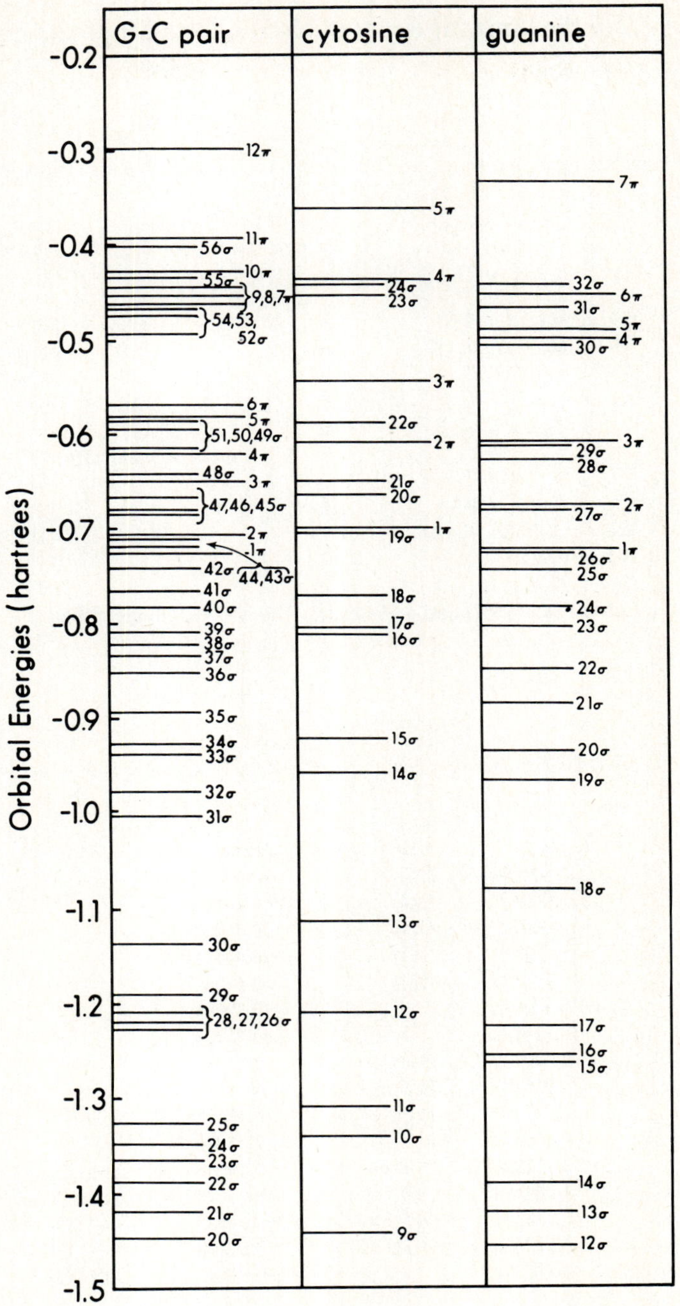

Figure 1-7. Orbital energy diagrams of the guanine-cytosine pair and of the separate components. The lowest-lying orbitals have been omitted [Clementi *et al.* (1971)].

through 1-11; their accuracy may be judged on the basis of the results, collected in Table 1-3, for the ionization potentials and dipole moments. Similarly, Fig.

Table 1-3. Ionization potentials and dipole moments for adenine, guanine, cytosine, and thymine

	Adenine	*Guanine*	*Cytosine*	*Thymine*
	Ionization Potential (eV)			
Clementi *et al.* (1969)	9.98	9.11	9.83	10.54
Mely and Pullman (1969)	9.41		9.79	10.16
Ladik *et al.* (1977)	9.29	8.21	9.77	9.60
Experimental*	8.91		8.90	9.43
	Dipole moment (D)			
Clementi *et al.* (1969)	2.61	6.94	6.40	3.29
Mely and Pullman (1969)	2.83		6.76	4.72
Experimental*	3.0			3.9

* As given by Lifschitz *et al.* (1967) and de Voe and Tinoco (1962), respectively.

Table 1-4. Energy (hartrees) and main character of the occupied orbitals of dimethylphosphate anion*

Energy	*Character*	*Energy*	*Character*
−78.89980	P	−0.49808	
−20.06082	O(4)	−0.46562	
−20.06074	O(5)	−0.46237	
−19.79346	O(1)	−0.44511	
−19.79313	O(3)	−0.41067	
−10.93501		−0.39044	
−10.93499		−0.26738	
− 7.12012	P	−0.25339	
− 4.95350	P	−0.23191	
− 4.95245	P	−0.17138	
− 4.95082	P	−0.16546	
− 1.13285	O(4),O(5)	−0.09600	O(1),O(3),P
− 1.08919	O(4),O(5)	−0.05574	O(1),O(3),P
− 0.91858	O(1),O(3)	−0.04933	O(1),O(3),P
− 0.85240	O(1),O(3)	−0.00468	O(1),O(3)
− 0.77137		+0.02440	O(1),O(3)
− 0.76811			

* As given by Perahia *et al.* (1975).

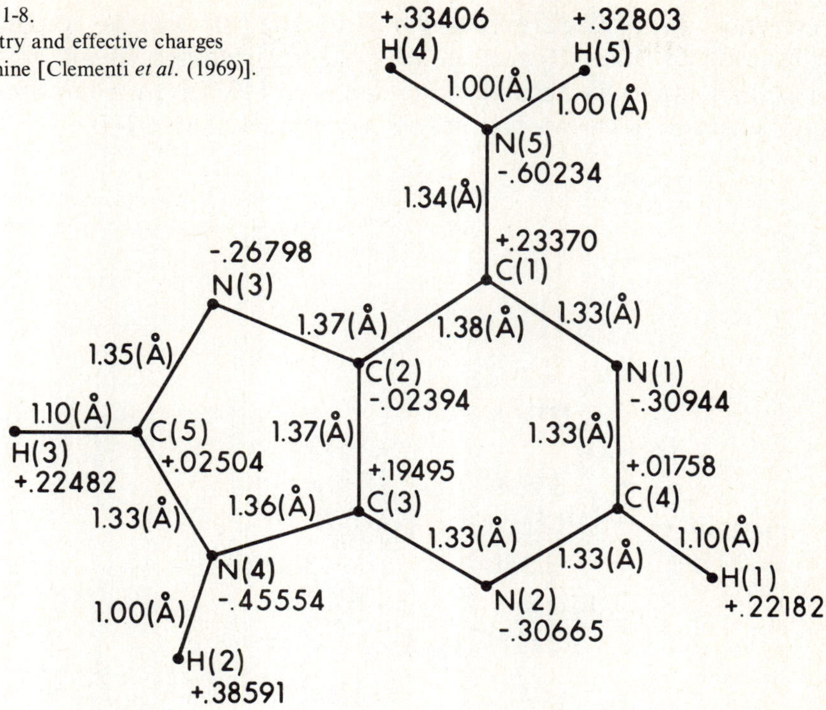

Figure 1-8.
Geometry and effective charges of adenine [Clementi *et al.* (1969)].

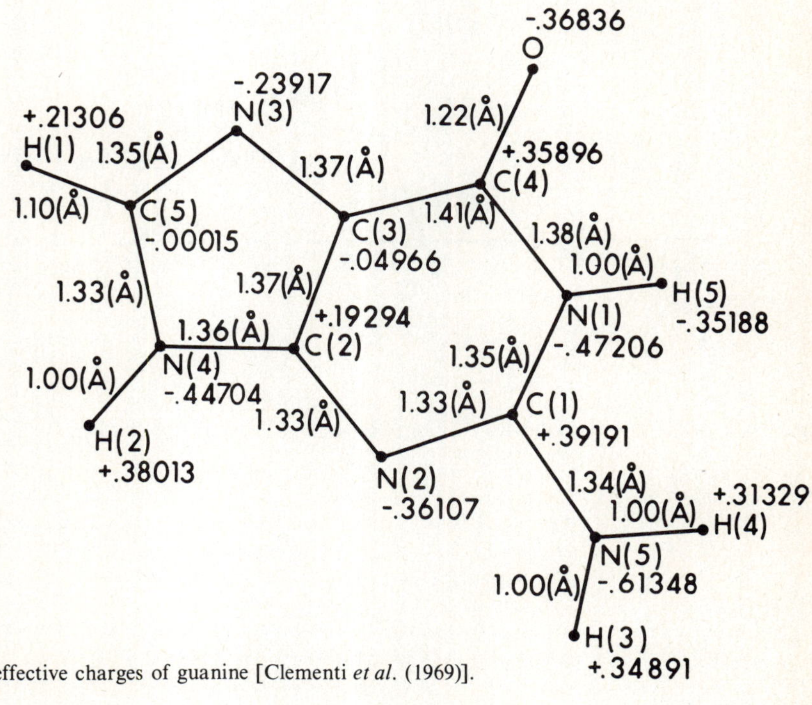

Figure 1-9.
Geometry and effective charges of guanine [Clementi *et al.* (1969)].

1-12 presents the effective charges for the phosphate group in dimethylphosphate anion (as a model system), obtained in an *a priori* calculation of comparable accuracy [Perahia *et al.* (1975)]; for completeness, the orbital energies of this system are also presented (Table 1-4).

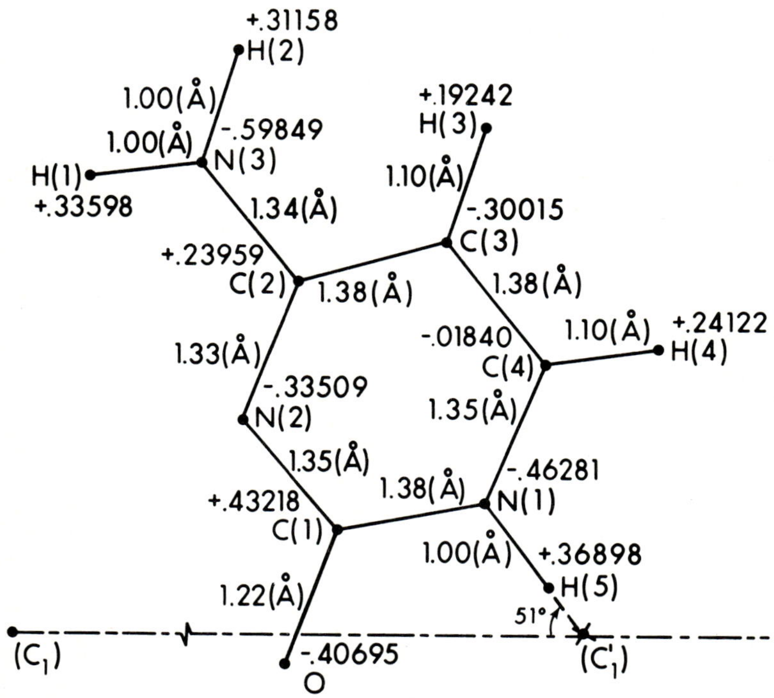

Figure 1-10. Geometry and effective charges of cytosine [Clementi *et al.* (1969)].

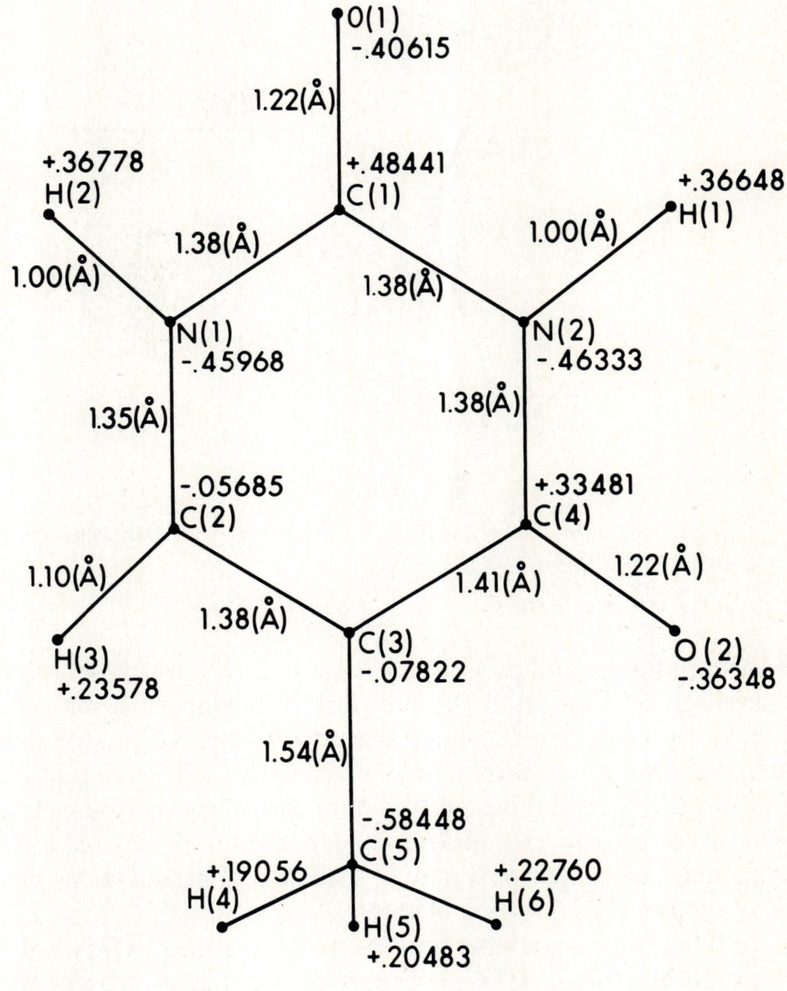

Figure 1-11. Geometry and effective charges of thymine [Clementi *et al.* (1969)].

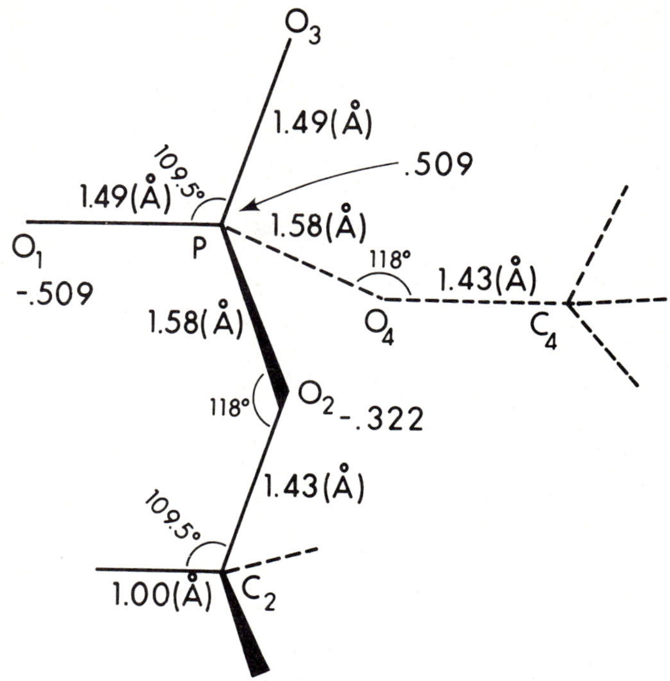

Figure 1-12. Geometry and effective charges of the phosphate group (in dimethylphosphate anion) [Perahia *et al.* (1975)].

Duplication and Transcription

The sequence of puric and pyrimidinic bases in DNA, which determines the genetic heredity, is transmitted through duplication (replication) of DNA. Prior to the mitosis, the double helix starts unwinding itself progressively and nucleotides present in the medium may become attached (by complementary recognition) to the two strands. Such a complementary pairing yields, upon completion of the process, two molecules identical to the original DNA; this pairing constitutes a problem that may be studied within the framework of the models discussed in the following chapters.

Tentative but rather suggestive arguments have been put forward for the interpretation of the initiation of the processes of cell division and replication of the DNA, based on the dipolar characteristics of the components of the cells which may exhibit collective dipolar oscillations [Frohlich (1968)]. The cell membranes, on one hand, consist of a strong dipolar layer, so that in local vibrations of a part of the membrane the positive and negative parts may vibrate against each other, resulting in an oscillating electric dipole; giant molecules (with dipolar properties due to hydrogen-bonds) and systems with non-localized electrons may also exhibit such oscillations. Certain processes may feed energy locally to some dipolar oscillations and long-range coulomb interactions result in a sharing of this energy with other dipoles which will tend to oscillate coherently, provided that the energy supplied is sufficiently large compared to the energy loss. Non-linear effects may reduce the loss, increasing therefore the excitation, with the possibility of the system reaching a metastable state in which the energy supplied to the dipolar constituents is channelled into a single longitudinal mode with long-range phase correlations. In this state, the mean value of the square of the electric polarization may be large and cause deformations and stresses, which may in turn lead to a relative stabilization of the excitation; a small supply of energy is then required to maintain it, while a continued higher energy supply will lead to an instability. Within the framework of this model (which fits the general conditions for the stability of the dissipative structures mentioned above), the deformations connected with the excitation could stimulate the cell division; the stimulus, proportional to the size of the cell, would be due not only to the vibrations inside the cell considered but also to the dipolar oscillations in other cells, except when densely packed, as experimentally observed in the contact inhibition. (The participation of non-localized electrons in these phenomena may lead to a qualitative change in all those cell activities that depend on the vibrations; a carcinogenic change would then be caused by an excessive density of non-localized electrons in a region of the cell and carcinogenic molecules should have the ability of transferring electrons to relevant regions of the cell.)

Strong local electric fields within the cell may determine a polarization of the mobile charges of the molecule, resulting in a repulsion between the charges at the same end of the two strands [Hoffman and Ladik (1964)]; on the other hand, approximate calculations [Seprodi *et al.* (1969)] have shown that a homogeneous electric field, in the plane of the base pairs, may cause the breaking of the hydrogen-bonds between them. These are two points that merit further investigation at a higher level of accuracy.

The complementary pairing involved in the formation of the two daughter molecules constitutes, on the other hand, a problem that may be studied within the framework of the methods discussed in the following chapters.

TRANSMISSION OF THE CODE

All the processes normally taking place in a cell constitute, in principle, a manifestation of the genetic code.

The immediate transmission of the code occurs, on one hand, in the duplication of DNA (mentioned above) and its transcription into the ribonucleic acids (RNA) involved in the synthesis of proteins. The processes involving the latter, which may function as, or form part of, enzymes, membranes and antibodies, represent the communication of the biological information.

Such a communication may be altered or disrupted, accidentally or on purpose, as evidenced in aging, carcinogenesis, and all those instances of interference with the reactions or the conditions of the communication processes or of interaction with the above systems.

Protein Synthesis

Proteins are biosynthesized from their components, the amino acids, in a process controlled by the cell nucleus and in which the nucleic acids act as templates.

The genetic message, corresponding to a given protein and contained in the DNA, is first transcribed into the (messenger) *m*-RNA which then selects, with the help of the corresponding (transfer) *t*-RNA, the component amino acids. The process takes place in the ribosomes and consists of four main steps: transcription of the *m*-RNA, binding of the amino acid to the *t*-RNA, reading of the code, and termination of the synthesis. Only those points corresponding to the proper transmission of information are considered here.

The *m*-RNA and *t*-RNA are formed by single helicoidal chains of ribonucleotides (mainly adenine-, guanine-, cytosine-, and uracil-nucleotides; see below for additional details). The sugar participating in these nucleotides is the *D*-ribofuranose (ribose), phosphorylated at positions 3′ and/or 5′; the position of the phosphate group at the beginning and end of the chain plays an important role in the synthesis of proteins. The chains of these RNA may be coiled to form loops at intervals with hydrogen-bonds between the bases.

A correlation exists between the sequence of nucleotides in *m*-RNA and the sequence of amino acids in the corresponding protein. It has been postulated that every amino acid is coded by a set of three consecutive bases, such a set being denoted as a codon. Assuming that the order of the bases plays a given role, sixty-four codons may be constructed, all of which are operative according to experimental information. Table 1-5 shows the correlation derived from experimental observations between codons and amino acids: the genetic code is degenerate since there are only twenty amino acids to be coded; with the exception of methionine and tryptophan, the remaining amino acids are coded by two, three, and even four synonyms.

The amino acids are transported to the *m*-RNA by the corresponding *t*-RNA, specific for the various amino acids. The specificity of each *t*-RNA depends on its sequence of nucleotides and, probably, on its secondary structure. The set of those nucleotides in a *t*-RNA, complementary to one of the codons in the *m*-RNA, is designated as a anticodon.

Table 1-5. Codons in *m*-RNA*

first base	second base: A	C	G	U	third base
A	lys	thr	arg	ile	A
	asn	thr	ser	ile	C
	lys	thr	arg	met[x]	G
	asn	thr	ser	ile	U
C	gln	pro	arg	leu	A
	his	pro	arg	leu	C
	gln	pro	arg	leu	G
	his	pro	arg	leu	U
G	glu	ala	gly	val	A
	asp	ala	gly	val	C
	glu	ala	gly	val[x]	G
	asp	ala	gly	val	U
U	term[+]	ser	term[+]	leu	A
	tyr	ser	cys	phe	C
	term[+]	ser	trp	leu[x]	G
	tyr	ser	cys	phe	U

* As given by Nirenberg (1969).
[x] Initiation of protein synthesis.
[+] Termination of protein synthesis.

The reading of a codon in *m*-RNA implies a codon-anticodon pairing. This pairing cannot be understood unless more than one *t*-RNA is postulated for those amino acids with degenerate codons. As an alternative, the 'wobble' theory [Crick (1966)] may serve to explain this paradox: it is accepted that the pairing $A \leftrightarrow U$, $G \leftrightarrow C$ is normal for the first two positions of the codon-anticodon association, while for the third position there are several ways of (codon-anticodon) pairing ($A \leftrightarrow U$, $G \leftrightarrow U$, $G \leftrightarrow C$, $U \leftrightarrow A$, $U \leftrightarrow G$, $C \leftrightarrow G$, $U \leftrightarrow I$, $C \leftrightarrow I$, $A \leftrightarrow I$, where I stands for inosine).

Inspection of the experimental information leads to some interesting conclusions. Presenting Table 1-5 in the form shown in Table 1-6 [Fraga and Valdemoro (1968)], it becomes evident that the existence of C and G at either or both of the first two positions plays, in most cases, a decisive role, the nature of the base (puric or pyrimidinic) appearing at the third position being irrelevant. On the other hand, the pairing $A \leftrightarrow U$, with only two hydrogen bonds, is weaker and the contribution of the third base may be needed in order to decide the specificity. This simple reasoning, however, fails to explain the different contribution of the second position, suggesting that the codon-

Table 1-6. Abbreviated representation of the codons in *m*-RNA*

first base		second base: A	C	G	U	third base
		A	C	G	U	
	A	lys	thr	arg	met[x]	pu
		asn		ser	ile	py
	C	gln	pro	arg	leu	pu
		his				py
	G	glu	ala	gly	val	pu
		asp				py
	U	term	ser	trp[+]	leu	pu
		tyr		cys	phe	py

*See Table 1-5 for details.
[x]Disregarding the coding of isoleucine by AUA.
[+]Disregarding the role of UGA as a terminator codon.

anticodon pairing, as considered so far, may not be the only factor. That is, other interactions may play an important role.

Although it has been mentioned above that the main deoxyribonucleosides were those of adenine, guanine, cytosine and thymine, and that the main ribonucleosides were those of adenine, guanine, cytosine and uracil, the fact is that a much larger number of both deoxyribonucleosides and ribonucleosides have been identified in both DNA and *t*-RNA [Hall (1970)]. Most of these nucleosides are related to the ones mentioned above, differing in some modifications.

Even if the majority of those modifications are rather simple (e.g., replacement of OH with S and of NH_2 with OH, attachment of a methyl group, etc.), they could cause significant changes in the structure and function of the nucleic acid molecule, the latter as a consequence of their effect on the interaction characteristics.

There are, in addition, some hypermodified nucleosides characterized by a relatively large side-chain with a functional group (hydroxyl, carboxyl, allylic double bond). These hypermodified nucleosides have only been found at the position adjacent to the 3′-end of the anticodon. Not all the *t*-RNA have modified nucleosides in that position, but such molecules are in a minority and in this connection two comments must be made: in some cases, the nucleoside at that position has not been identified (perhaps because of its unusual structure); in others, adenosine appears in that position, suggesting the possibility of the existence of a very labile substituent, not identified so far. Examples of hypermodified nucleosides are *N*-(nebularin-6-

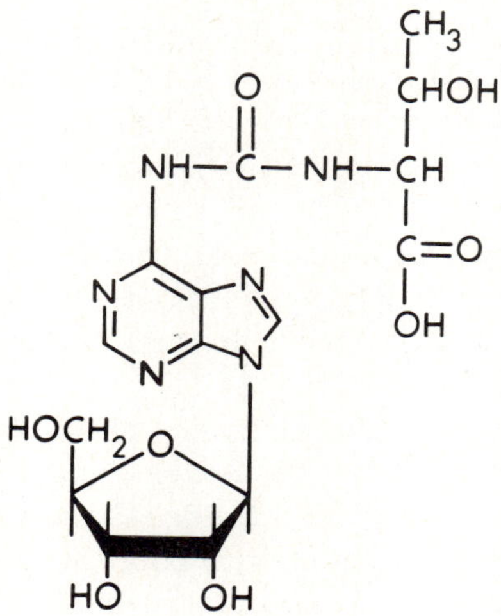

(I)

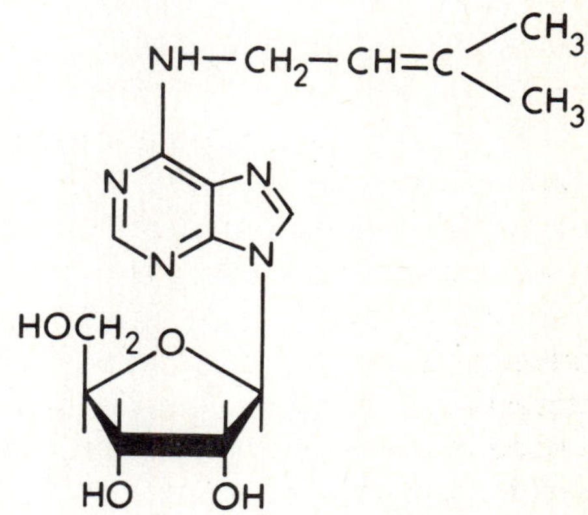

(II)

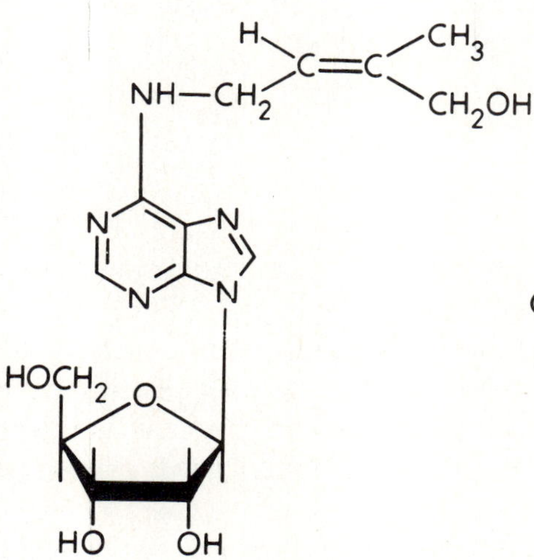

(III)

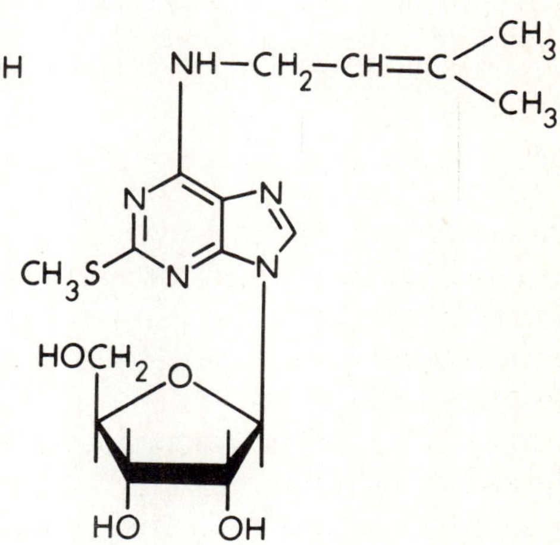

(IV)

Figure 1-13. Formulas of N-(Nebularin-6-ylcarbamoyl)threonine (I), N^6-(Δ^2-isopentenyl)adenosine (II), N-(*cis*-4-hydroxy-3-methylbut-2-enyl)adenosine (III), and N^6-(Δ^2-isopentenyl)-2-methylthioadenosine (IV) [Hall (1970)]. (Reprinted with permission. Copyright by Academic Press, Inc.)

ylcarbamoyl)threonine, N^6-(Δ^2-isopentenyl)adenosine, N^6-(cis-4-hydroxy-3-methylbut-2-enyl)adenosine, and N^6-(Δ^2-isopentenyl)-2-methylthioadenosine (Fig. 1-13).

The strategic location of the hypermodified nucleosides in the *t*-RNA molecule and the existence of a functional group in the side-chain suggests a number of roles that they may play: The carboxyl group, easily ionizable, could enhance the stability of the transfer-messenger-ribosome complex through electrostatic interaction with the basic ribosomal proteins [Schweizer *et al.* (1969)]; the isopentenyl group, with its reactive double bond, might facilitate the association-dissociation process during the amino acid transfer and translocation steps [Hall (1970)], providing, in addition, the basis for a rapid mechanism of activation-deactivation of the *t*-RNA specific to certain protein molecules, thus controlling their synthesis [Hall (1970)].

A rather fundamental function is implicitly postulated in the above statement for the isopentenyl group: an effect on the codon-anticodon interaction. In fact, three possibilities may be considered: it distorts the codon-anticodon interaction, thus permitting abnormal recognition; it contributes to the fidelity of such an interaction; and, consequently, as a result of an alteration in its structure, it may prevent such an interaction [Hall (1970)]. In any case, but especially in the first two instances, the active presence of such a hypermodified nucleoside implies a tetranucleoside sequence for the anticodon and it is evident that a variety of possibilities may be postulated for the correspondence between codon (anticodons), the amino acids and the proteins.

Sequences and Conformations

A polypeptide chain may fold spontaneously to the three-dimensional structure of a native protein, believed [Anfinsen and Scheraga (1975)] to be the one in which the Gibbs free energy of the whole system is minimum with respect to all the degrees of freedom. That is, the native conformation is determined by all the possible inter- and intra-molecular interactions. In a given environment, therefore, the preferred conformations of polypeptide chains will depend on their amino acid sequence.

The basic problem to be solved regarding the relationship between sequence and conformation is then the determination of the preferred conformation (i.e., that one with lowest energy) for a given sequence of amino acids. It is evident that the considerations outlined above for polynucleotides apply just as well to the case of polypeptides. That is, it is not realistic to think of carrying out accurate *a priori* calculations for such systems and approximate methods must be used; and, in fact, semiempirical methods have met with considerable success. Their detailed discussion, from a general point of view, is postponed to the following chapters and here only the methodological foundation will be considered.

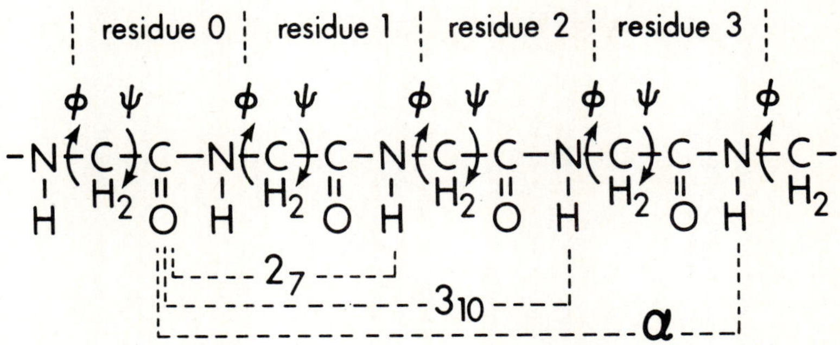

Figure 1-14. Schematic representation of part of the polyglycine chain showing the rotation angles (ϕ, ψ) and the hydrogen-bond scheme for the 2_7, 3_{10}, and α-helices [Morokuma (1971)].

Table 1-7. Comparison of results for some conformations of polyglycine*

Geometry	α	3_{10}	2_7	β
ϕ	−48°	−49°	−75°	−140°
ψ	−57°	26°	70°	135°
O-H	1.84(3)	1.72(2)	1.93(1)	
Energies				
E_{00}	−45.53197	−45.52293	−45.54222	−45.55420
E_{01}	− 1.02784	− 1.02840	− 1.04267	− 1.02741
E_{02}	− 0.00092	− 0.03930	− 0.00040	+ 0.00049
E_{03}	− 0.02802	− 0.00271	− 0.00048	− 0.00019
Densities				
C(1)	3.9591	3.9621	3.9595	3.9615
H(2)	0.9944	0.9890	0.9871	0.9824
H(3)	0.9853	0.9741	0.9855	0.9843
C(4)	3.6357	3.6373	3.6515	3.6541
O(5)	6.3873	6.3969	6.3767	6.3656
N(6)	5.2010	5.2172	5.1986	5.1822
H(7)	0.8372	0.8264	0.8411	0.8698

* As given by Morokuma (1971). The O-H distance is given in A; the number in parentheses indicates the segment number of the hydrogen atom which forms a hydrogen bond with the oxygen atom of segment 0. The energies are given in hartrees; the subscripts indicate the segments involved in the interaction. For the electron densities, the atoms given correspond to the groups CH_2, CO, and NH, respectively.

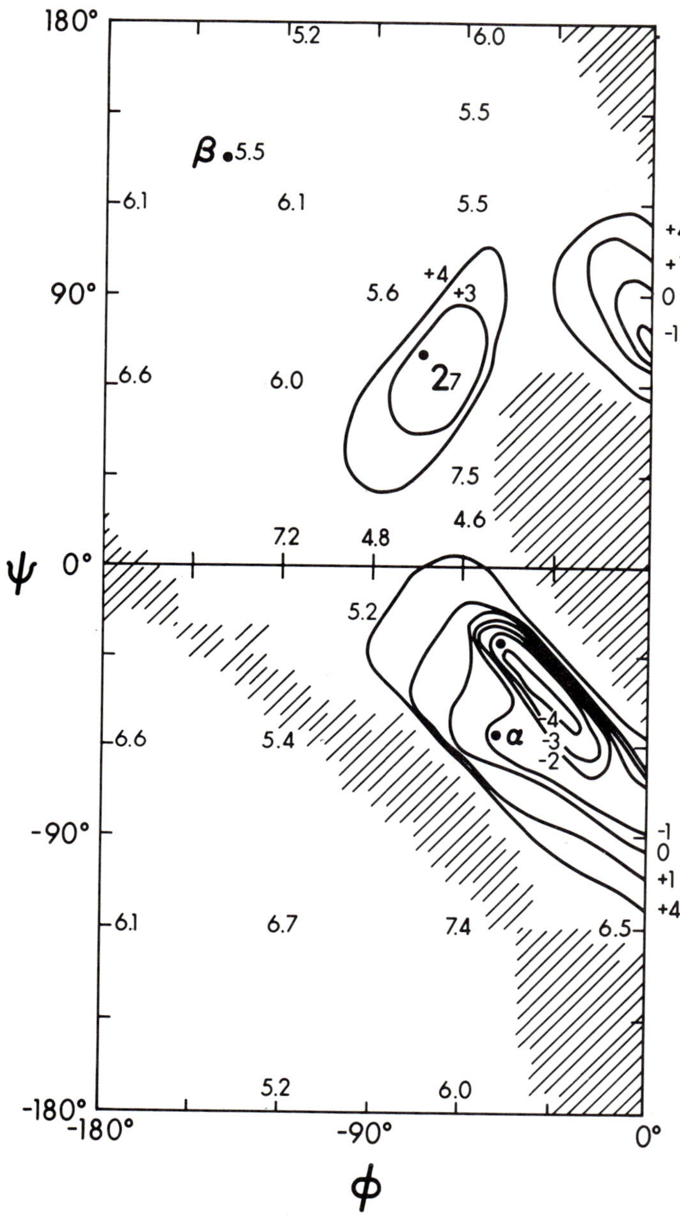

Figure 1-15. Conformational energy map of polyglycine (with energies referred to an arbitrary origin). Contours have been hand-interpolated for energies up to 4 kcal/mole, while higher values are given as discrete entries. The shaded area is expected to be extremely unstable [Morokuma (1971)].

As a starting point, one may consider the results obtained for polyglycine (Fig. 1-14) using a Hartree-Fock method (including all the valence electrons) for linear polymers [Morokuma (1971)]. The general features of the conformation energy map (Fig. 1-15) are in qualitative agreement with those found in empirical treatments, the main difference being that the minimum is found around the conformation ($\phi = -30°$, $\psi = 45°$) resembling a 3_{10} helix ($-49°$, $-26°$) (not shown explicitly in the figure), while the empirical maps give the minimum near the α-helix conformation ($-48°$, $57°$); the reason for the difference may be found in an overestimation of the hydrogen-bonding contribution.

The main conclusion (Table 1-7) is that the intraresidual interactions (i.e., those within a single amino acid residue) play, by far, the dominant role, although interactions between the various residues contribute to the stabilization of the chain; the latter are due, primarily, to hydrogen-bonding (with an increased polarity in the atoms involved).

Therefore, as a first approximation, one might consider the possibility of building up a polypeptide using the results obtained for independent residues. The 'dipeptide' model (Fig. 1-16), although neglecting the interresidual interactions, includes the effect of the formation of the peptidic bonds, so that the corresponding results will be superior (for the purpose in mind) to those

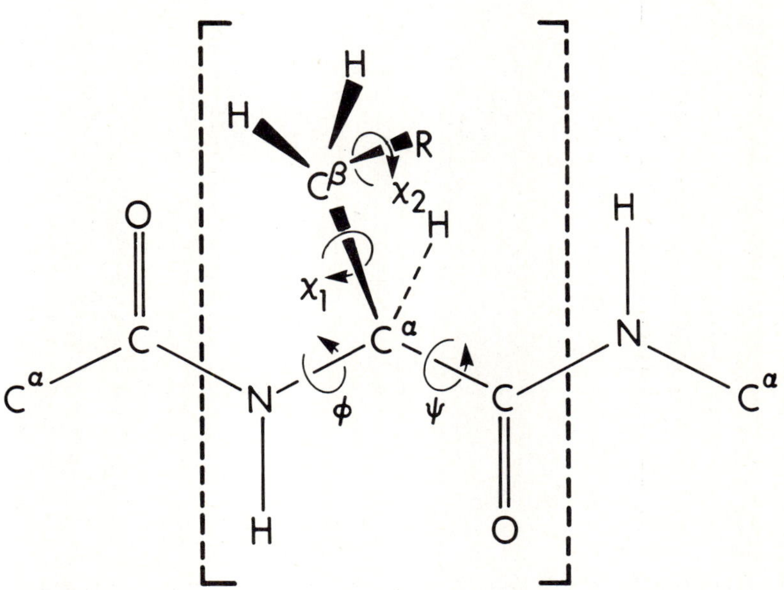

Figure 1-16. Schematic representation of the 'dipeptide' model, with the amino acyl residue limited by the dashed lines [Pullman *et al.* (1970)].

obtained for independent amino acids. The importance attached, at this moment, to this model arises from the fact that extensive *a priori* results are available. The calculations were carried out, in all cases, for the *N*-acetyl-*N'*-methyl amino acylamide with the peptide bond taken in the planar *trans* conformation ($\omega = 0°$) and the two backbone terminal methyl groups fixed so that the N-H and C = 0 terminal bonds are each eclipsed by one of the C-H bonds of these groups, with additional simplifications or assumptions depending on the case (see Chapter 4 for full details). Conformational ϕ, ψ

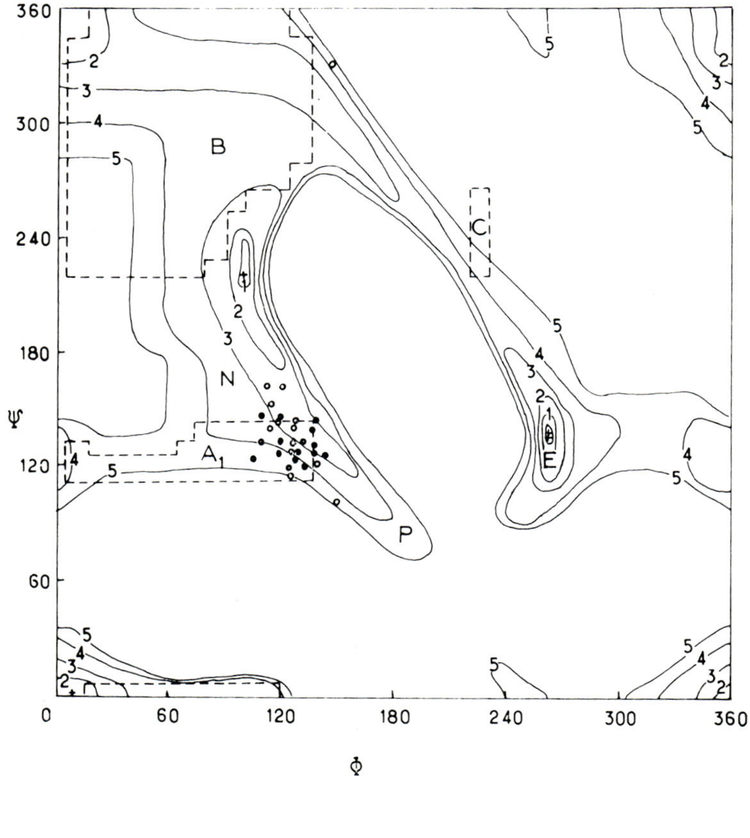

Figure 1-17. Conformational energy map for N-acetyl-N'-methylalanylamide. The notation used is: +, local minima; double cross, global minimum; dashed lines, contours from semiempirical calculations; circles, values found in crystalline lysozyme; shaded circles, values found in myoglobin. The contours are in kcal/mole, referred to the absolute minimum of the calculation [Maigret *et al.* (1971)].

maps, i.e., representations of the molecular energy as a function of ϕ and ψ, were obtained for each residue: e.g., first, conformational ϕ, ψ maps were prepared for different values of the dihedral angles χ of the side-chain; then, the complete map was constructed by superposition (i.e., taking the lowest energy points) of the preceding submaps. A representative example (for *N*-acetyl-*N'*-methylalanylamide) is presented in Fig. 1-17, and a complete summary is given in Chapter 4.

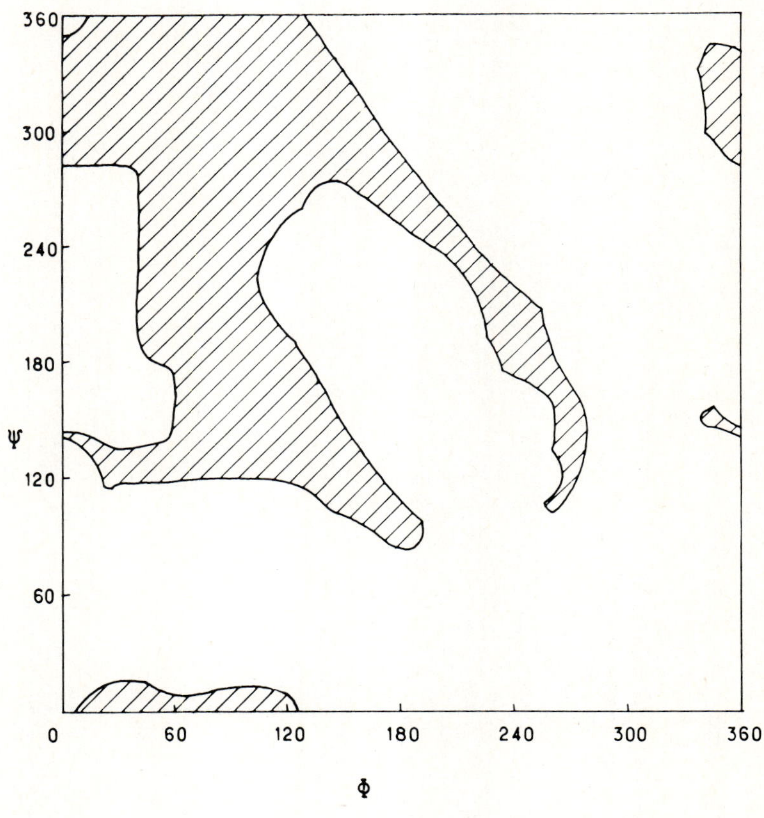

Figure 1-18. Lower limits of the allowed conformational space for the four aliphatic residues (within 5 kcal/mole above the global minima) [Maigret *et al.* (1971)].

Figures 1-18 through 1-21 present the upper and lower limits of the allowed fraction of the conformational ϕ, ψ map within the contour corresponding to an energy of 5 kcal/mole above the global minima, for the four aliphatic (alanyl, isoleucyl, leucyl and valyl) and the four aromatic (histidyl, phenylalanyl, tryptophanyl and tyrosyl) residues. In these figures, the upper limit encloses that part of space where there is a probability of finding any one

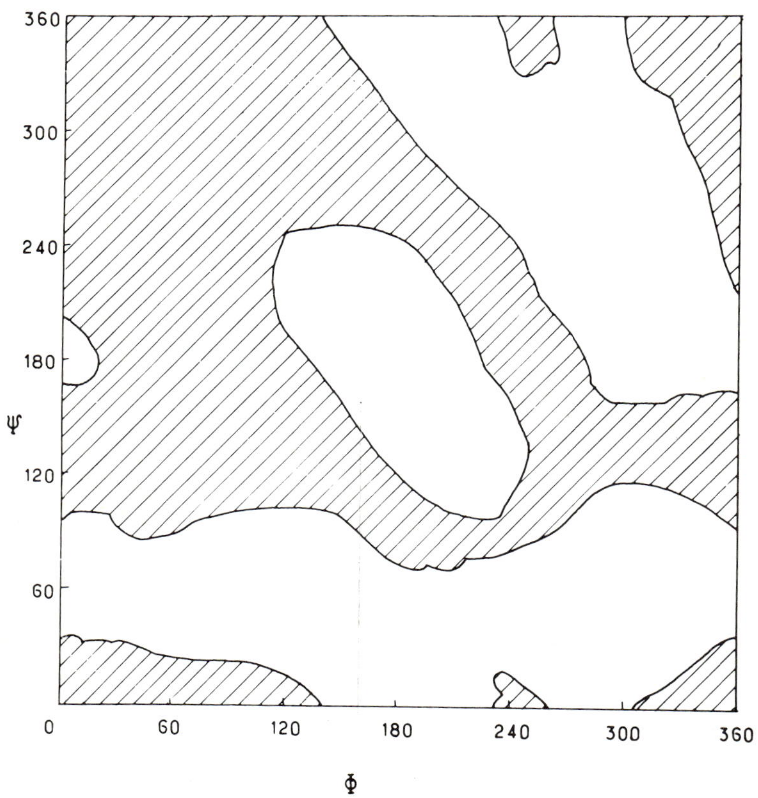

Figure 1-19. Upper limits of the allowed conformational space for the four aliphatic residues (within 5 kcal/mole above the global minima) [Maigret *et al.* (1971)].

of the four residues under consideration; the lower limit, on the other hand, encloses the part of the conformational space common to the four residues. The minimum conformational space is common to both groups, but the maximum space is smaller for the aliphatic residues than for the aromatic residues.

The problem, however, with the dipeptide model is that intraresidual interactions allow several conformations and that, if a single preferred

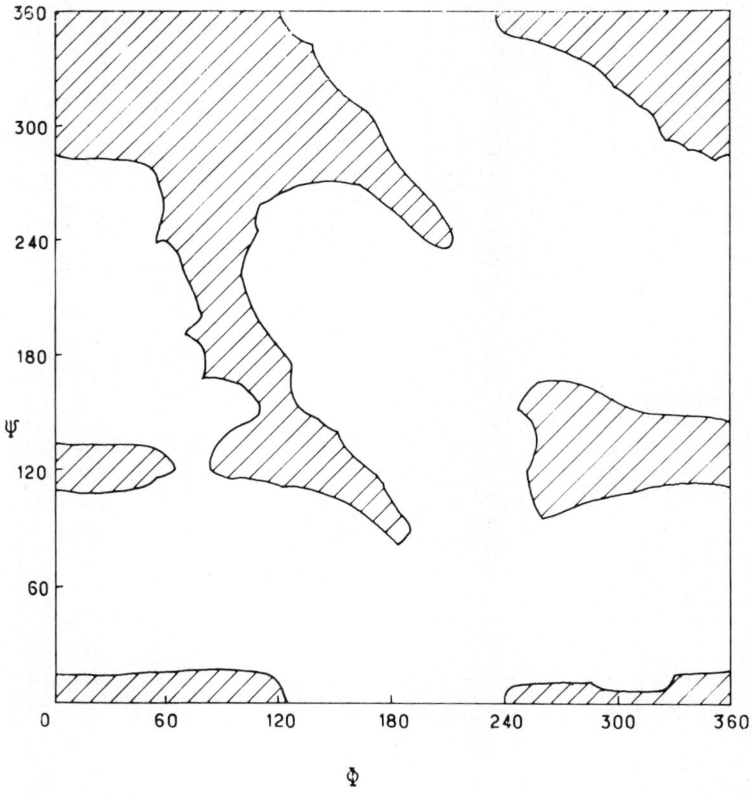

Figure 1-20. Lower limits of the allowed conformational space for the four aromatic residues (within 5 kcal/mole above the global minima) [Maigret *et al.* (1971)].

conformation is to be obtained, it will be necessary to take into consideration the interresidual interactions.

The results obtained in a semiempiricial calculation for nonapeptides show the difference between medium- and long-range interresidual interactions [Ponnuswamy *et al.* (1973)]. The molecule of egg-white lysozyme was divided into 14 nonapeptide segments and the energies of the 14 sets of tri-, penta-, hepta- and nona-peptides within these segments minimized with respect to all the backbone and side-chain dihedral angles of the same central residue within

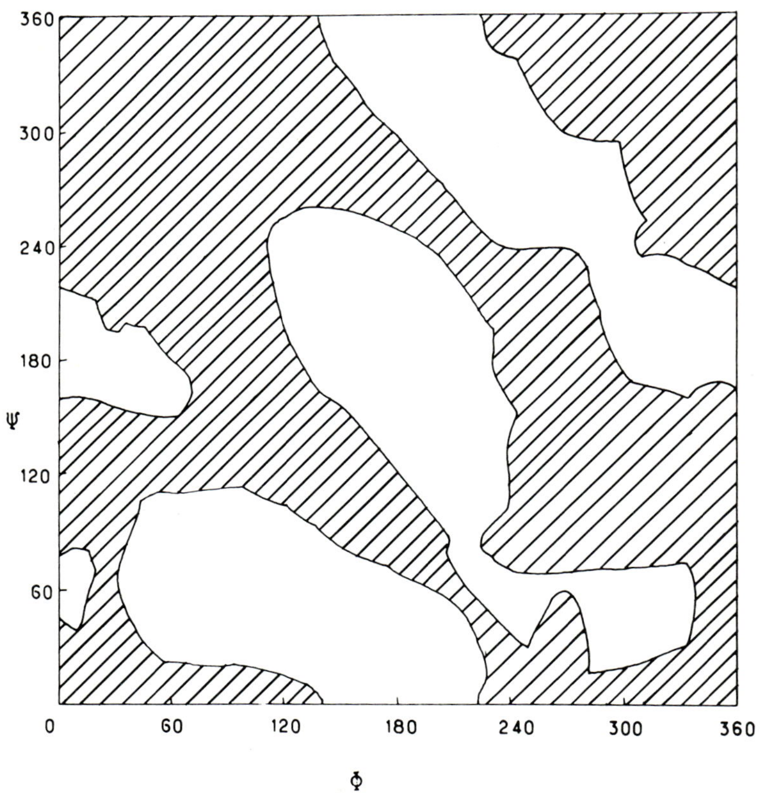

Figure 1-21. Upper limits of the allowed conformational space for the four aromatic residues (within 5 kcal/mole above the global minima) [Maigret *et al.* (1971)].

each set (Fig. 1-22), maintaining fixed the conformations of all the other residues. The starting backbone conformations were a right-handed α-helix, a left-handed α-helix, two in the β region (β_1 and β_2), an equatorial, seven-membered hydrogen-bonded ring (C_7) and the one corresponding to the X-ray structure (N). The results (Fig. 1-23) show that, as the length of the oligopeptide increases, the conformation obtained by minimization of the conformation N tends to become the preferred one.

The conclusion is that medium-range interresidual interactions, at least within a nonapeptide, determine (in conjunction with the intraresidual interactions) the conformation of a given residue in a protein in most cases, although not in all. Therefore, it would be expected that a rather satisfactory protein conformation might be obtained using the results obtained for amino acyl residues in appropriate nonapeptides, but such a proposal cannot be considered as very realistic and it will be necessary to consider a simplified model.

It has been observed that most of the turns (denoted as U-turns) present in proteins or in cyclic or linear oligo- and polypeptides are stabilized by an intramolecular hydrogen-bond involving the residues *i* and *i* + 3 (Fig. 1-24). Model *a priori* calculations were carried out for 'tripeptides' CH_3CO-X-Y-$NHCH_3$ (Fig. 1-25), X always being an *L*-alanyl residue and Y being either an *L*-alanyl, a *D*-alanyl, or a glycyl residue [Maigret and Pullman (1974)] and the competition between C_5, C_7, C_{10} rings and open structures was investigated through mapping of the whole (ϕ, ψ) conformational space and energy minimization. The results show that the stable 'tripeptide' conformations are simply combinations of 'dipeptide' stable conformations, with the exception of the C_{10} rings (which cannot exist in the dipeptide).

It is believed [Maigret and Pullman (1974)] that such calculations reflect

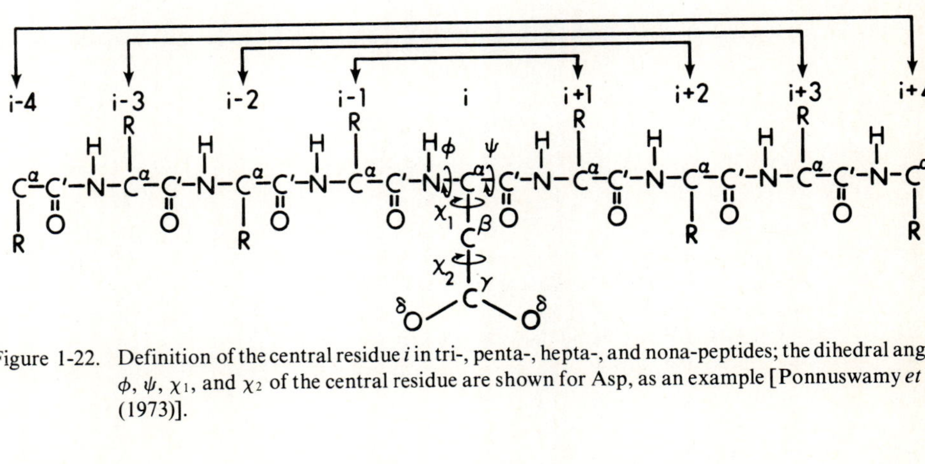

Figure 1-22. Definition of the central residue *i* in tri-, penta-, hepta-, and nona-peptides; the dihedral angles ϕ, ψ, χ_1, and χ_2 of the central residue are shown for Asp, as an example [Ponnuswamy *et al.* (1973)].

satisfactorily the short- and medium-range interresidual interactions in polypeptides and, therefore, one would tend to suggest that the results obtained for amino acyl residues in tripeptides could be used in the

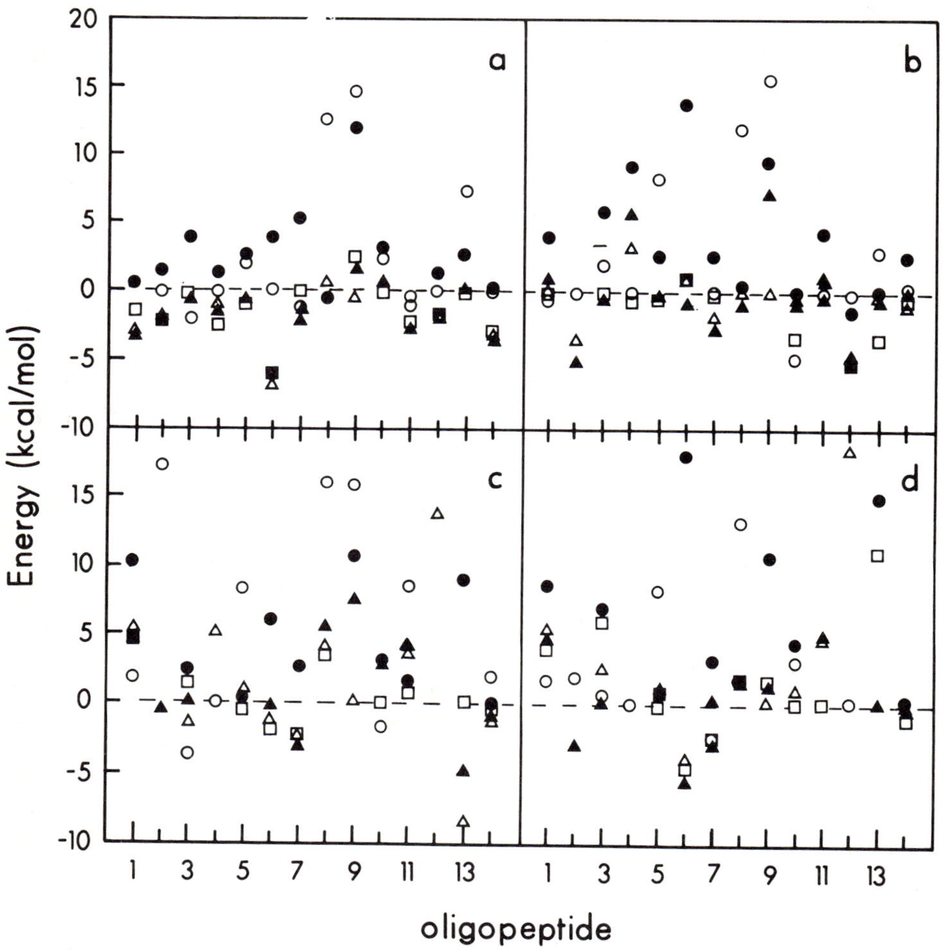

Figure 1-23. Relative minimized energies of oligopeptides for six conformations of the central residue *i* in 14 segments of lysozyme. The results in each box correspond to tri-, penta-, hepta-, and nona-peptides, respectively. The local conformations in the energy minimization were the right-handed α-helix (circle), left-handed α-helix (shaded circle), β_1 (triangle), β_2 (shaded triangle), C_7 (square), and N (dashed line) [Ponnuswamy *et al.* (1973)].

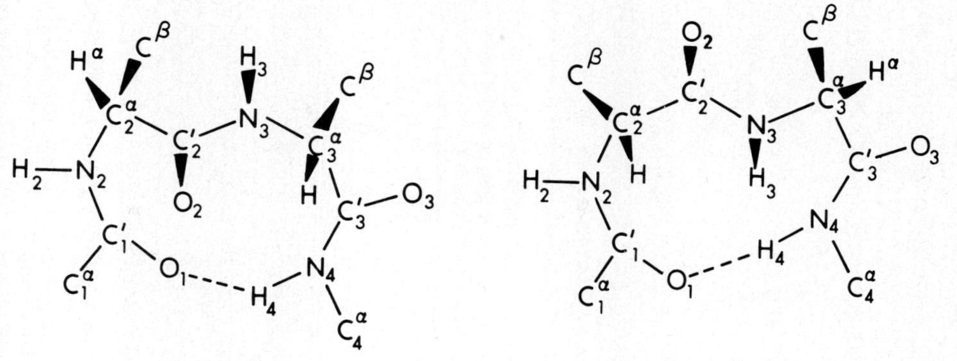

Figure 1-24. Types of U-turns in an *LL* sequence with (i/i +3) hydrogen-bonding corresponding to the conformation C_{10} [Maigret and Pullman (1974)].

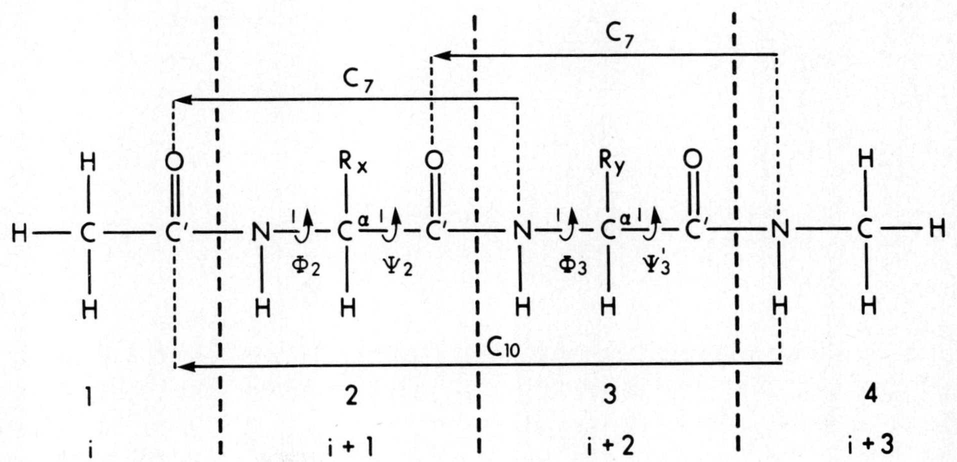

Figure 1-25. Schematic representation of the 'tripeptide' model [Maigret and Pullman (1974)].

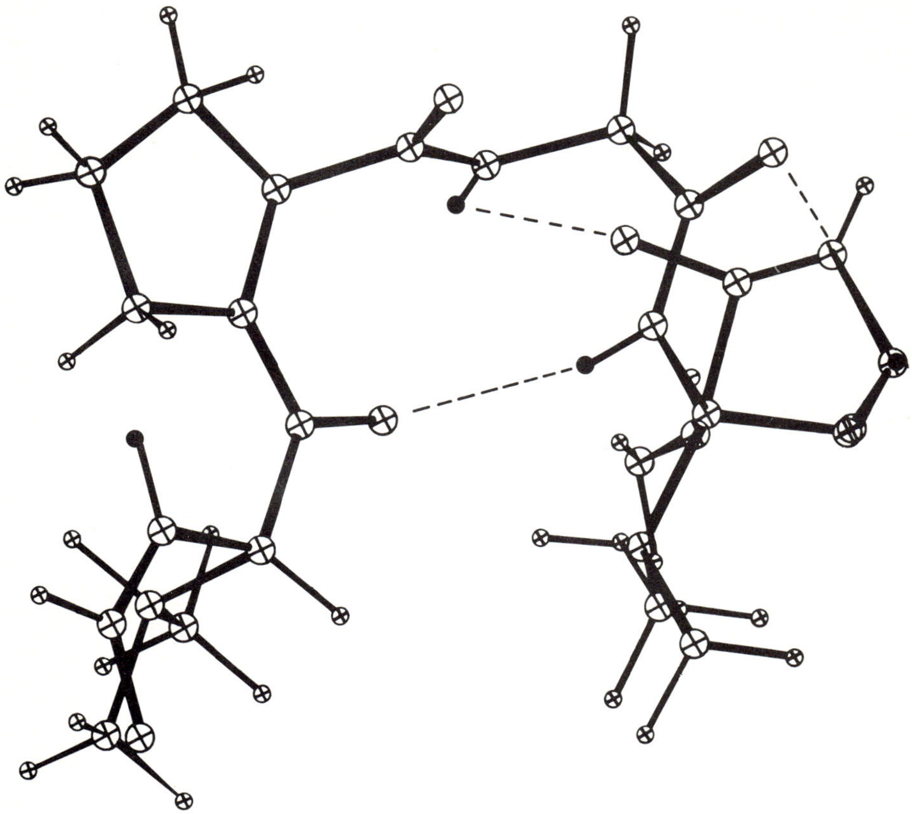

Figure 1-26. Preferred folded conformation of the system N-formyl-*L*-val-*L*-pro-gly-*L*-val-gly-OCH_3 [Renugopalakrishnan and Urry (1976)].

construction of polypeptides. There are, however, two problems associated with this suggestion: on one hand, the C_{10} ring that simulates the U-turn is not the most stable conformation, although it represents the most probable structure; on the other hand, it is doubtful that the conformations obtained for a given residue in different tripeptides would be identical, so that an 'averaging' would be required.

In any case, the resulting conformation must be refined by consideration of the long-range interactions in the polypeptide considered. Therefore, it would seem pertinent, after all, to use the results obtained for the amino acyl residues in dipeptides.

Such is the approach adopted in a semiempirical study of the repeat unit *L*-val-*L*-prol-gly-*L*-val-gly (blocked in the calculations by *N*-formyl and OCH_3 groups) of elastin [Renugopalakrishnan and Urry (1976)]. The net charges used for the residues were those from *a priori* calculations on *N*-acetyl *L*-valine amide, *N*-formyl *L*-proline amide, and *N*-acetyl glycine *N*-

methylamide (taking care of preserving the electrical neutrality of the assembled polypeptide). In the calculations, the dihedral angles were optimized systematically, starting from the fully extended conformation. It is interesting to note that the predicted preferred conformation (Fig. 1-26), in qualitative agreement with the experimental information, presents a C_{10} structure, with a non-linear hydrogen-bond between the two valyl residues.

INFORMATION AND RECOGNITION

Information has emerged, from the discussion given in this Chapter on the experimental observations and the theoretical interpretation on the development, evolution, and operation of the genetic code, as the fundamental concept underlying all the biological processes, whether generative or operative. The information grows during the evolution and, in each generation, is transmitted and utilized, all the processes implying a recognition of information.

The goal ahead is the prediction of the workings of the biological information. Three distinct points require attention in this consideration: what are the criteria for the measurement of recognition processes? What is the formulation associated to such criteria? How is the prediction carried out in practice?

These are the subjects of the three following chapters — recognition processes, interaction energies, and computational simulations — within the general framework of molecular information theory.

Molecular Information Theory

'Many are now racing to see who can be the first to calculate correctly the three-dimensional structure of an enzyme from its sequence.'

D.F. Bradley

The outcome of a recognition process is based on energetic considerations, the probability of recognition depending on the values of the stabilization energies of the various receptor-carrier complexes. In aqueous solution, the binding process is influenced by the solvent molecules through the contribution of the solvation energies and by binding to the receptor in competition with the information carriers.

Molecular information theory embodies all the pertinent theoretical formulations for the qualitative interpretation and the quantitative prediction of recognition processes. The practical details presented and the numerical information collected are those specifically related to recognition processes involving biopolymers in an aqueous environment.

The computational simulation of interactions and reactions, with the help of fast, large-capacity computers and interactive display terminals is already operative. More distant, but within reach, are such undertakings as the design of new enzymes, polypeptidic antibiotics, membranes, . . .

Chapter 2

Recognition Processes

Chemical (and biochemical) processes occur through the formation of molecular associations, due to the action of intermolecular forces.

At the lowest level, the formation of a molecular association involving two molecules may be considered to represent a simple recognition process, whereby the information stored in a molecule (carrier) is read by the other (receptor). In such a case, the efficiency of the reading may be correlated with the stability of the associate.

In the case when several carriers form simultaneously molecular associations with a given receptor (complex recognition), the definition must be changed in order to take into account the existence of such competing processes. The adoption of a given convention for the description of the overall process constitutes, therefore, the first step towards the development of a recognition theory for the quantitative prediction of complex recognition processes.

DEFINITIONS AND CONVENTIONS

The recognition process between two systems, R (the receptor) and C (the carrier), may be represented by the reaction

$$R + C \leftrightarrow RC$$

assuming a one-to-one stoichiometry. If the receptor may recognize more than one carrier, the possible processes (always assuming one-to-one stoichiometry) may be represented similarly by the equations

$$R + C_l \leftrightarrow RC_l$$

where the index l labels the various carriers.

The development of a method by which to measure the relative selectivity of the receptor towards the various carriers should be based on both thermodynamic and kinetic arguments. The difficulties are, of course, staggering and, consequently, it is necessary to look for possible simplifications. In this connection, it is worthwhile to take into account the existence of two types of recognition processes. In certain cases, the formation of the associate constitutes the end of the process (e.g., association antigen-antibody), but there are other instances where the molecular association represents only an intermediate step (e.g., enzymatic reactions where the substrate is recognized by the enzyme), followed by its dissociation into the receptor and the products of the reaction. These two kinds of processes, labelled as associative and reactive respectively, may be represented by the reactions (assuming one-to-one stoichiometry)

$$R + C_l \leftrightarrow RC_l \text{ (associative process)}$$
$$R + C_l \rightarrow R + \text{products (reactive process)}$$

with the added assumption that the associate, RC_l, constitutes the transition state in the reactive process, that is,

$$R + C_l \leftrightarrow RC_l \rightarrow R + \text{products}$$

The change in potential energy associated with either process may be 'represented' as an n-dimensional surface (hypersurface) in an (n+1)-variable space, the remaining n variables corresponding to the separation and relative orientations of the systems involved as well as the internal parameters defining their conformations.

It will be assumed here, in order to simplify the development of the theory, that equilibrium is reached in the formation of all the associates and that their dissociation (with equal probability in both directions in the case of reactive processes) is the rate-determining step in each case. Under such assumptions it is appropriate to define a thermodynamic probability for the associative processes and a kinetic probability for the reactive processes. The main characteristic of this convention is that both probabilities will be seen to be identical, given in terms of the equilibrium constants for the formation of the associates.

Probability of Recognition

In the absence of competing processes, there is a certainty that if the reaction

$$R + C \leftrightarrow RC$$

occurs, a recognition process characterized by the equilibrium constant

$$K = [RC]/[R]\,[C]$$

has taken place; [R], [C] and [RC] denote the activities of the sytems.

Whenever several processes are to be considered, two scales for the measurement of the selectivity may be used, depending on whether the corresponding processes are considered separately or simultaneously (that is, in competition).

For the processes

$$R + C_l \leftrightarrow RC_l$$

the corresponding equilibrium constants

$$K_l = [RC_l]/[R]\,[C_l]$$

could again be taken as a measure of the corresponding probabilities

$P_l = K_l$

of independent recognition. It must be taken into account, however, that for a given concentration of the receptor, the extent of the recognition will depend on the initial activity of the carrier considered, $[C_{l\,0}]$. Therefore, the definition of the (thermodynamic) probability of recognition should be changed to

$P_l = K_l\,[C_{l\,0}]$

Whenever the recognition processes take place in simultaneous competition, it is more appropriate to define a relative probability

$p_l = P_l/\Sigma P_q$

where the summation extends to all the carriers q; the above definitions imply a normalization of the probability, so that

$\Sigma p_q = 1$

Because of the difficulties in dealing with activities, hereafter they will be approximated by the corresponding concentrations (for which the same notation will be used). This approximation is certainly valid for dilute solutions and whenever systems of similar structure and concentrations are considered and ratios of activities are involved. In such a case, the expression

$p_l = K_l\,[C_{l\,0}]/\Sigma K_q\,[C_{q\,0}]$

may be simplified whenever it may be assumed that the initial concentration of every carrier is much larger than the concentration of the receptor. Under such an assumption one can write

$[C_{q\,0}] = [C_q]$ (for any q)

so that one obtains

$p_l = K_l\,[C_l]/\Sigma K_q\,[C_q] = [RC_l]/\Sigma[RC_q]$

(Hereafter, no distinction will be made between $[C_{q\,0}]$ and $[C_q]$ and, for simplicity, the latter will be used in all cases.)

By analogy with the discussion given above for the thermodynamic probabilities, one may define a (relative) kinetic probability of recognition by

$p'_l = k_l\,[C_l]/\Sigma k_q\,[C_q]$

where k_q stands for the rate constant of the overall reactive process.

Within the transition state theory, the rate constant, k_q, is proportional to the equilibrium constant, K_q, for the formation of the corresponding associate. Therefore, the above probability transforms (after simplification) into

$p'_l = K_l\,[C_l]/\Sigma K_q\,[C_q]$

which is identical to the expression for the thermodynamic probability, p_l. Consequently, the labels 'thermodynamic' and 'kinetic' will be dropped and the simplified designation 'probability of recognition' used thereafter.

Stabilization Energy

The expressions developed for the probabilities of recognition provide the foundations for an analysis of the conditions for a correct recognition as well as for the numerical prediction of such a process.

These objectives are more easily attained if those expressions are recast in terms of the energy changes that accompany the process. Such a transformation is based on the relationship between the equilibrium constants and the changes in free energy

$$\Delta G^{\circ}_{l} = -RT \ln K_{l}$$

where ΔG°_{l} represents the change in the standard Gibbs free energy associated with the reaction and R is the universal gas constant (not to be confused with the receptor R).

The change in the standard free energy, ΔG°_{l}, may be approximated by

$$\Delta G^{\circ}_{l} = -RT \ln(f(RC_{l})/f(R)\ f(C_{l})) + N\ \Delta E_{l}$$

where

$$f(RC_{l}) = \Sigma \exp(-E(RC_{l};i)/kT)$$
$$f(R) = \Sigma \exp(-E(R;i)/kT)$$
$$f(C_{l}) = \Sigma \exp(-E(C_{l};i)/kT)$$

represent the partition functions of the three systems, with the summations extending to all their respective states, including separate terms for all those (degenerate) states with the same energy; the energy levels, $E(RC_{l};i)$, $E(R;i)$, $E(C_{l};i)$ are referred to a common zero, which in this case corresponds to the energy of the ground state of the associate. N is the Avogadro number, k is the Boltzmann constant, and

$$\Delta E_{l} = E(RC_{l};0) - (E(R;0) + E(C_{l};0))$$

represents the so-called stabilization energy of the molecular association, evaluated from the ground state molecular energies ($i = 0$); it corresponds to the change in potential energy accompanying the formation of the associate from its components.

The *a priori* evaluation of the partition functions, with consideration of the translational, rotational, vibrational and electronic states is really not feasible in practice, with the possibility of errors in all the terms [Alagona *et al.* (1972)]. Therefore there is no recourse but to make use of approximations.

The usual approximation consists of assuming all the partition functions to be unity, in which case the expression of $\Delta G°$ becomes

$$\Delta G°_l = N\ \Delta E_l$$

and an extimate of

$$K_l = \exp(-N\ \Delta E_l / RT) = \exp(-\Delta E_l / kT)$$

will be available once ΔE_l has been evaluated. Substitution of the expressions for the equilibrium constants in the definition of the probability of recognition yields

$$p_l = K_l\ [C_l] / \Sigma K_q\ [C_q] = [C_l]\ \exp(-\Delta E_l / kT) / \Sigma [C_q]\ \exp(-\Delta E_q / kT)$$

which can be rewritten as

$$p_l = 1/(1 + \kappa_l)$$

with

$$\kappa_l = \Sigma'([C_q]/[C_l])\ \exp((\Delta E_l - \Delta E_q)/kT)$$

(with $q \neq l$).

It is possible, however, as a consequence of the structure of the expression for the recognition probability to arrive at the same result by an argument that obviates the need of invoking such a drastic approximation. First of all, the partition function for the receptor may be eliminated, as it appears in all the terms of both the numerator and denominator. On the other hand, whenever the carriers are all very similar (see below), it may be assumed that their partition functions are nearly identical; the partition functions of all the receptor-carrier associations, in such a case, would also be essentially identical. Elimination of all the partition functions leads then to the expression given above for p_l.

For a discussion on the influence of the molecular structure on the recognition process, it is convenient to rewrite the above expression in an alternate way. Taking into accound the definition of the stabilization energy, one can write

$$\begin{aligned}\Delta E_l - \Delta E_q &= E(RC_l;0) - E(R;0) - E(C_l;0) - E(RC_q;0) + E(R;0) + E(C_q;0)\\ &= E(RC_l;0) - E(RC_q;0) - E(C_l;0) + E(C_q;0)\end{aligned}$$

which simplifies to

$$\Delta E_l - \Delta E_q = E(RC_l;0) - E(RC_q;0)$$

whenever the molecular energies of the groundstates of all the carriers

$$E(C_l;0) \simeq E(C_q;0) \text{ (for any q)}$$

may be assumed to be identical. In such a case, one can write

$$\kappa_l = \Sigma'([C_q]/[C_l]) \exp((E(RC_l;0) - E(RC_q;0))/kT)$$

which reduces to

$$\kappa_l = \Sigma' \exp((E(RC_l;0) - E(RC_q;0))/kT)$$

when the concentrations of all the carriers are identical.

CONDITIONS FOR RECOGNITION

Because of the competitiveness of the processes, it is not possible to decide on the correct recognition process in terms of the probabilities just defined. In a simple approach, one might expect that the correct process would be that one involving the carrier C_l for which p_l is higher. This condition may be appropriate in some cases, but one should remember that the possibility

$$p_l < \Sigma' p_q$$

(with $q \neq l$) may occur, in which case one is bound to consider that the recognition process involves all the competing carriers.

Therefore, if the carrier C_l is to be 'recognized' by the receptor (i.e., if it is the 'correct' partner in the recognition process), the condition to be satisfied is

$$p_l > \Sigma' p_q$$

which can be written as

$$p_l > \kappa_l/(1 + \kappa_l)$$

taking into account the normalization condition for the probabilities. Recognition of the carrier C_l will be said to occur whenever

$$1 < p_l < 1/2.$$

The numerical prediction of a competitive recognition process may be based on the above expression and it implies the evaluation of the corresponding stabilization energies.

The technical details that by necessity play an important role in the formulation for such a prediction may obscure some points of interest. Consequently, it is convenient, first of all, to present a qualitative discussion regarding the molecular characteristics and the influence of the concentrations.

Characteristics of the Interacting Sites

When carrying out a calculation for a recognition process, the complete interaction between the two molecules must be taken into account (at the level of approximation considered). There is no doubt, however, that steric conditions, especially when biopolymers are involved, determine that the main contribution to the interaction energy will be made by specific regions in R and C, denoted hereafter as recognition and combining sites, respectively, or interacting sites when referring simultaneously to both of them. Although all types of interactions will be in operation, the stabilization of the associate at the equilibrium separation will be considered here to be mainly due to hydrogen-bonds between the interacting sites. A (hydrogen-bond) complementarity of both sites will be required for recognition as hydrogen-bonding is geometry-dependent. This consideration constitutes the basis for the study of the molecular characteristics of the interacting sites.

There are several possibilities, depending on the size of the receptor and the carriers. Usually, the receptor is a polymer, but the carriers may be either momomers or polymers. For the theoretical discussion, when considering the interaction between polymers, it is convenient to distinguish two cases depending on whether the interacting sites have the same size or not. In the former case, one may also consider the possibilities of equal or different number of monomers in the recognition and the combining sites; when the number of monomers is the same, there is the added distinction of equal or different numbers of hydrogen-bonding centers per monomer.

In the case of equal size of the interacting sites with the same number of monomers per site, there is a one-to-one correspondence between the coding units (monomers) and the theoretical study will yield information on the size (number of units) of the sites. In all other cases, the discussion must be based on the total number of hydrogen-bonds formed, as the problem is complicated by geometry considerations, specific for the systems under consideration.

The information carriers may be completely different systems or have similar structures, differing in such a case in the combining sites. The difference in the latter case may be due to the fact that different monomers are involved in the combining sites of the various carriers or to their sequence, if all the carriers have the same monomers in the combining site.

Given a sequence of monomers (coding units) in the recognition site, each characterized by the corresponding number of centers capable of forming hydrogen-bonds, the recognition site may then be defined by the sequence $[n_1, n_2, \ldots, n_p, \ldots, n_N]$, where n_p defines the maximum number of bonds that may be formed by the p-th unit of the recognition site, and N′ is the number of monomers in the site. Therefore, the total number of bonds that may be formed is then

$$B = \Sigma n_p$$

If a given carrier, l, is similarly characterized by the sequence $[n'_1, n'_2, \ldots, n'_p, \ldots, n'_N]$, the actual number of bonds formed between the receptor and the carrier (assuming the appropriate complementarity) will be

$$b_l = \Sigma \min(n_p, n'_p)$$

(where the summation runs over the number of units). For example, if the combining sites of the carriers are formed by the same three units, A, B, and C, differing in the sequence in which they appear, the possible number of carriers is $(N')! = 3! = 6$, characterized by the sequences ABC, ACB, BAC, BCA, CAB, CBA. If n'_1, n'_2, n'_3 denote the maximum number of bonds that A, B, and C may form, the bonds formed between the receptor and each carrier are

$$b_1 = \min(n_1, n'_1) + \min(n_2, n'_2) + \min(n_3, n'_3)$$
$$b_2 = \min(n_1, n'_1) + \min(n_2, n'_3) + \min(n_3, n'_2)$$
$$b_3 = \min(n_1, n'_2) + \min(n_2, n'_1) + \min(n_3, n'_3)$$
$$\ldots$$
$$b_6 = \min(n_1, n'_3) + \min(n_2, n'_2) + \min(n_3, n'_1)$$

In order to use this information in the evaluation of the recognition probabilities, it must be assumed that the energy per hydrogen-bond is independent of the hydrogen-donor and -acceptor involved. One can then approximate the interaction energy between the two systems by

$$\Delta E_l = -b_l \, \Delta e$$

where Δe represents the energy (in absolute value) per hydrogen-bond. Substitution in the expression for κ_l yields

$$\kappa_l = \Sigma'([C_q]/[C_l]) \exp((b_q - b_l) \Delta e/kT)$$

Assuming the concentrations of all the carriers to be identical, one can then write

$$p_l = (1 + \Sigma' \exp((b_q - b_l) \Delta e/kT))^{-1}$$

which can be used for the determination of the allowed differences in the number of bonds formed by the various competing carriers and the size of the interacting sites.

First of all, one can see that whenever

$$b_l = b_q \text{ (for any q)}$$

one has

$$p_l = 1/N$$

where N is the total number of carriers considered. That is, when all the carriers form the same number of bonds, the recognition of any carrier is equally probable at any temperature.

On the other hand, for a more general discussion, taking into account that it must be

$$p_l > 1/2$$

one can write

$$(1 + \Sigma' \exp((b_q - b_l)\, \Delta e / kT))^{-1} > 1/2$$

or, alternatively,

$$\Sigma' \exp((b_q - b_l)\, \Delta e / kT) < 1$$

Inspection of this expression shows that in order for the carrier C_l to be the correct partner, it must be

$$b_l > b_q \text{ (for any q)}$$

That is, the correct carrier is the one capable of forming the higher number of bonds with the receptor.

Restricting the discussion to carriers differing in the sequence of the combining sites, formed by the same set of N′ units, and classifying the incorrect carriers into groups according to the number of bonds that they form, one can write

$$\Sigma\, N_{-t} = N - 1 = (N')! - 1$$

where N_{-t} denotes the number of carriers that form t bonds less than the correct carrier; the summation over *t* runs from 1 to $b_l - 1$. Then, the above inequality can be written (by changing the summation over *q* to a summation over *t*) as

$$\Sigma\, N_{-t} \exp(-\, t\, \Delta e / kT) < 1$$

which shows an interesting feature of the recognition process. On one hand, the stability of the association between the receptor and the carrier increases with the number of hydrogen-bonds, or, equivalently, for the case considered, with the number of monomers in the interacting sites. On the other hand, as the number of monomers increases, so does the number of incorrect carriers, with a corresponding increase in the possibility of reading errors. The above expansion expresses mathematically the compromise between the two conditions.

Such an expression may be used in the estimation of the optimal size of the interacting sites. Assuming, for example, that all the incorrect carriers form a single group with all of them forming *t* bonds less than the correct carrier, one has

$$N_{-t} \exp(-\, t\, \Delta e / kT) < 1$$

or

$$N_{-t} = (N')! - 1 < \exp(t\, \Delta e / kT)$$

so that

$$(N')! < 1 + \exp(t\, \Delta e / kT)$$

Adopting $R \simeq 2$ cal/°K mole, at 300°K one obtains

$$(N')! < 1 + \exp(t\ \Delta e/600)$$

which yields the limiting values of N′ in Table 2-1 for various choices of *t* and Δe. Thus the values given for Δe = 5 kcal/mole indicate that the interacting site could have a maximum of 6, 10, or 13 monomers depending on whether all the incorrect carriers would form 1, 2, or 3 bonds less than the correct partner.

Table 2-1. Estimates of the size (N′) of the interacting sites*

Δe (kcal/mole)	t = 1	t = 2	t = 3
2	4	6	7
3	5	7	9
4	6	9	12
5	6	10	13

* The values in each column have been obtained assuming that all the incorrect partners form t bonds less than the correct carrier.

One can proceed in a similar way when there are several types (in terms of number of bonds formed) of incorrect carriers, in the same or different ratios. For example, if half of the incorrect carriers form one hydrogen-bond less than the correct carrier and the other half two bonds less, than one can write

$$N_{-1} \exp(-\Delta e/600) + N_{-2} \exp(-2\ \Delta e/600) < 1$$

Taking into account that in this case

$$N_{-1} = N_{-2} = (1/2)\ ((N')! - 1)$$

one obtains

$$(N')! < 1 + 2/(\exp(-\Delta e/600) + \exp(-2\ \Delta e/600))$$

so that with Δe = 5 kcal/mole, one has

$$(N')! < 8318$$

and N′ = 7; that is, the combining site should be formed by seven units.

The main point, in this connection, is that values obtained from the above condition fall within the ranges determined experimentally. Therefore, it may be interesting to try and apply it to a more relevant case, although at this point the biochemical consequences will not be considered.

The case to be investigated is one where the interacting sites are formed by three monomers (with the possibility of repetition), chosen from the set A, B, C, and D; A and D recognize each other, forming two bonds, and B and C recognize each other with formation of three hydrogen-bonds, but there is no recognition between the members of the two subsets. Given a sequence XYZ (where X, Y, Z can be any of A, B, C, D) for the recognition site, the theoretical calculation implies the evaluation of the recognition probabilities for the 64 possible carriers that may be formed by the permutations (with repetition) of the four monomers taken three at a time. From the values of the probabilities one can decide in each case which one (if any) is the correct partner, as well as the corresponding recognition error. In each case, the calculations yield, for the combining site, the monomer sequence complementary (with maximum number of bonds) to that of the recognition site: for example, the recognition site ABC recognizes the sequence DCB. The results corroborate the initial hypothesis of a relationship between correct recognition and high number of bonds formed between the receptor and the carrier.

Unfortunately, even the smallest change in the conditions originally assumed disrupts profoundly such a symmetric and well-behaved picture. For example, a possible case of interest is that where D, in addition to binding A, may also form two bonds with C. In such a case, it is observed that the monomers B and C are not equivalent any longer, B retaining its decisive character insofar as recognition is concerned. The second point is that no recognition is achieved whenever the monomer D is present; a receptor sequence, say ABD, would recognize equally the combining sequences DCA and DCC. The results indicate that, under the assumptions adopted, a recognition code based on sequences of these four monomers will not work, in the sense that a given receptor will recognize more than one carrier. It may be worthwhile to emphasize at this point that one of the most crucial assumptions made here is that one relating the recognition exclusively to the interaction (involving only hydrogen-bonding) between the recognition and combining sites without any consideration to all the remaining interactions (particularly those involving the regions adjacent to the interacting sites).

Although a more general analysis of the problem of the size of the interacting sites may be, in principle, of interest [Simon (1968, 1972)], it will not be pursued in this work, more concerned with the correlation between recognition and stabilization energy and the evaluation of the latter.

Influence of the Concentrations

The assumptions made above regarding the concentrations represent a rather artificial constraint in some cases. In fact, concentrations play in some instances (e.g., recognition of drugs) a rather decisive role and, consequently, their effect must be investigated.

For simplicity one may consider the case of only two carriers, l and m, with the assumption that their activity coefficients are similar, so that the ratio of their activities may be approximated by the ratio of their concentrations. The condition for correct recognition of carrier l may be rewritten as

$$[C_m]/[C_l] < \exp((b_l - b_m)\,\Delta e/kT)$$

or

$[C_m]/[C_l] < \exp(t\ \Delta e/kT)$

where t is the difference of the number of bonds in the associate RC_l minus the number of bonds formed in the associate RC_m.

Table 2-2 presents the upper bounds for the ratio $[C_m]/[C_l]$, corresponding to $p_l = 1/2$, with $R \simeq 2$ cal/°K mole and T = 300° K, for various values of Δe and t, while Table 2-3 collects the upper bounds corresponding to various degrees of recognition. It becomes evident from these results that the formation of just one bond less by C_m is a sufficient condition for the correct recognition of C_l, unless C_m is present in extremely large concentrations.

Table 2-2. Upper bounds of the ratio $[C_m]/[C_l]$ for correct recognition*

Δe(kcal/mole)	*t = 1*	*t = 2*	*t = 3*
2	2.80(1)	7.85(2)	2.20(4)
3	1.48(2)	2.20(4)	3.27(6)
4	7.85(2)	6.17(5)	4.85(8)
5	2.08(3)	1.73(7)	7.20(10)

* The numbers in parentheses denote the powers of ten.

Table 2-3. Upper bounds of the ratio $[C_m]/[C_l]$ for various values of the recognition probability*

recognition probability (%)	*ratio*
50	2080
60	1664
75	1040
80	832
90	416

* At 300° K, with $\Delta e = 5$ kcal/mole and $R \simeq 2$ cal/°Kmole, for the case $t = 1$.

The situation is different, of course, when both C_l and C_m form the same number of bonds with the receptor. The recognition condition with $t = 0$ becomes

$[C_l] > [C_m]$

and the recognition is governed absolutely by the concentrations.

THE DETERMINATION OF STABILIZATION ENERGIES

As discussed above, the quantitative prediction of a recognition process requires a knowledge of the stabilization energies for the associations of the various carriers with the receptor, to be evaluated from the corresponding ground state molecular energies for the preferred conformations.

That is, one is faced with the study of the corresponding potential energy hypersurfaces, involving their determination and the search for their stationary points. The pertinent formulation for the evaluation of the potential energy values and the practical details of such calculations (including the necessary minimization techniques) are presented in Chapters 3 and 4 respectively, but first some comments of a general nature are in order. The points to be considered are the role of the solvent, the reference states, and the topological characteristics of the hypersurface.

Solvation Effects

The solvating medium affects the recognition probabilities in two ways: on one hand, there is an explicit dependence taken into account through the inclusion of the concentration of the systems involved in the expressions of the probabilities; on the other hand, solvation affects the values of the probabilities through the stabilization energies. The influence of the concentrations has already been discussed and now the general problem of solvation must be examined.

In this connection there are again two distinct manifestations of the influence of solvation. The direct dependence of the stabilization energies on the solvation manifests itself in the following way: the stabilization energies must be evaluated from the energies of the solvated receptor-carrier aggregate, the solvated receptor, and the solvated carrier. This is a point that figures preeminently in the following chapters, both in terms of methods (Monte Carlo techniques) and practical details (expansions for the evaluation of the interaction of biomolecules of interest with water).

There is, however, an additional involvement of the solvent in the recognition process: the solvent itself may be considered as an information carrier, insofar as solvation of the receptor implies a recognition of the solvent molecules. Therefore, the solvent should be considered, for the evaluation of the recognition probabilities, as one more carrier competing for recognition by the receptor.

This participation of the solvent may account for the distinction between correct and wrong information carriers. In a simple interpretation, one may assume that such a distinction may arise simply because the stabilities of the various associations may be in the order solvated receptor-correct carrier > solvated receptor > solvated receptor-wrong carrier. That is, the solvent is displaced from its association with the receptor by the correct carrier, but, on the other hand, the solvent displaces the wrong carriers [Bradley (1970)].

It is evident, therefore, that the solvent makes, through different roles, a rather decisive contribution to recognition processes and must be taken into account in comparative studies. It is in this connection that the concept of binding energy must be introduced through the definition of reference states.

Reference States

The use of stabilization energies in terms of energies of the solvated species is adequate, both formally and numerically, when dealing exclusively with the quantitative prediction of recognition processes (i.e., with the evaluation of recognition probabilities). There are cases, however, when a distinction between the proper binding energy and other concurrent effects may be necessary; such is the case when the comparative binding characteristics of structurally related drugs are studied.

It has been assumed implicitly throughout the discussion that the receptor and carrier molecules exist and interact with the solvent. There are cases, however, where the receptor is embedded in the lipoidal membrane phase and the carrier exists in both an aqueous and the lipoidal phase; in such cases, the hypothetical reference states, insofar as binding is concerned, may be chosen as to correspond to the lipid-solvated molecules in the non-aqueous phase, unsolvated over the relevant surfaces [Davies (1977)].

For a discussion of the binding energies, independent of the energy changes associated with solvation effects, the following hypothetical process (Fig. 2-1) may be considered [Davies (1977)]:

a) the carrier (drug) is transferred from the aqueous to the lipoidal phase, with an energy change ΔE_p;

b) the lipid is removed from the carrier relevant surface, leaving a cavity or small set of cavities in the lipoidal phase, with an energy change ΔE_{ph};

c) the lipid is removed from the receptor relevant surface with an energy change ΔE_{RSh};

d) the lipid removed in the preceding step is introduced into the matching cavities created in step *b*, with an energy change ΔE_{SSh};

e) the carrier molecule is bound to the receptor surface with an energy change ΔE_{DRh}.

Therefore, the observed stabilization energy referred to the aqueous phase is given by

$$\Delta E_{obs} = \Delta E_p - \Delta E_{ph} + \Delta E_{DRh} - \Delta E_{RSh} + \Delta E_{SSh}$$

Assuming that the last two terms cancel each other and rearranging one can write

$$\Delta E_{DRh} = \Delta E_{obs} - \Delta E_p + \Delta E_{ph}$$

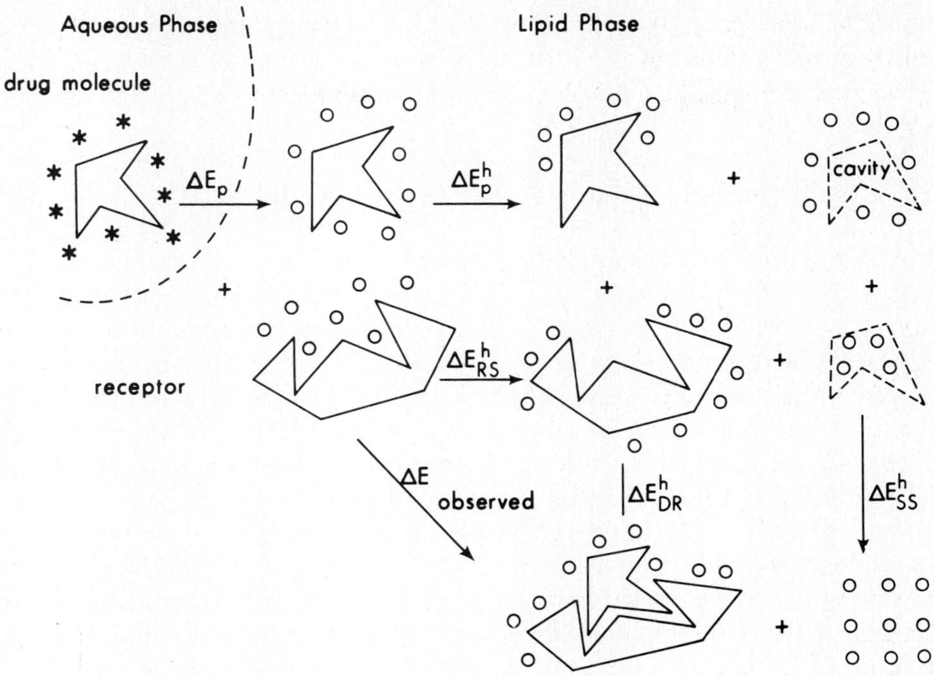

Figure 2-1. Schematic representation of drug-receptor binding. The superscript 'h' is used to indicate hypothetical energies of the process [Davies (1977)].

Considering two structurally related carriers, *l* and *m*, the difference between their binding energies to the receptor may be written as

$$\Delta(\Delta E_{DRh}) = \Delta(\Delta E_{obs}) - \Delta E_{p\,l} + \Delta E_{ph\,l} + (\Delta E_{p\,m} - \Delta E_{ph\,m})$$

which transforms into

$$\Delta(\Delta E_{obs}) = \Delta E_{p\,l} - \Delta E_{p\,m} - \Delta E_{ph\,l} + \Delta E_{ph\,m}$$

whenever their binding energies are essentially identical.

If their combining surfaces are similar, it may be assumed that

$$\Delta E_{ph\,l} = \Delta E_{ph\,m}$$

so that the difference in the observed binding energies

$$\Delta(\Delta E_{obs}) = \Delta E_{p\,l} - \Delta E_{p\,m}$$

is simply a function of the partitioning effect (between the aqueous and the lipoidal phases): if the two carriers differ in some substituent groups outside the combining areas, the change in the observed binding energy depends essentially on the solvation effects of those groups.

On the other hand, if the two carriers differ in their combining sites (e.g., by substitution in that region), it is not possible to give an interpretation of the changes in the binding energy, as $\Delta E_{ph\,l}$ will not necessarily be equal to $\Delta E_{ph\,m}$ any longer.

Topological Characteristics of Energy Hypersurfaces

Although an N-dimensional hypersurface cannot be visualized (except in the case of N = 2), it is possible to obtain its representation in one dimension [Krivoshey and Sleta (1976)]. Each separate contour (component) of an equipotential level of a hypersurface is ascribed a point in the one-dimensional representation, and the points ordered according to the values of the energy; the resulting figure is a 'tree'. In such a representation, all the differential properties are ignored but the topological properties of the hypersurface are retained. A tree may consist of several end-points and branch-points, their vertex orders (i.e., the number of branches intersecting at one point) being 1 and $\geqslant 3$, respectively. The stationary points of the hypersurface (local maxima and minima for the conformations of the reactants and the products, saddle points for the activated complexes) are represented by end-points (maxima and minima) and branch-points (saddle points). Representative examples of trees are presented in Figs. 2-2 and 2-3 for the ortho-para hydrogen transition

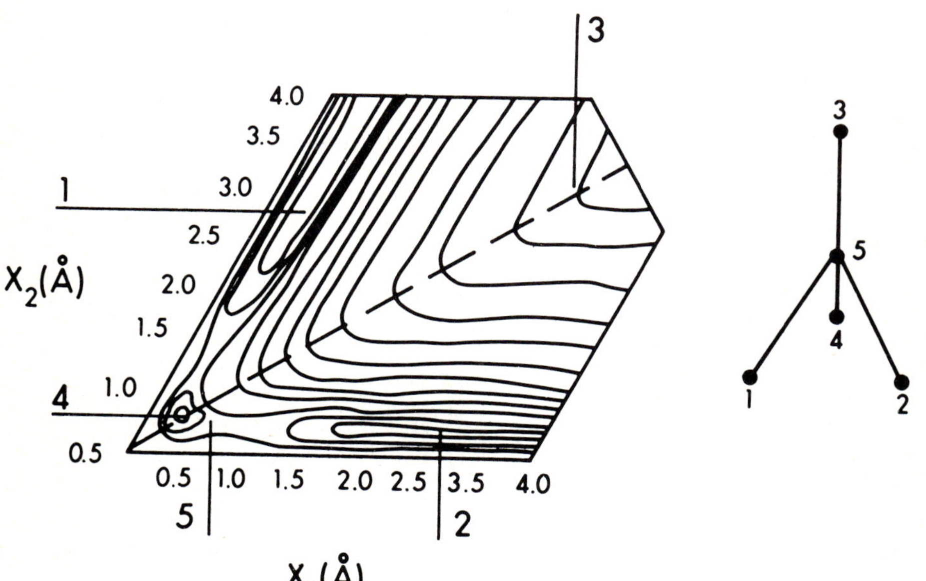

Figure 2-2. Potential energy surface and tree representation for the ortho-para hydrogen transition [Krivoshey and Sleta (1976)].

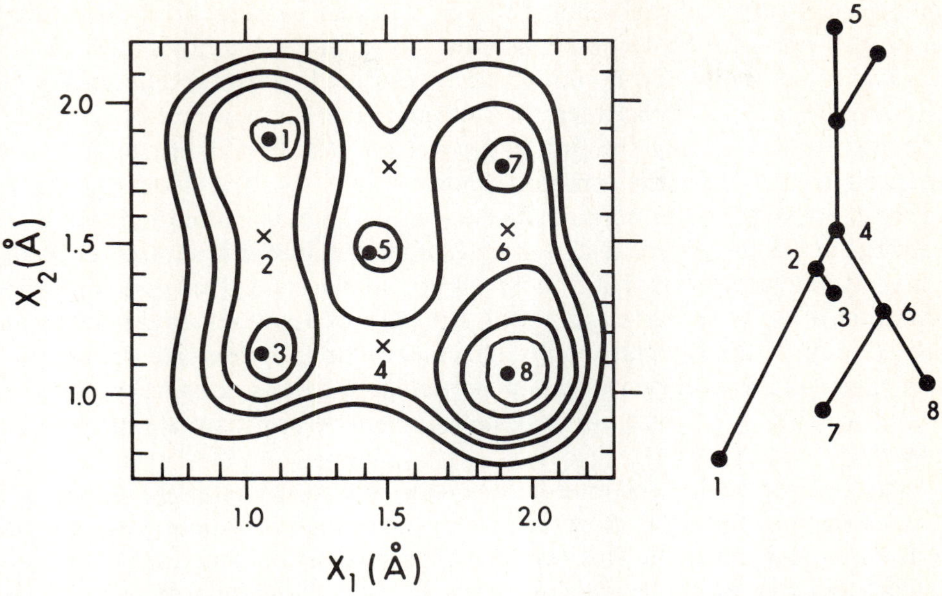

Figure 2-3. Potential energy surface and tree for the tautomeric rearrangement of the guanine-cytosine pair [Krivoshey and Sleta (1976)].

and the tautomeric rearrangement of guanine-cytosine, respectively.

In a chemical reaction, the representative point moves on the hypersurface along a path between the original conformation of the reactants and the final conformation of the products. All the points of this path belong to constant-level contours and, since every component of each level is represented on the tree, every path along the potential energy hypersurface will have its corresponding path from one tree point to another.

The value of the tree representation arises from the possible savings in computation work, in connection with the expensive evaluation of energy values and the difficulties in the minimum search. An inspection of the guanine-cytosine tree, with comparison with the corresponding two-dimensional map, leads to some interesting conclusions:

a) the path between points 3 and 1 follows the branches of the tree;

b) the path between points 3 and 8 goes directly from 3 to 4 and then to 8, without going through points 2 or 6;

c) the path between points 3 and 7 goes from 3 to 4, then to 8, and then proceeds along the branches, going through 6.

Generalization from these observations leads to the following procedures for the determination of the paths to be followed in order to get from one end-point to another:

1) end-points separated by only one branch-point: a) go from the first end-

point to the common branch-point along the branch joining them; b) go from the branch-point to the second end-point along the branch joining them;

2) end-points separated by more than one branch-point: a) determine the intermediate end-points between the two end-points considered; b) go to the branch-point connecting the first end-point and the first intermediate end-point, directly or along the common branch (if it exists); c) go from the above branch-point to the first intermediate end-point, directly or along the common branch (if it exists); d) repeat steps *b* and *c* until the final end-point is reached.

It is evident, therefore, that if the tree were known for a given case (e.g., the formation of a molecular association), the path of approach could be found without any difficulty. Although the construction of the tree requires a knowledge of the energy values (in particular, of the stationary points), a limited number of calculations may be sufficient to approximate its general shape.

Given two molecules, the interaction energy is a function, $E=E(R, \theta, \phi, \alpha, \beta, \gamma)$, of the position (R, θ, ϕ) of the cartesian system of coordinates associated with the second molecule with respect to the system associated with the first molecule, as well as of the orientation (α, β, γ) of the second molecule with respect to its associated system. At large separations, the effect of the orientation defined by α, β, γ, will be small, so that, for a given value of R, it is sufficient to determine the intervals of existence of θ and ϕ corresponding to a given value of the energy. Those intervals define the components of the energy level and, therefore, points on different branches of the tree representation. Repetition of the process at various distances would provide sufficient information for a rough representation of the tree.

The analytical determination of those ranges would constitute the most satisfactory procedure, but because of the difficulties to be encountered, a numerical approach may be more appropriate in practice. It is realized that, in such a case, one is carrying out simply a two-variable mapping, but supplemented with the conceptual simplicity of the tree representation. This approach might be used in order to decide the most appropriate paths that would lead from the initial situation to the molecular aggregate.

Chapter 3

Interaction Energies

The quantitative prediction of a recognition process requires, as discussed in the preceding Chapter, the evaluation of the corresponding stabilization energy, ΔE. Such calculations, in accurate treatments, are extremely complicated but predictive results may be obtained at various levels of approximation. A complete discussion is therefore convenient in order to obtain a proper perspective of the difficulties and possibilities of theoretical treatments, as well as of the validity of their results. But before proceeding with such a discussion, the general problem must be outlined so that the reasons for the procedure to be followed will be understood.

Two comments must be made at this point. An extrapolation of past experience in the field of theoretical calculations (see Chapter 1) supports the contention (pessimistic as it may be) that accurate calculations for large biomolecules and for their interactions will not be possible. Nevertheless, for completeness, one might feel tempted to present the formulation of those methods that some day might be useful in this connection. But, for simplicity and brevity and with practical considerations in mind, it has been decided to restrict the discussion to those points of relevance in the foreseeable future. Theoretical data for the separate systems will be required for the application of the resulting formulation and, in principle, it would be necessary to describe the methods by which they are obtained. However, because the theoretical treatments for such systems have been repeatedly discussed in the literature, those details are omitted here in order to avoid an unnecessary repetition, and a knowledge of quantum-mechanical concepts at the theoretical and practical levels will be assumed. It must be emphasized, however, that such a knowledge is not a *sine qua non* condition for the practical application of the formulation, to be presented here, to biochemical problems. [In any case, the reader is referred to the works of Carbo and Riera (1978) and Czismadia (1976) for details on the general theory, formulation of the self-consistent field method (at the various levels of approximation), practical considerations, representative examples of calculations, and characteristics and availability of existing programs.]

GENERAL OUTLINE

The designation 'recognition process' has been used, so far, with a connotation of a one-to-one relationship, even when considering competing process; and the evaluation of ΔE has been understood within the same context. This implication will be retained for simplicity in the wording throughout the text, but consideration must be taken of the fact that biochemical processes occur in aqueous solution where interactions between large numbers of molecules exist.

Traditional quantum-mechanical methods are not adequate for such a situation, and other procedures (e.g., Monte Carlo techniques) must be used. The Monte Carlo procedure makes use of data (interaction energy) for bimolecular aggregates. And, therefore, the complete study of a recognition process will involve the evaluation of the potential energy for bimolecular associates (involving the receptor, the carriers, and the water) and the determination of the potential and internal energies of the receptor, the carriers, and the various receptor-carrier systems in water clusters (in an approximation to the real situation in aqueous solutions). These three steps, to be discussed below, represent the levels of approximation, that are possible.

BIMOLECULAR ASSOCIATES

General Theory

The theoretical study of the formation of the molecular association implies the determination of the corresponding energy hypersurface as a function of the conformational variables: that is, the intermolecular separation and relative orientation of the two systems as well as the internal variables, in order to allow for their complete freedom in adopting the optimum conformation at each point. The energy hypersurface may be defined in terms of the molecular energy for the complete system or of the interaction energy

$$\Delta E = \langle AB|H_{AB}|AB\rangle - \langle A|H_A|A\rangle - \langle B|H_B|B\rangle = E_{AB} - E_A - E_B$$

where $|A\rangle$, $|B\rangle$, and $|AB\rangle$ represent the electronic ground state functions of the systems and of their association, respectively, and E_A, E_B, and E_{AB} are the corresponding molecular energies. The molecular hamiltonian operators for the independent units, H_A and H_B, and for the complete system, H_{AB}, are given by

$$H_A = -(1/2)\Sigma\Delta_i - \Sigma\Sigma Z_a/R_{ia} + (1/2)\Sigma\Sigma' 1/R_{ij} + (1/2)\Sigma\Sigma' Z_a Z_{a'}/R_{aa'}$$

$$H_B = -(1/2)\Sigma\Delta_i - \Sigma\Sigma Z_b/R_{ib} + (1/2)\Sigma\Sigma' 1/R_{ij} + (1/2)\Sigma\Sigma' Z_b Z_{b'}/R_{bb'}$$

$$H_{AB} = -(1/2)\Sigma\Delta_i - \Sigma\Sigma Z_a/R_{ia} - \Sigma\Sigma Z_b/R_{ib} + (1/2)\Sigma\Sigma' 1/R_{ij} + (1/2)\Sigma\Sigma' Z_c Z_{c'}/R_{cc'}$$

where the summations over i, j extend to all the electrons in the system considered, those for *a* and *b* run over the nuclei, with nuclear charges Z_a and

Z_b, of A and B respectively, and c and c' stand for a and/or b. The terms in these hamiltonians correspond to the kinetic energy of the electrons, the nuclear attraction energy of the electrons by the nuclei, the electronic repulsion energy, and the nuclear repulsion energy; Δ_i denotes the kinetic energy operator for the i-th electron and R_{ij}, R_{ia},... stand for the distance between the two particles involved.

The evaluation of $|AB>$ and E(AB) should be carried out by a configuration interaction (CI) or multi-configuration Hartree-Fock (MCHF) treatment so that for a sufficiently large number of configurations ΔE would approach the exact value (when equivalent CI or MCHF values are used for E_A and E_B). Until such extensive calculations will be feasible, the self-consistent field (SCF) method at the monoconfigurational level within the expansion approximation may be used. In any case, but especially so for accurate calculations (with large basis sets), the practical problem of the determination of an extremely large number of integrals must be faced unless A and B have a small number of electrons.

An important factor to be taken into account in calculations of this type is that of the accuracy of the results. Normally, the values of the electronic energies will be obtained with, say, eight significant figures. Consequently, for sufficiently large systems, the energies (in hartrees) may have only two or three significant decimals and the error in E may be of the order of 1 kcal or larger. This condition should be examined in each case in order to ascertain whether the results have any meaning at all.

In addition, other factors dependent on the level of approximation affect considerably the results, even at the Hartree-Fock level. One can write (not including the nuclear repulsion terms for simplicity)

$$\Delta E_X = \Delta E_{HF} + \Delta E_C$$

where ΔE_X, ΔE_{HF}, and ΔE_C are the differences in the exact electronic, Hartree-Fock, and correlation energies of AB and the isolated molecules, respectively. Therefore, the validity of the discussion given on the basis of the Hartree-Fock results for all the systems depends on the magnitude of ΔE_C. Even so, this situation might be acceptable but, in practice, one is still a long way from having Hartree-Fock results, especially for such large systems as those of biological interest. In most cases, one can hope to have, at best, *a priori* SCF results obtained with limited basis sets with the corresponding uncertainty as to whether the errors in E_{AB}, E_A and E_B are comparable or not.

From these comments it becomes evident that, except for very accurate treatments such as CI or MCHF, the results of even expensive calculations may not always be absolutely reliable. It is for that reason that one should consider whether more economical approaches may be, in some cases, both valid and useful.

Whenever the associate must be considered as a true molecule, there would seem to be no alternative, if accurate results are required, but to carry out a treatment as discussed above, preferably and if possible at a level of approximation higher than Hartree-Fock. There are situations (at larger separations), however, where the components of the aggregate retain to a considerable extent their identity; in such cases a perturbational approach within the framework of the self-consistent field theory may be adequate. In fact, as it will be seen below, the resulting formulation may be modified for its application in approximate treatments to short and medium separations and even to the conformational analysis of biopolymers.

The basic approximation, implicit in the above statement, is that the total function of the associate may be constructed as the product of the functions of the individual molecules, mutually perturbed by the interaction with each other. The total electronic energy may then be expressed in terms of the (perturbed) electronic energies of the separate molecules (expressed in terms of their perturbed orbitals) and a term corresponding to the expectation value, for those functions, of the operator

$$- \Sigma\Sigma Z_b / R_{ib} - \Sigma\Sigma Z_a / R_{ja} + (1/2)\Sigma\Sigma 1 / R_{ij}$$

where the summations extend over the electrons, *i* and *j*, and nuclei, *a* and *b*, of A and B, respectively.

The SCF formulation for the evaluation of the interaction energy is straightforward and the application of perturbation theory to the resulting eigenvalue equation does not offer any difficulty either. The question to be considered, and which has already been partially discussed above, concerns the practical applicability of the resulting formalism. There is no doubt of its usefulness for small systems, that is, molecules with limited numbers of electrons and nuclei. But computational times for real biomolecules (especially biopolymers) would impose too heavy a load on any computing facility. If anything is to be done with some chance of practical feasibility, a simplifying approximation must be introduced.

Electrostatic Molecular Potentials

Retaining the separability of the total function, as introduced above, the most logical and immediate approximation consists of neglecting the perturbation of each molecule on the other.

That is, the electronic interaction energy corresponds then to a coulombic interaction between the unperturbed charge densities of A and B and may be evaluated by

$$\Delta E'_{AB} = \int\int \rho(A;\mathbf{R}_1)\, \rho(B;\mathbf{R}_2)\, (1/|\mathbf{R}_1 - \mathbf{R}_2|)\, dv_1\, dv_2 \\ - \Sigma Z_a \int \rho(B;\mathbf{R}_2)\, (1/|\mathbf{R}_2 - \mathbf{R}_a|)\, dv_2 - \Sigma Z_b \int \rho(A;\mathbf{R}_1)\, (1/|\mathbf{R}_1 - \mathbf{R}_b|)\, dv_1$$

where, e.g., $\rho(A;\mathbf{R}_1)$ represents the total electronic density, at the point defined by the position vector $\mathbf{R}_1$, evaluated from the orbitals of the unperturbed molecule A. Introducing the nuclear repulsion energy term, the resulting expression may be rearranged [Bonaccorsi *et al.* (1971)] as

$$\Delta E_{AB} = \int\int(\rho(A;\mathbf{R}_1) - \Sigma Z_a\ \delta(\mathbf{R}_1 - \mathbf{R}_a))\ V(B;\mathbf{R}_1)\ dv_1$$

where

$$V(B;\mathbf{R}_1) = \int(\rho(B;\mathbf{R}_1) - \Sigma Z_b\ \delta(\mathbf{R}_2 - \mathbf{R}_b))\ (1/|\mathbf{R}_1 - \mathbf{R}_2|)\ dv_2$$

represents the electrostatic molecular potential at the point defined by the position vector $\mathbf{R}_1$, due to the electronic distribution and the nuclear charges of the unperturbed molecule B.

The results of a calculation within the framework of this approximation may then be corrected by consideration of the polarization of the electron densities. Restricting the discussion to the case of a molecule in the presence of a (positive) charge, Q, at a distance R, the polarization energy, P_R may be expressed as

$$P_R = E_R - E - V_R$$

where E_R represents the energy of the molecule in the presence of the charge, E is the energy of the unperturbed molecule and V_R denotes the electrostatic interaction energy. Assuming an approximate group additivity for the polarization energy, an analytical expression has been proposed [Bonaccorsi *et al.* (1976)] for its evaluation in terms of the polarizabilities associated with the various groups; the formulation is simplified if the partition adopted is made in terms of bonds and lone pairs.

The 1/R Expansion

Unfortunately, it may happen (e.g., when the interacting systems are large, such as biopolymers) that the unperturbed orbitals, and consequently the unperturbed electronic distributions, are not available, in which case the preceding formulation must be simplified further. The basic assumption for such a simplification consists of approximating the molecular charge distributions (corresponding to the electronic distributions and the nuclear charges) by model distributions of localized point charges.

For simplicity in the development of the formulation, it is more appropriate to consider first the interaction of a molecule with an external charge Q. When the charge lies on the *z*-axis at a distance R from the center of coordinates, the hamiltonian operator for the interaction may be written (Appendix A6) as

$$\begin{aligned} H &= -Q\Sigma Q_i/|\mathbf{R}_i - \mathbf{R}| + Q\Sigma Z_a/|\mathbf{R}_a - \mathbf{R}| \\ &= Q' V - \mu_z F_z \\ &- (1/3)(Q_{xx} F_{xx} + Q_{yy} F_{yy} + Q_{zz} F_{zz}) \\ &+ Q\Sigma Z_a/|\mathbf{R}_a - \mathbf{R}| \ldots \end{aligned}$$

where Q′ denotes the total electric charge of the molecule; V stands for the potential at the origin, F_z is the *z*-component of the field at the origin, and F_{xx}, F_{yy}, F_{zz} represent components of the field gradient at the origin, all of them due to the external charge. The operators (in atomic units)

$$\mu_z = -\Sigma Q_i z_i$$
$$Q_{pq} = (1/2)\Sigma(3 p_i q_i - R^2_i \delta_{pq})$$

(where *p,q* stand for any of the cartesian coordinates) correspond to the components of the dipole and quadrupole moments of the electronic distribution.

The interaction energy may be written for a fixed position and orientation [according to Buckingham (1967), from whose work the general expressions in this section have been taken], as

$$\begin{aligned} \Delta E = Q' V &- \mu_{z0} F_z - (1/2)\Sigma\Sigma\alpha_{tu} F_t F_u \\ &- (1/6)\Sigma\Sigma\Sigma\beta_{tuv} F_t F_u F_v \\ &- (1/24)\Sigma\Sigma\Sigma\Sigma\gamma_{tuvw} F_t F_u F_v F_w \\ &- (1/3)\Sigma\Sigma Q_{tu0} F_{tu} \\ &- (1/3)\Sigma\Sigma\Sigma A_{v\,tu} F_v F_{tu} \\ &- (1/6)\Sigma\Sigma\Sigma\Sigma B_{tu\,vw} F_t F_u F_{vw} \\ &- (1/6)\Sigma\Sigma\Sigma\Sigma C_{tu\,vw} F_{tu} F_{vw} \ldots \end{aligned}$$

in terms of the polarizability and hyperpolarizability tensors α, β, γ, A, B, C,... (see below) and where μ_{z0} and Q_{tu0} stand for the electric dipole and quadrupole moments of the unperturbed molecule; the summations over *t,u,v,w* extend over *x,y,z*.

For a spherically symmetric system, such as an atom, without dipole or quadrupole moments and for which β and A vanish, the remaining polarizability tensors are isotropic, that is

$$\alpha_{tu} = \alpha\,\delta_{tu}$$
$$\gamma_{tuvw} = (1/3)\,\gamma\,(\delta_{tu}\delta_{vw} + \delta_{tv}\delta_{uw} + \delta_{tw}\delta_{uv})$$
$$B_{tu\,vw} = B((3/4)(\delta_{tv}\delta_{uw} + \delta_{tw}\delta_{uv}) - (1/2)\,\delta_{tu}\delta_{vw})$$
$$C_{tu\,vw} = C((1/2)(\delta_{tv}\delta_{uw} + \delta_{tw}\delta_{uv}) - (1/3)\,\delta_{tu}\delta_{vw})$$

Taking into account (Appendix A6) that the only non-vanishing components of the external field and field gradient at the origin are F_z and $F_{zz} = -2\,F_{yy} = -2\,F_{xx}$, the interaction energy may be written as

$$E = Q' V - (1/2)\,\alpha F^2 - (1/24)\,\gamma F^4 - (1/4)\,B F^2 F' - (1/4)\,C (F')^2 \ldots$$

where the convention

$$\alpha = (1/3)\Sigma\alpha_{tt}$$
$$\gamma = (1/5)\Sigma\Sigma\gamma_{tu\,tu}$$
$$B = (2/15)\Sigma\Sigma B_{tu\,tu}$$
$$C = (1/5)\Sigma\Sigma C_{tu\,tu}$$

has been adopted and the notation for the field and field gradient, F_z and F_{zz} has been simplified to F and F′, respectively. The above constants are known as dipole polarizability (α), dipole hyperpolarizability (γ), uniform field quadrupole polarizability (B), and field gradient quadrupole polarizability (C). Finally, with

$$V = Q/R$$
$$F = -Q/R^2$$
$$F' = 2\,Q/R^3$$

one can write

$$\Delta E = Q'\,Q/R - (1/2)\,\alpha\,Q^2/R^4 - C\,Q^2/R^6 - (1/2)\,B\,Q^3/R^7 - (1/24)\,\gamma\,Q^4/R^8 \ldots$$

The evaluation of the interaction between two molecules may now be obtained in a rather simple way using the preceding developments if their charge distributions are approximated by point charges, Q_a and Q_b, centered at the nuclei. In such a case one can write

$$\Delta E = \Sigma\Sigma(Q_a\,Q_b/R_{ab} - (1/2)\,\alpha_a\,Q^2_b/R^4_{ab} - C_a\,Q^2_b/R^6_{ab} - (1/2)\,B_a\,Q^3_b/R^7_{ab} - (1/24)\,\gamma_a\,Q^4_b/R^8_{ab} \ldots)$$

for the interaction of molecule A (as a set of atoms, characterized by the parameters Q_a, α_a, γ_a, B_a, C_a, ...) and the external charges Q_b at the positions of the atoms of molecule B. Therefore, the total interaction may be written as

$$\Delta E = \Sigma\Sigma(a_{ab}/R_{ab} + b_{ab}/R^4_{ab} + c_{ab}/R^6_{ab} + d_{ab}/R^7_{ab} + e_{ab}/R^8_{ab} \ldots)$$

with

$$a_{ab} = Q_a\,Q_b$$
$$b_{ab} = -(1/2)\,(\alpha_a\,Q^2_b + \alpha_b\,Q^2_a)$$
$$c_{ab} = -(C_a\,Q^2_b + C_b\,Q^2_a)$$
$$d_{ab} = -(1/2)\,(B_a\,Q^3_b + B_b\,Q^3_a)$$
$$e_{ab} = -(1/24)\,(\gamma_a\,Q^4_b + \gamma_b\,Q^4_a)$$

In practice, the first problem to be solved is the determination of the values of the above coefficients. Theoretical values may be available for the charges and the dipole polarizabilities, but in most cases not even approximate values are known for C, B, and γ. Consequently, it has been necessary to develop semiempirical rules for their estimation (see Chapter 4, where the computational details of this type of calculations are discussed). Alternatively, although it represents a rather drastic simplification, one could approximate the interaction energy by the first term

$$\Delta E = \Sigma\Sigma Q_a\, Q_b / R_{ab}$$

of the above expansion; the charges used in practical calculations are either effective charges (Appendix A7), determined in separate treatments of the individual molecules, or some others corresponding to a model distribution (see below), in which case the charges are not necessarily centered on the atoms of the molecules.

Secondly, it should be emphasized that the above expression is only valid when the conditions stated are satisfied: that is, when the overlap between the density distributions of the two systems is negligible. For smaller separations, the formulation must be modified in order to correct for the neglect of the overlap; usually the term included for this purpose has the form

$$g_{ab} / R^{12}_{ab}$$

although an exponential term has sometimes been used [Kitaigorodsky (1961), Sundaram (1974), Caillet *et al.* (1978)].

Hydrogen-bonding

It may happen that at shorter distances the possibility of hydrogen-bonding will exist. Such an interaction is not accounted for by the above expansion and the corresponding correction must be included.

A number of empirical hydrogen-bond potential energy functions, developed over the years, have been obtained by fitting of experimental data on hydrogen-bond lengths and energies. One such function, for a bond X-H . . . Y [Lippincott and Schroeder (1955), Schroeder and Lippincott (1957)] is

$$\begin{aligned}\Delta E = {} & D\,(1 - \exp(-n\,(r - r_0)^2/2\,r)) \\ & - D'\exp(-n'\,(R - r - r'_0)^2/2\,(R - r)) \\ & + A\exp(-b\,R) \\ & - (1/2)\,A\,(R_0/R)^m \exp(-b\,R_0)\end{aligned}$$

where R is the distance between X and Y, R_0 is the distance between X and Y at the minimum of the potential function, *r* is the length X-H when the hydrogen-bond is formed, D and D′ are the strengths of the bonds X-H and H-Y, with normal lengths r_0 and r'_0, respectively, and *n*, *n′* are related to the ionization potentials of the atoms; *m* may take values between 1 and 6. The constant A may be otained from the condition $\delta\Delta E/\delta R = 0$ and the parameter *b* may be evaluated through the use of a known hydrogen-bond energy. The first two terms represent the stretching energies of the X-H and H-Y bonds, the second term corresponds to a van der Waals repulsion, and the fourth term gives the electrostatic attraction.

This expression was modified initially [Moulton and Kromhout (1956)] by inclusion of a directional correction; the first term was multiplied by $\cos^2\theta'$ and the second term by $\cos^2\theta$, where θ' is the angle between the bond X-H and some normal bonding direction for the atom X and θ is the angle between the bond H-Y and some normal bonding direction for the atom Y.

Such an expression has been applied [Scott and Scheraga (1966)] to calculations of N-H ... O bonds in polypeptides with appropriate modifications:

a) the first term may be omitted if the bonds are held rigid;

b) the normal bonding direction for the oxygen atom is assumed to be such that a normal hydrogen-bond, along the direction of one of the lone pairs, would form an angle HOC of 120°; as there are two lone pairs, two terms, similar to the second term of the above expression, must then be included.

The new expression is then

$$\begin{aligned}\Delta E = &-D' \cos^2 \theta_1 \exp(-n' (R - r - r'_0)^2 / 2 (R - r)) \\ &- D' \cos^2 \theta_2 \exp(-n' (R - r - r'_0)^2 / 2 (R - r)) \\ &+ A \exp(-b R) \\ &- (1/2) A (R_0/R)^m \exp(-b R_0)\end{aligned}$$

where θ_1 and θ_2 are the angles which the hydrogen atom makes with the directions of the lone-pair orbitals of the oxygen atom, anywhere on a cone. Considering that the oxygen atom may present sp^2 hybridization, the angular factor $\cos^2\theta$ (for both θ_1 and θ_2) may be replaced [Ooi *et al.* (1967)] with

$$((1 + (6)^{1/2} \cos \theta_1)/(1 + (6)^{1/2})^2$$

With $A = 4.941\ 10^6$, $D' = 82.4$, $n' = 13.15$, $r = 1.01$ A, $r'_0 = 0.97$ A, $R_0 = 2.85$ A, a maximum hydrogen-bond strength of −3.25 kcal/mole is obtained for $R = 2.70$ A, $\cos \theta_1 = 1$, and $\cos \theta_2 = 0$.

It is believed [Ooi *et al.* (1967)] that the above expression provides a reasonable representation of hydrogen bonding in conditions favourable for hydrogen-bond formation, not being appropriate for the borderline cases.

An alternate method based on theoretical calculations has also been proposed [McGuire *et al.* (1971)] and been widely used for conformational calculations of polypeptides. The fitting in this method involves several steps. First, calculations (at whatever level is possible) are carried out for linear hydrogen-bond systems. Then, using empirical energies for all the interactions except for that between the proton and the proton-acceptor, the function for the hydrogen-bond interaction may be determined by fitting the empirical expression for the energy of formation of the dimer to the quantum-mechanical expression in the region of the attractive part of the potential. The resulting expression may then be refined by comparison with experimental data.

The function

$$E = f_{ab} / R^{10}_{ab} + g'_{ab} / R^{12}_{ab}$$

was adopted as an extension of the terms for the non-bonded interactions (other than the electrostatic charge interaction) and because of its convenience for numerical calculations; R_{ab} denotes the separation between the hydrogen atom and the proton-acceptor, and the expansion coefficients depend, naturally, on the type of hydrogen-bonding considered. For the evaluation of the hydrogen-bonding contribution to the total interaction energy, the terms corresponding to all the possible hydrogen-bonding interactions must be included.

This fitted function, that reproduces the properties of the hydrogen-bonding interaction without the explicit introduction of an angular-dependent term, depends on the quantum-mechanical method used for the calculations as well as on the expression adopted for all the remaining interactions. That is, such an expression cannot be used by itself for the study of a given hydrogen-bond, but must be used in conjunction with all the other terms in order to predict the total interaction, the only quantity that has a meaning.

Practical Approach

Assuming for the sake of the argument that the required data (unperturbed orbitals of the molecules and the expansion coefficients discussed above) are available and that the computing facilities will permit one to tackle the problem at any level of approximation, one is, in principle, in the position of carrying out the study of any molecular association.

There are, however, three points that must be taken into account when facing reality. First, there are the difficulties mentioned above regarding the accuracy of the results, especially when dealing with large systems, Second, the assumptions just stated are far too optimistic regarding both the imput data and the computing facilities. Third, the calculations would be extraordinarily expensive. So that practical considerations on what to do and how to do it must be brought into play.

The goal of the study of the interaction between two systems consists in determining the path of approach and the stability of their association. It requires, as stated at the beginning of this section, a knowledge of the energy hypersurface. Two questions arise in this connection: is it necessary to determine, as completely as possible, the energy hypersurface? Should one always use the most accurate (and, therefore, most expensive) method?

The practical distinction between long and short range interactions provides the answer to the second question and, implicitly, to the first question as well. At long distances, it is sufficient to carry out approximate calculations, while at short ranges it is imperative to use more accurate methods. Therefore, in practice, the calculations could be carried out in a number of steps, the scope of each one being determined by the preceding step.

First, the interaction energy is determined at the zero-th order level and within the point-charge approximation for a wide range of relative positions at large separations of the two molecules. The results, in the general case, will show the directions of preferred approach (see Chapter 2). It must be mentioned that this type of calculation may be carried out at two levels of quite different accuracy. First, of course, there is the approximation of the interaction energy by the electrostatic interaction between point charges, using the effective charges on the atoms or any other model charge distribution. The latter is far more accurate but it is not so widely used because of the difficulties in the selection of the model distributions, particularly for large molecules. This approximation may be used, in a reverse application, for the determination of analytical expressions for interaction energies, obtained in accurate calculations; such expressions are extremely important because of the consequent reduction of the calculation time for the study of polymolecular systems within the framework of Monte Carlo techniques. Evidently, if available, they should be used over the complete range of distances (for which they are valid) with preference to any other approximation (except for more accurate calculations).

Refined calculations may now be carried out for those directions of interest for different separations between the molecules. The zero-th order approximation may still be used, but at a higher level of accuracy if the unperturbed densities of the two systems are available; if not, the $(1/R)$ expansion must be used. At short range separation, it will be necessary to switch to SCF calculations at the level of accuracy permitted by the practical restrictions in the computing facilities.

The outlined procedure corresponds to an ideal situation. One must remember, in fact, that the reason for the development of the $1/R$ expansion, corrected for the possibility of hydrogen bonding, is that it will be only possible to carry out approximate calculations. Therefore, it will be assumed hereafter that interaction energies are evaluated using a $1/R$ expansion valid for the complete range of separations between the interacting systems. The practical details regarding the expansion coefficients are discussed in Chapter 4.

The above comments describe a procedure that might be used for the evaluation of points in an energy hypersurface but the problem remains of how to determine the true hypersurface (or, at least, its region of interest), and in this connection it is necessary to distinguish between associative and reactive processes.

In the cases of associative processes, calculations such as those outlined above should yield the necessary information. The lowest minimum in the hypersurface should correspond to the aggregate.

For reactive processes, however, it must be remembered that the intermediate aggregate will be characterized by a saddle point, so that additional work may be required. The calculations should lead to a situation where the conformation that might be expected for the aggregate would correspond to the beginning of a 'pass' in the hypersurface, facing a fast-rising ridge for shorter separations. The existence of a saddle point, corresponding to the aggregate, would require confirmation by means of calculations for the systems in the right-hand side of the reaction.

In all instances there is the possibility of a refinement. The calculations described imply the adoption of a given geometry or conformation for the interacting molecules. Usually the experimentally determined geometry is used, but a consistent study should include a conformational analysis in order to determine the preferred geometry (corresponding to a minimum energy) under the conditions considered.

In fact, $1/R$ expansions have been used very successfully for conformational analyses of biopolymers, although correction terms are required in such a case in order to include the torsional energy (not accounted for completely by the other interactions) corresponding to the rotation around a bond.

The time consuming calculations (of interactions and conformations) for large systems may be simplified by the neglect of the interactions between distant atoms or of, at least, some of the terms of their interaction. Within the formulation discussed here, such a neglect is simply based on the practical fact that for given coefficients in the above expansion, the contribution from any or all of the terms may be smaller than a threshold value for distances beyond a certain range.

In conformational analysis, the neglect of atoms that are far away from a rotating group may be justified by application of the integral Hellman-Feynman theorem [Kier (1968)]. Given two conformations of the molecule, with corresponding molecular hamiltonian operators, H_1 and H_2, and exact ground state functions, Ψ_1 and Ψ_2, the energy difference between the two conformations is given by

$$\Delta E = E_1 - E_2 = \langle \Psi_1 | \Delta H | \Psi_2 \rangle / \langle \Psi_1 | \Psi_2 \rangle$$

where ΔH, which represents the difference between the two hamiltonians, contains only nuclear-dependent terms (i.e., nuclear attraction and nuclear repulsion terms).

Assuming that the rotation takes place with bond lengths maintained fixed, and that m of the n nuclei of the molecule are involved in the rotation, the hamiltonian difference can be expressed as

$$\Delta H = \Sigma\Sigma Z_a Z_{a'} (1/R_{aa'1} - 1/R_{aa'2}) - \Sigma\Sigma Z_a (1/R_{ia\,1} - 1/R_{ia\,2})$$

where $R_{aa'1}$ denotes the distance between nuclei a and a' in conformation 1 and $R_{ia'1}$ represents the distance of the i-th electron from nucleus a in conformation 1; the summations in the first term run from 1 to m and from $(m + 1)$ to n, respectively (with the nuclei labelled so that the first m are those involved in the rotation), and the summations in the second term run over all the electrons and all the nuclei in the molecule. The energy difference may then be written as

$$\Delta E = \Sigma\Sigma Z_a Z_{a'}(1/R_{aa'1} - 1/R_{aa'2}) \\ - \Sigma Z_a < \Psi_1|\Sigma(1/R_{ia\,1} - 1/R_{ia\,2}|\Psi_2 >/< \Psi_1|\Psi_2 >$$

with the same summations as in the previous case.

This energy difference may now be evaluated approximately using functions in terms of localized orbitals. Knowing that localized orbitals with centroids of density far enough from the rotating groups are not affected by the rotation, the sets of s orbitals may be partitioned into two groups respectively: the first r orbitals, ϕ, are those that remain constant and are, therefore, common to both conformations; the remaining orbitals, from $(r + 1)$ to s, are labelled ϕ' and ϕ'', respectively, for each conformation. Taking into account that the orbitals of each conformation are orthonormalized but that the sets ϕ' and ϕ'' are not necessarily orthogonal, the nuclear attraction part of ΔE may be approximated (for closed-shell configurations) by

$$- 2\Sigma Z_a\Sigma < \phi_r|1/R_{a\,1} - 1/R_{a\,2}|\phi_r >$$

plus other terms involving only orbitals of the sets ϕ' and ϕ''; the summations in this expression extend over all the nuclei and the orbitals of the common set and $1/R_{a\,1}$ denotes the distance between the nucleus a and the electron for which the integration is being carried out. Assuming that the centroids of those localized orbitals (which correspond closely to bond orbitals) lie at the bond midpoints and taking into account that these points are very far from the rotating nuclei, the above integration may be written as

$$- 2\Sigma\Sigma Z_a\ (1/R_{ka\,1} - 1/R_{ka\,2})$$

where the summations extend to the nuclei and the bonds far from the rotating nuclei and $R_{ka\,1}$ represents the distance between the nucleus a and the midpoint of bond k. Taking into account that $2/R_{ka\,1}$ (and similarly for the second conformation) is, for large distances, nearly equal to the sum of the inverse distances of the two nuclei forming bond k from nucleus a, the contribution to ΔE from the nuclei far from the rotating group is essentially cancelled.

CLUSTERS

The second step towards the study of recognition processes in aqueous solution implies the consideration of clusters of water molecules around the molecule(s) under consideration. The most appropriate technique for such a case is that embodied in the Monte Carlo method.

Monte Carlo Techniques

The Monte Carlo method is based on a stochastic process which generates a Boltzmann-weighted chain of configurations of a given N-body system.

Using the canonical ensemble, the equilibrium value of any quantity of interest, Q, is given by

$$Q = (\int Q \exp(-\Delta E/kT)\, dV) \;/\; (\int \exp(-\Delta E/kT)\, dV)$$

where ΔE denotes the potential energy of the system and dV is the volume element for the configuration space (taking into account that the interactions between the components of the system are independent of the velocity). The Monte Carlo method comes into being in connection with the evaluation of the many-dimensional integrals in the above expression, with the integration carried out over a random sampling of points instead of over a regular array of points.

A normal Monte Carlo method would imply the selection of a random configuration and its weighting by $\exp(-\Delta E/kT)$, where ΔE represents the potential energy for such a configuration. Such a method is not appropriate for close-packed configurations and consequently a modified Monte Carlo technique must be used [Metropolis *et al.* (1953)], in which the configurations are chosen with probabilities $\exp(-\Delta E/kT)$ and are then weighted evenly.

This modified technique, originally developed for a periodic system considered formed by squares, involves the following steps:

a) Place the N bodies contained in the square in any given configuration.

b) Move one of the bodies according to

$$x \rightarrow x + \kappa\, d$$
$$y \rightarrow y + \lambda\, d$$

where d is the maximum (arbitrary) allowed displacement and κ and λ are random numbers between -1 and $+1$ (generated, in this particular case, by a middle-square process); if the move takes the body outside the square, it is assumed that it reenters the square from the opposite side.

c) Evaluate the change, $\delta(\Delta E)$, in potential energy due to the move.

d) Decide whether the move is to be allowed or not on the basis of the following conditions:

if $\delta(\Delta E) < 0$, allow the move and place the body in its new position;

if $\delta(\Delta E) > 0$, allow the move and place the body in the new position if $\exp(-\delta(\Delta E)/kT) > h$ (h being a random number between 0 and $+1$); otherwise, if $\exp(-\delta(\Delta E)/kT) < h$, return the body to the old position.

e) Having attempted to move a body, proceed similarly with the next one.

For averaging purposes, the configuration resulting from every move, whether it is a new or the old configuration, is considered as a new configuration. The average value of the quantity Q is given by

$$Q = \Sigma Q_i / n_c$$

where Q_i denotes the value of Q after the *i*-th move and the summation over the moves runs from 1 to n_c, the number of configurations.

It can be shown [Metropolis *et al.* (1953)] that the method is ergodic (i.e., any state can be reached from any other, even if it implies several moves) and that if the ratio of the number of systems in state *r* to the number of systems in state *s* is greater than $\exp(-\Delta E_r / kT) / \exp(-\Delta E_s / kT)$, more systems move from state *r* to state *s*. Therefore, the ensemble must approach the canonical distribution, although how long it takes to reach equilibrium will depend on the value of *d* : if it is too large, more moves will be forbidden and, if it is too small, the configuration will not change enough.

This method has been extended to the case of cubic [Barker and Watts (1969)] and spherical cells [Abraham (1974)], with dimensions chosen according to the experimental conditions; for example, the volume of the constraining spherical boundary, V_c, has been chosen [Abraham (1974)] for water clusters as

$$V_c = 5 \cdot V_b$$

where V_b is the volume for bulk water of N molecules at the temperature T considered.

In the case of spherical boundaries it is necessary to introduce an additional restriction: the cluster is defined by the criterion that all molecules must always lie within the constraining sphere, centered at the center of mass of the cluster. Such a condition ensures the constant density of the cluster and the periodicity of the system.

For clusters of molecules within a spherical boundary, the steps of the procedure [Barker and Watts (1969) and Abraham (1974)] are:

a) Select an initial configuration (see below).

b) Select at random a molecule of the cluster, a rotation axis (x, y or z), as well as displacements (Δx, Δy, Δz) and a rotation angle between $-\alpha$ and $+\alpha$, both from uniform random distributions.

c) If the movement corresponding to the displacement and the rotation considered takes the center of mass of the molecule outside the spherical boundary (centered at the new center of mass of the cluster), the new configuration is rejected and the initial configuration retained.

d) If the new configuration is acceptable on the basis of the preceding condition, evaluate the change, $\delta(\Delta E)$, in the potential energy due to the move and apply the Metropolis tests (Fig. 3-1).

The values of α and Δ are to be chosen in each case: for example, α may be fixed, say, at 0.2 radians and Δ chosen so that approximately half the attempted moves are actually made, the choice governed by the size and temperature of the cluster [Abraham (1974)].

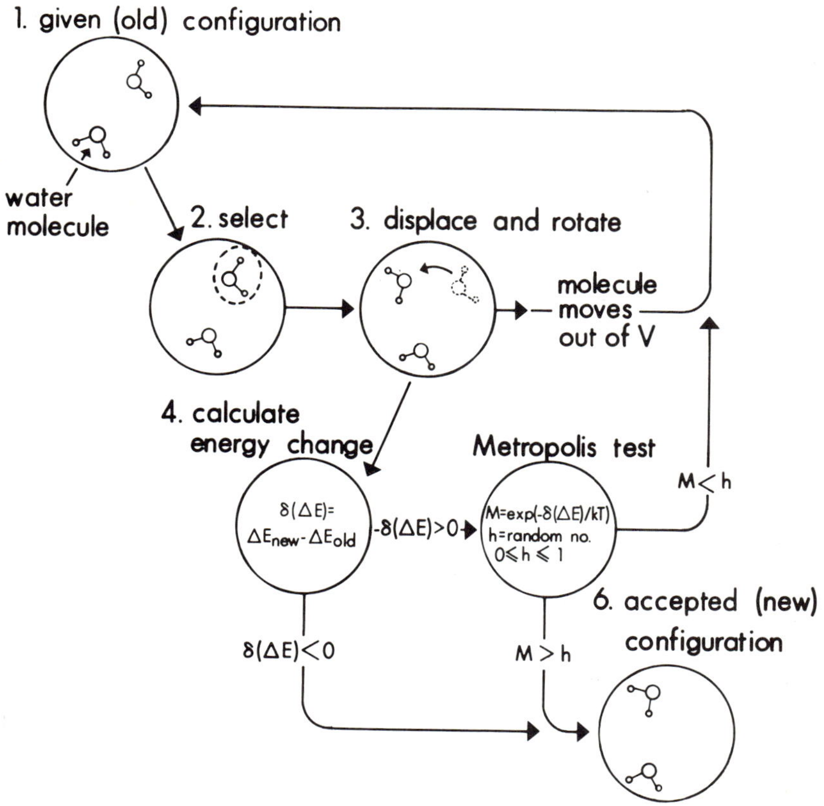

Figure 3-1. The Monte Carlo algorithm for the evaluation of canonical averages for water clusters [adapted from Abraham *et al.* (1976)].

After equilibrium has been reached, an additional number (say, 500,000) of configurations are generated in order to evaluate the potential and internal energy

$$< \Delta E > = \Sigma \, \Delta E_i / n_c$$
$$U = 3kT \, (N - 1/2) + < \Delta E >$$

of the cluster.

The evaluation of the potential energy, after each move, could be carried out, in principle, by an accurate method. The computation time, for N and particularly n_c so large, would be prohibitive and an approximate method must be used. Within the framework of an accepted approximation (with consideration only of two-body interactions), the potential energy is written

$$\Delta E = \Sigma\Sigma' \Delta E_{pq}$$

where the summations extend over all the molecules in the system (with $q > p$) and ΔE_{pq} represents the interaction between the molecules *p* and *q*. The evaluation of ΔE_{pq} by an accurate method may still be prohibitive and recourse must be made of another approximation. In general, the $1/R$ expansion may be appropriate but, in those cases where the information is available, fitted analytical expansions (mentioned above) should be used. Such is the case for water molecules, discussed below before considering a cluster of water around a given system.

Water Clusters

Water being a small molecule (with only ten electrons), the interaction between two molecules may be treated, without any difficulty, by accurate methods. Among other treatments, Hartree-Fock calculations have been carried out [Popkie *et al.* (1973)] for an extensive number of conformations of water dimers, leading to a comprehensive description of the Hartree-Fock potential for this system.

Fitting of the data has yielded, on the basis of an adopted charge distribution,

$$\begin{aligned} \Delta E = {} & Q^2 (1/R_{13} + 1/R_{14} + 1/R_{23} + 1/R_{24}) \\ & + 4Q^2/R_{78} \\ & - 2Q^2 (1/R_{18} + 1/R_{28} + 1/R_{37} + 1/R_{47}) \\ & + a_1 \exp(-b_1 R_{56}) \\ & + a_2 (\exp(-b_2 R_{13}) + \exp(-b_2 R_{14}) + \exp(-b_2 R_{23}) + \exp(-b_2 R_{24})) \\ & + a_3 (\exp(-b_3 R_{16}) + \exp(-b_3 R_{26}) + \exp(-b_3 R_{35}) + \exp(-b_3 R_{45})) \end{aligned}$$

of the potential energy, where (in atomic units)

$$\begin{aligned} a_1 &= 582.277054 \\ a_2 &= 0.143789 \\ a_3 &= 5.470184 \\ b_1 &= 2.520593 \\ b_2 &= 1.221756 \\ b_3 &= 1.936626 \\ Q^2 &= 0.449387 \end{aligned}$$

and a distance of 0.436 bohrs between O_5 and M_7 or O_6 and M_8 (see Fig. 3-2). This fit is satisfactory for the attractive regions but less accurate for the repulsive regions.

The above expansion has been used [Abraham (1974)] for the evaluation of ΔE_{pq}, discussed above, in a Monte Carlo treatment of water clusters. In fact, the actual expression used for the evaluation of the potential energy was

$$\Delta E = \Sigma\Sigma' (\Delta E_{HF\,pq} + \Delta E_{S\,pq} + \Delta E_{L\,pq})$$

where the additional terms, given by

$$\Delta E_{S\,pq} = -c_S \exp(-\,d_S\, R_{Op\,Oq})$$
$$\Delta E_{L\,pq} = -c_L / R^6_{Op\,Oq}$$

represent corrections for short and long distance effects; $R_{Op\,Oq}$ denotes the separation between the oxygen atoms of the molecules *p* and *q* and

$c_S = 0.1030181$
$c_L = 49.23446$
$d_S = 0.9753$

(in atomic units); the long-range potential is only operative for distances $R_{Op\,Oq} \geqslant 7.56$ bohr.

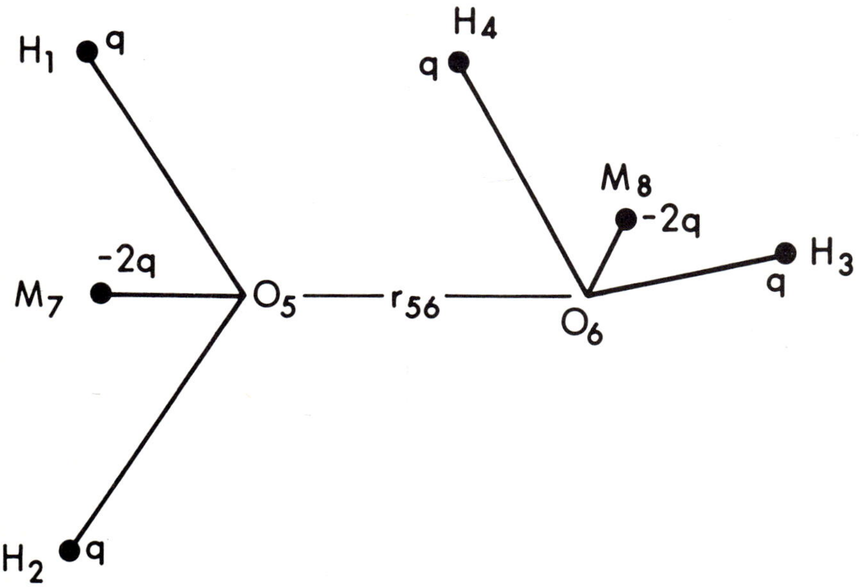

Figure 3-2. Point-charge model used in the derivation of the analytical expression for the water-water interaction potential [Popkie *et al.* (1973)].

Monte Carlo calculations were carried out [Abraham (1974)] for clusters of 8, 27, and 64 molecules. The results show that, as expected, the internal energy decreases with decreasing temperature and with increasing size of the cluster (as a consequence of the smaller influence of the surface on the internal energy for larger clusters) and that the thermal expansion and shape distortion increase with increasing temperature. These two effects are observed in the snapshot pictures of a cluster of 64 molecules at 263.2°K and 298.0°K presented in Figs. 3-3 and 3-4, respectively. On the other hand, Fig. 3-5 provides a stereoscopic viewing of the cluster of 27 molecules at 298.0°K.

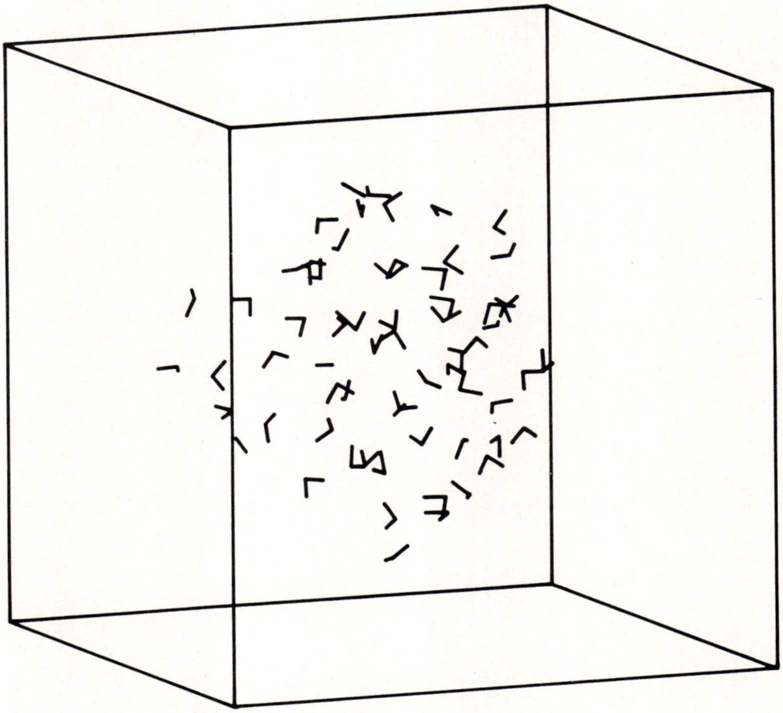

Figure 3-3. Snapshot picture of a cluster of sixty-four water molecules at 263.2°K [Abraham (1974)].

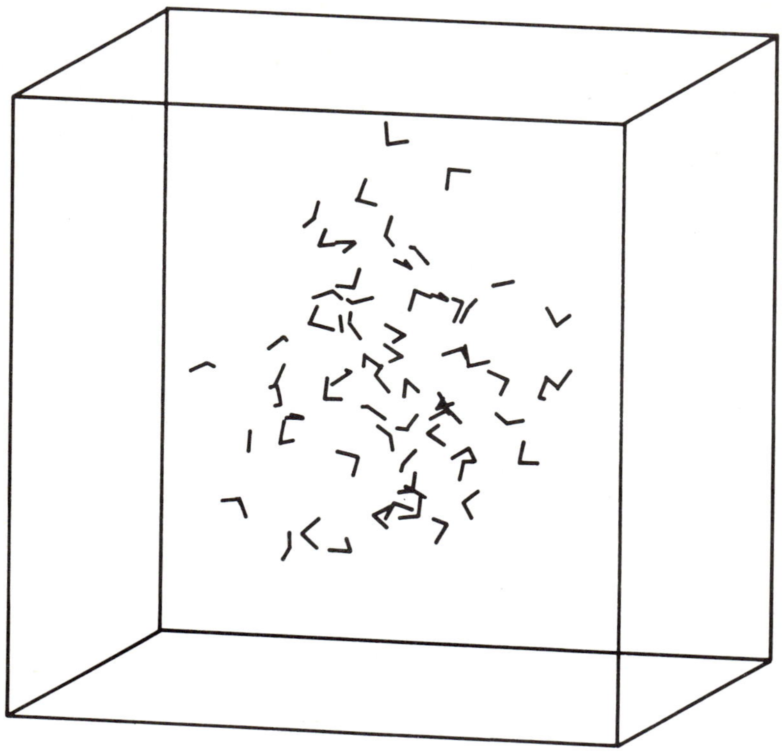

Figure 3-4. Snapshot picture of a cluster of sixty-four water molecules at 298.0°K [Abraham (1974)].

Figure 3-5. Snapshot picture of a cluster of twenty-seven water molecules at 298.0°K drawn for stereoscopic viewing [Abraham (1974)].

Hydrated Ions

A hydrated ion represents the simplest case of a system surrounded by a cluster of water molecules. A gas-phase situation still exists if the number of water molecules is small (microcluster), but nevertheless a theoretical treatment of such a system represents a considerable step forward.

The data needed are the values (or analytical expressions) for the interaction energy between two water molecules and between a water molecule and the ion. The Monte Carlo procedure must be modified in order to take into account the presence of the ion, which will be placed at the center of the spherical boundary.

The first calculation of this type has been carried out [Abraham *et al.* (1976)] for the case of hydrated alkali and halide ions. The water-water interactions were evaluated by the formula

$$\Delta E = \Delta E_{HF} + \Delta E_{dispersion}$$

where ΔE_{HF} corresponds to the Hartree-Fock interaction (being of the same form as that considered above) and $\Delta E_{dispersion}$, which accounts for the dispersion effects, is taken as

$$\Delta E_{dispersion} = c_1 / R^6_{56} - c_2 / R^8_{56} + c_3 / R^{10}_{56}$$

with coefficients

$c_1 = 922.781$ kcal A^6/mol
$c_2 = 17283.5$ kcal A^8/mol
$c_3 = 24119.7$ kcal A^{10}/mol

The parameters, in the improved fitting of the Hartree-Fock interaction,

$a_1 = 71533.4$ kcal/mol
$a_2 = 779.885$ kcal/mol
$a_3 = 4084.02$ kcal/mol
$b_1 = 3.96994\ A^{-1}$
$b_2 = 3.12544\ A^{-1}$
$b_3 = 3.91443\ A^{-1}$

and the value

$Q^2 = 139.272$ kcal A/mol

are such that the formula includes now the correlation effects.

The Hartree-Fock interaction between the ions and the water was approximated by

$$\Delta E = Q^2\,(1/R(I,H_1) + 1/R(I,H_2) - 2/R(I,M)) - Q^2_M\,(1/R(I,M_1) + 1/R(I,M_2) - 2/R(I,O)) + a_1\,(\exp(-b_1\,R(I,H_1)) + \exp(-b_1\,R(I,H_2))) + a_2\,\exp(-b_2\,R(I,O))$$

for Li^+ and Na^+ (see Fig. 3-6) and by

$$\Delta E = Q^2\,(1/R(I,H_1) + 1/R(I,H_2) - 1/R(I,R_1) - 1/R(I,R_2)) - Q^2_M(1/R(I,M_1) + 1/R(I,M_2) - 2/R(I,O)) + a_1\,(\exp(-b_1\,R(I,H_1)) + \exp(-b_1\,R(I,H_2))) + a_2\,\exp(-b_2\,R(I,M))$$

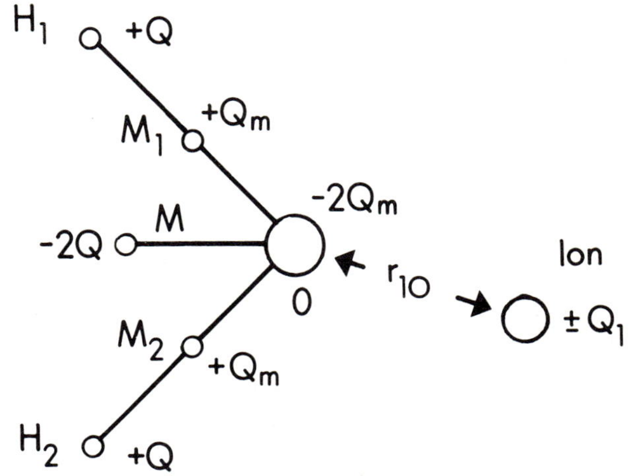

Figure 3-6. Charge model for the interaction of water with an ion (Li^+, Na^+) [Abraham *et al.* (1976)].

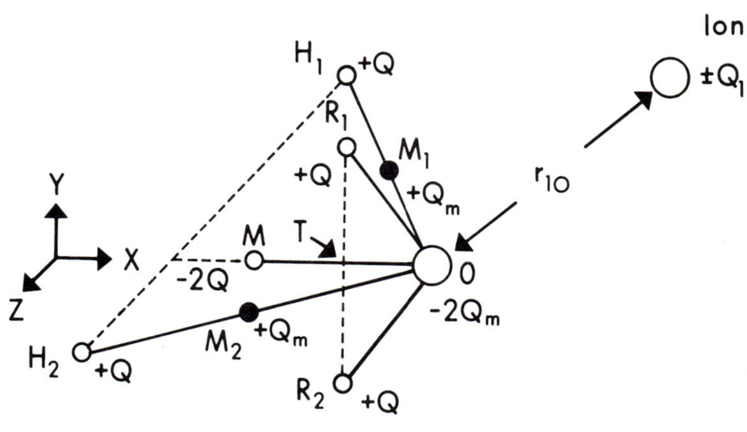

Figure 3-7. Charge model for the interaction of water with an ion (K^+, F^-, Cl^-); the dotted lines are parallel to the respective axes [Abraham *et al.* (1976)].

for K^+, F^-, and Cl^- (see Fig. 3-7) [Kistenmacher *et al.* (1973a,b,c, 1974)], the corresponding values of the parameters (for the ions of interest) being given in Table 3-1; in these equations I denotes the ion.

Table 3-1. Parameters for the analytical expression of the ion-water potential energy (in atomic units)*

	Na^+	K^+	Cl^-
Q	0.53885	0.59564	−0.61465
Q_M	−0.22106	−0.05095	0.12698
a_1	164.80064	4.45471	6.06616
a_2	166.32249	373.78784	69.61395
b_1	3.814561	2.64194	1.63899
b_2	2.330438	2.16348	1.74914
OM		0.07975	0.28188
OT		0.35249	0.46735
TR_1		0.013476	0.21344

* As given by Kistenmacher *et al.* (1973c).

The change in the Helmholtz free energy, $\Delta F_{N-1\,N}$, for the reaction

$$\text{ion} \cdot (N-1)\, H_2O + H_2O(\text{vapour}) \leftrightarrow \text{ion} \cdot N\, H_2O$$

is given [Abraham *et al.* (1976)] by

$$\Delta F_{N-1\,N} = \int < \Delta E(H_2O;\lambda) > d\lambda + kT \ln N - \int p\, dV$$

where $< \Delta E(H_2O;\lambda) >$ denotes the canonical average of the potential energy of the water molecule, which is being added to the cluster, and the integration limits of the first integral are 0 (when the monomer does not interact and behaves as an ideal gas) and 1 [when the monomer is undistinguishable from the other (N − 1) water molecules]; the integration limits in the second integral are the volume corresponding to 1 atm and the volume of the constraining boundary. [See the work of Abraham *et al.* (1976) for details of the integration procedure.] The remaining thermodynamic quantities are then given by

$$\Delta H_{N-1\,N} = \Delta U_{N-1\,N} - kT$$
$$\Delta U_{N-1\,N} = < \Delta E(N;\lambda = 1) > - < \Delta E(N-1;\lambda = 1) >$$
$$\Delta G_{N-1\,N} = \Delta F_{N-1\,N} + \Delta(PV) = \Delta F_{N-1\,N} - kT$$

(for the standard state, at 298°K and 1 atm).

The calculations show that the enthalpy change for the addition of one more water molecule becomes more positive as the cluster increases in size. This behavior is due to a decrease in the ion-water attraction and (e.g., for Na^+) an increase in the water-water repulsion, as observed in Table 3-2.

Table 3-2. Monte Carlo values (kcal/mol) of the average ion-water and water-water potential energies per water molecule*

N	$Na(H_2O)_N^+$		$K(H_2O)_N^+$		$Cl(H_2O)_N^-$	
1	−24.32	0.00	−16.92	0.00	−10.93	0.00
2	−24.24	0.79	−16.66	0.37	−10.73	0.08
3	−24.02	1.82	−16.58	1.10	−10.58	0.11
4	−23.65	2.74	−16.02	1.39	−10.08	−0.19
5	−22.82	3.24	−14.81	0.98	− 9.44	−0.53
6	−21.53	3.18	−13.75	0.57	− 9.18	−0.74

* As given by Abraham *et al.* (1976).

RECOGNITION IN AQUEOUS SOLUTION

The study of recognition processes in aqueous solution implies the consideration of the receptor, the carrier, and the receptor-carrier aggregate in the presence of water molecules and involves, therefore, three sets of separate calculations such as described above, with some appropriate modifications. [See also the review of Scheraga (1978) for other approaches.]

The interaction energies, that must be evaluated, are those between two water molecules and between the solvated system and a water molecule. It is hoped that, in due time, fitted analytical exapansions (from *a priori* SCF calculations) will be available for the interactions between systems of interest and a water molecule (see Chapter 4), in which case they should be used, together with the expansion described above for the water-water interaction. Until that time, however, in order to avoid a mixture of two levels of approximation, it may be more appropriate to use 1/R expansions for all the interactions.

In order to take into account the concentration dependence at a quantitative level, the number N of water molecules to be considered within the spherical boundary should be that corresponding to the concentration. It is in this connection that two different difficulties may arise, depending on whether the value for the ratio of the number of molecules of the solvated system to the number of water molecules is small or large.

If such a ratio is small, the solution may be approximated by consideration of a single molecule of the solvated system within the constraining boundary. Depending on the value required for N, the Monte Carlo calculation may be feasible or not. In the second case there is always the possibility of carrying out the calculation for a smaller value of N, although as high as practically possible in order to obtain the best approximation.

On the other hand, the problem encountered when the ratio is larger is of different nature. For single-phase systems, the number of molecules to be included in the Monte Carlo calculation must be, at least, of the order of several hundreds and therefore, in order to satisfy the value of the above ratio, it may be necessary to consider more than one molecule of solvated system within the constraining boundary. That is, the relative positions of the molecules of the solvated system must then be taken into consideration, which implies a set of Monte Carlo calculations (as well as the availability of the fitted analytical expansion for the interaction between two such molecules, when operating at that level of approximation).

In order to reproduce the periodicity of the solution, it is appropriate in this case to revert to cubic constraining boundaries (as in the original Monte Carlo formulation). The calculations will involve the following steps:

a) Determine the volume, V_c, of the constraining cube, corresponding (e.g., as specified above) to the number, N, of water molecules to be considered in the calculation.

b) Divide this volume in *n* cubes.

c) Position the *n* solvated molecules in space so that each one is at the center of one such cube and carry out (in gas-phase conditions) an optimization, for minimum energy, of their relative orientations.

d) Carry out a Monte Carlo calculation for a cluster of N / n water molecules in a small cube (defined in step *b*), with a molecule of the solvated system at the center of the cube.

e) Position the equilibrated miniclusters, determined in the previous step, within the original constraining volume, so that the molecules of the solvated system have the relative orientations determined in step *c*.

f) Carry out a Monte Carlo calculation for the total cluster.

g) Optimize the relative orientations of the molecules of the solvated system [Intuitively, one would expect the decrease in energy corresponding to this optimization to be small and, therefore, it might be omitted, taking into account that it will involve new Monte Carlo calculations.]

Chapter 4

Computational Simulations

The study of recognition processes, as discussed in Chapters 2 and 3, implies conformational analyses and a knowledge of interactions and reactions, requiring the determination of energy hypersurfaces or, at least, of parts of them.

Two distinct problems must be dealt with in this connection: the determination of the points of the hypersurface and the study of its characteristics (i.e., the positions of its minima and saddle points).

The difficulties in the evaluation of the points of energy hypersurfaces have been examined in Chapter 3, where it was concluded that rigorous quantum-mechanical calculations are not practically feasible, especially when dealing with large biomolecules and when recognition processes in aqueous solution are considered. It was suggested, at that point, that 1/R expansions could be appropriate for such calculations in all instances, that is, both for gas-phase calculations and in conjunction with the Monte Carlo calculations for systems in water clusters. The general theoretical details have been discussed, and it remains to examine the practical considerations, such as the atomic parameters and molecular data required in such expansions.

On the other hand, the study of the characteristics of hypersurfaces require special techniques of minimization, in order to determine the preferred conformations and paths of approach. Many optimization procedures have been developed, but not all are appropriate or economically applicable and, for that reason, a detailed examination of the problem is required.

In summary, the computational simulation of biopolymers and of their interactions requires

a) the adoption of a potential energy expression
b) the values of the corresponding parameters
c) data for the units involved

and demands the availability of very efficient programs for

a) the evaluation of the interaction energies
b) the determination of the minima in the energy hypersurface
c) the Monte Carlo simulation of the aqueous environment.

ENERGY EXPRESSIONS AND DATA BANKS

The data needed for a calculation of the type of those envisaged here depend, evidently, on the formulation used.

Energy Expressions

Restricting the discussion to interaction energy expressions of the type 1/R (Chapter 3), the possible cases (using a unified notation) are:

a) Evaluation of intermolecular interactions (recognition processes)

The interaction energy is given by

$$\Delta E = \Sigma\Sigma(a_{ab}/R_{ab} + b_{ab}/R^4_{ab} + c_{ab}/R^6_{ab} + d_{ab}/R^7_{ab} + e_{ab}/R^8_{ab} + \ldots)$$

where the summations extend to the atoms *a*, *b* of the two systems A, B, respectively, R_{ab} denotes the distance between two such atoms and

$$a_{ab} = Q_a Q_b$$
$$b_{ab} = -(1/2)(\alpha_a Q^2_b + \alpha_b Q^2_a)$$
$$c_{ab} = -(C_a Q^2_b + C_b Q^2_a)$$
$$d_{ab} = -(1/2)(B_a Q^3_b + B_b Q^3_a)$$
$$e_{ab} = -(1/24)(\gamma_a Q^4_b + \gamma_b Q^4_a)$$

where Q, α, γ, B, and C are the charge, dipole polarizability, dipole hyperpolarizability, uniform field quadrupole polarizability, and field gradient quadrupole polarizability, respectively.

The first three terms make the predominant contribution to the interaction energy and correspond to a short-range (R^{-6}) dispersive attraction, a medium-range (R^{-4}) electrostatic-induced dipole attraction, and a long-range (R^{-1}) electrostatic charge interaction. In such a case, only Q, α, and C are required in order to evaluate the expansion coefficients: the effective charges may be obtained (for small molecules) in an SCF calculation or may be adapted (for large molecules) from the results of calculations for their fragments or related systems and the dipole polarizabilities are available for all the neutral atoms and ions of interest [Fraga *et al.* (1976)] and interpolated values might be used (Fig. 4-1) in correspondence with the charge of the atom. Unfortunately, such is not the case for the field gradient quadrupole polarizabilities, with the result that the corresponding coefficients must be approximated, usually by means of the Slater-Kirkwood expression [Pitzer (1959)]

$$c_{a\,b} = -(3/2)(eh/2\pi m^{1/2})\,\alpha_a\,\alpha_b/((\alpha_a/n_a)^{1/2} + (\alpha_b/n_b)^{1/2})$$

where *e* and *m* are the charge and the mass of the electron, *h* is Planck constant and n_a, n_b denote the effective numbers of valence electrons (see Table 4-39).

The above expression must be modified by the inclusion of an additional term (R^{-12}) that accounts for the very short range overlap repulsion. Usually, in addition to this term, only those for R^{-1} and R^{-6} are retained, so that the expression used is

$$\Delta E = \Sigma\Sigma(a_{ab}/R_{ab} + c_{ab}/R^6_{ab} + g_{ab}/R^{12}_{ab})$$

An approach that has been adopted [Clementi (1976)] for the determination of

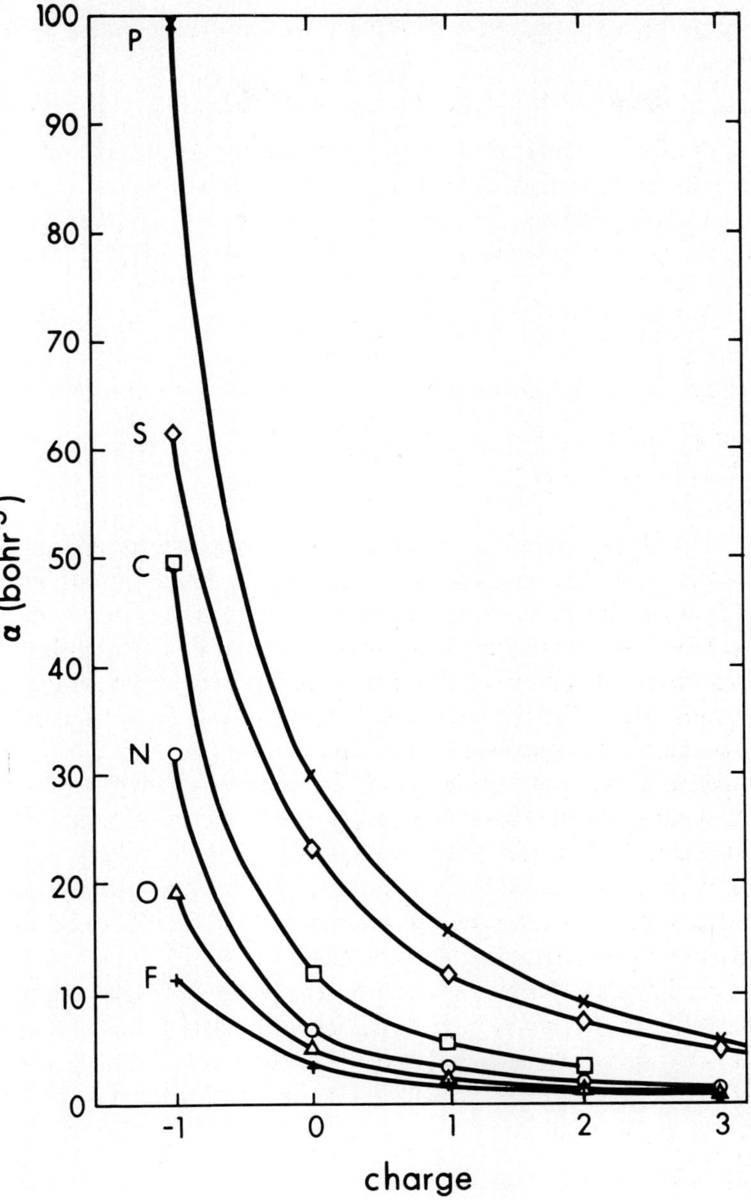

Figure 4-1. Representation of the atomic dipole polarizabilities as a function of the state of ionization.

the corresponding expansion coefficients makes use of the fitting of the *a priori* results obtained for

$$\Delta E = E_{AB} - (E_A + E_B) = \Sigma\Sigma(a'_{ab}\, Q_a\, Q_b / R_{ab} + c_{ab} / R^6_{ab} + g_{ab} / R^{12}_{ab})$$

for a large number of separations and orientations of the two systems. In this connection, it has been found necessary to distinguish the atoms not only by their nuclear charges (i.e., H, C, N, O, etc.) but also according to their environments, so that several classes of a same atom (say, H) may exist in a molecule.

b) Evaluation of intramolecular interaction energies (conformational analysis)

An expression widely used in conformational analysis consists of the terms

$$\Delta E = \Sigma\Sigma(a_{ab} / R_{ab} + c_{ab} / R^6_{ab} + g_{ab} / R^{12}_{ab}) \\ + \Sigma\Sigma(f_{ab} / R^{10}_{ab} + g'_{ab} / R^{12}_{ab}) \\ + \Delta E_{torsion}$$

where the first three terms correspond to the electrostatic charge and non-bonded interactions, the next two terms give the hydrogen-bond interactions, and $\Delta E_{torsion}$ corresponds to the torsional energy for rotation about a given bond; the first set of summations include the 1-4, 1-5 and higher order interactions (where 1-4 denotes the interaction between two atoms separated by three bonds, etc.), and the second set extends to the hydrogen and proton-acceptors that may be involved in hydrogen-bonding.

The corresponding coefficients are determined as follows:

i) The charges in the coefficients a_{ab} are those (effective charges) obtained in theoretical calculations for the corresponding residues; the electrostatic interaction is to be corrected by inclusion (in the denominator) of an effective dielectric constant with a value 2.0 whenever charges, obtained in semiempirical, all-valence-electron calculations are used.

ii) The coefficients c_{ab} are evaluated by the expression given above;

iii) The coefficients g_{ab} have been usually evaluated in terms of the minimum for the non-bonded interactions [Scott and Scheraga (1966)]; for example [Momany *et al.* (1974)], the value of the interaction energy ϵ_{ab} at such a minimum, $R_{ab\,min}$, is related by

$$\epsilon_{ab} = (1/2)\, c_{ab} / R^6_{ab\,min} = -g_{ab} / R^{12}_{ab\,min}$$

to the coefficients c_{ab} and g_{ab}, and this expression may be used for the evaluation of g_{ab} if the position of the minimum is known from experimental information. These repulsive non-bonded interactions are corrected by a factor with a value of 0.5 for 1-4 type interactions and a value 1.0 for all the other interactions.

iv) The coefficients f_{ab} and g'_{ab} for the hydrogen-bonding interactions may be determined by comparison with theoretical results (see Chapter 3).

In connection with the torsional energies for internal rotations in amino acids, one may distinguish between rotations about backbone bonds, to be treated in the same way for all the amino acids, and rotations about bonds of the side chains to be considered separately for each system [for more details see the work of Momany *et al.* (1975) on which the following outline has been based].

Regarding the rotation about backbone bonds, it is assumed, first of all, that there is no intrinsic torsional contribution to rotation about the N-C(α) and C(α)-C′ bonds, such rotations being represented adequately by the interatomic interactions across the bonds. The torsional energy corresponding to the rotation about the peptide bond, on the other hand, is given by

$$\Delta E(\omega) = (20.0/2)(1 - \cos 2\omega)$$

where ω is the angle of rotation (the six atoms which constitute the peptide group lying on a plane when $\omega = 0°$ or $180°$).

In most of the cases for the rotation about bonds of the side chains, there is an intrinsic torsional term in addition to all the other interactions. The parameters to be used in each case in the expression

$$\Delta E(\chi_i) = (U_{i0}/2)\ (1 \pm \cos n\chi_i)$$

are determined by fitting of the difference between the experimental (or theoretical) values and the remaining contributions to the barrier. The general values (except as indicated below) are n = 3 and

$U_{i0} = 2.7$ kcal/mole for rotation about C-C (saturated) bonds;
$U_{i0} = 0.6$ kcal/mole for rotation about C-O (hydroxyl) bonds;
$U_{i0} = 1.8$ kcal/mole for rotation about C-N (amine) bonds.

The torsional energy terms to be used in each case are:

amino acid	*variation of dihedral angle*	*expression to be used*
alanine	χ_1	general expression
arginine	χ_1, χ_2, χ_3	general expression
	χ_4	no contribution
	χ_5	$(18.0/2)\ (1 - \cos 2\chi_5)$
	χ_{61}	$(20.0/2)\ (1 - \cos 2\chi_{61})$
	χ_{62}	$(20.0/2)\ (1 - \cos 2\chi_{62})$
asparagine	χ_1	general expression
	χ_2	$(15.0/2)\ (1 - \cos 2\chi_2)$
aspartic acid	χ_1	general expression
	χ_2	no contribution
	χ_{32}	$(8.0/2)\ (1 - \cos 2\chi_{32})$

cysteine	χ_1	general expression
	χ_2	$(1.5/2)\,(1 + \cos 3\chi_2)$
cystine		see discussion below
glutamic acid	χ_1, χ_2	general expression
	χ_3	no contribution
	χ_{42}	$(8.0/2)\,(1 - \cos 2\chi_{42})$
glutamine	χ_1, χ_2	general expression
	χ_3	$(15.0/2)\,(1 - \cos 2\chi_3)$
glycine		no side chain
histidine	C(α)-C(β)	general expression
	C(β)-C(γ)	no contribution
hydroxyproline		see discussion below
isoleucine	χ_1, χ_{21}, χ_{22}, χ_{31}	general expression
leucine	χ_1, χ_2, χ_{31}, χ_{32}	general expression
lysine	χ_1, χ_2, χ_3, χ_4, χ_5	general expression
methionine	χ_1, χ_2	general expression
	χ_3	$(2.0/2)\,(1 + \cos 3\chi_3)$
	χ_4	$(2.0/2)\,(1 + \cos 3\chi_4)$
ornithine	χ_1, χ_2, χ_3, χ_4	general expression
phenylalanine	C(α)-C(β)	general expression
	C(β)-C(γ)	no contribution
proline		see discussion below
serine	χ_1, χ_2	general expression
threonine	χ_1, χ_{21}, χ_{22}	general expression
tryptophan	C(α)-C(β)	general expression
	C(β)-C(γ)	no contribution
tyrosine	C(α)-C(β)	general expression
	C(β)-C(γ)	no contribution
	χ_6	$(3.5/2)\,(1 - \cos 2\chi_6)$
valine	χ_1, χ_{21}, χ_{22}	general expression

The above tabulation requires some additional comments for some special cases:

arginine	the expressions for χ_5, χ_{61} and χ_{62} are also valid for the uncharged guanidyl group
asparagine	the amide group is taken to be planar
glutamine	the amide group is taken to be planar
valine	the variations in χ_{21} and χ_{22} are strongly coupled through hydrogen-atom interactions and must be carried out simultaneously

In addition, it must be noticed that the expression for χ_{32} of aspartic acid [or χ_{42} of glutamic acid] may also be used for the α-carboxyl end-group of polypeptides; for carboxylates there is no torsional contribution for χ_2 (in aspartic acid) or for χ_3 (in glutamic acid).

In the case of cystine it is necessary to include the two contributions

$$\Delta E(\chi_{SS}') = A\,(R_4 - R_{40})^2 \text{ with } A = 10 \text{ kcal/mol A}^2$$
$$\Delta E_{loop} = B\,(R_i - R_{i0})^2 \text{ with } B = 100 \text{ kcal/mol A}^2,\ i = 1,2,3$$

corresponding, respectively, to rotation about the S-S bond and the loop-closing energy; the distances R_i correspond to the separations S-S′, C(β)-S, S-C′(β), and C(β)-C′(β) (where the prime denotes atoms in one of the half-cystine residues), and R_{i0} are the equilibrium distances when $\chi_{SS}' = 90°$.

In proline and hydroxyproline there is the torsional potential for the backbone peptide bond and an internal energy change associated with the form (up, down or planar) of the pyrrolidine ring; the corresponding energies are given in Table 4-1. The torsional potential for the rotation of the hydroxyl group in hydroxyproline is the same as that in serine and threonine.

Table 4-1. Internal conformational energy (kcal/mole) of the fixed part of a proline residue*

Puckering Conformation	*Peptide Bond Conformation*	*Internal Energy*
down	cis	0.000
up	cis	0.202
planar	cis	0.920
down	trans	0.336
up	trans	0.550
planar	trans	1.366

* The designations down, up and planar correspond to the values (−75.0°, +18.67°), (−67.6°, −6.11°), and (−57.6°, 0.0°) for the angles that define the position of C(γ). The energies, which include the nonbonded and electrostatic contributions, are referred to the down-cis conformation with a value of 12.355 kcal/mole [Momany *et al.* (1975)]. (Reprinted with permission. Copyright by The American Chemical Society.)

Atomic and Molecular Data

The data required are

a) molecular geometry

This information may be available in the simple form of cartesian coordinates, structural data (bond lengths and bond and dihedral angles), or crystallographic results, from which the cartesian coordinates must be obtained (see Appendix A3).

b) effective charges

These values will not be available for biopolymers and must be, therefore, approximated. The most usual approximation consists of using the charges obtained in *a priori* calculations (if possible) for the molecules, whose residues form the polymer. In such an adaptation, care must be taken to conserve the electrical neutrality (or charge) of the system, introducing the appropriate changes.

c) expansion coefficients

Theoretical values of α and C may be used for long-range interactions, but in all other cases they must be obtained by semiempirical methods.

In due time, data will be available for more and more systems, but at this moment only a restricted set of data exists:

a) molecular geometry and effective charges (obtained in *a priori* calculations) for the naturally occurring amino acids, the puric and pyrimidinic bases of interest (adenine, guanine, cytosine, thymine and uracil), the phosphate group and ribose (see Tables 4-2 through 4-27 and Figs. 1-6, and 1-8 through 1-12);

Table 4-2. Geometry and effective charges of alanine*

Atom	*Class*	*x*	*y*	*z*	*Charge*
O	10	8.302	1.971	6.866	−0.42
O	9	5.076	4.360	8.391	−0.55
N	11	7.473	3.218	2.030	−0.55
C	5	6.389	3.299	6.575	0.50
C	8	5.434	3.765	3.884	−0.13
C	6	3.131	2.137	3.306	−0.59
H	2	4.967	5.767	3.690	0.21
H	1	8.101	1.384	2.190	0.26
H	1	8.939	4.472	2.307	0.25
H	3	3.635	0.157	3.342	0.20
H	3	1.663	2.449	4.695	0.20
H	3	2.394	2.579	1.453	0.22
H	4	5.851	3.929	9.952	0.41

* The coordinates (in bohrs) have been taken from the work of Lehmann *et al.* (1972) [as given by Clementi *et al.* (1977)] and the effective charges from the work of Clementi *et al.* (1977). The notation for the atoms is: O(1), O(2), N, C′, C(α), C(β), H(α), H(1), H(2), H(β_1), H(β_2), H(β_3), H(O_2).

b) expansion coefficients (obtained by fitting of *a priori* results) for the long-range interaction of amino acids and bases with water (see Tables 4-29 through 4-31);

c) expansion coefficients (obtained by semiempirical fitting) for electrostatic charge, non-bonded, and hydrogen-bonded interactions as well as torsional energy data, for amino acids (see Tables 4-32 through 4-41);

d) conformational maps (obtained by *a priori* calculations within the framework of the dipeptide model) for amino acyl residues (see Figs. 4-2 through 4-39);

e) conformational maps (obtained in *a priori* calculations) for ribose and deoxyribose (Figs. 4-40 and 4-41) and β-nucleosides (Figs. 4-42 through 4-45).

The details of the conformational analyses for the amino acyl residues and the β-nucleosides are:

Preferred conformations of the amino acyl residues

1) alanyl residue [Maigret *et al.* (1970a,b, 1971c), Pullman *et al.* (1970)]

The calculations were carried out for N-acetyl-N′-methylalanylamide, with

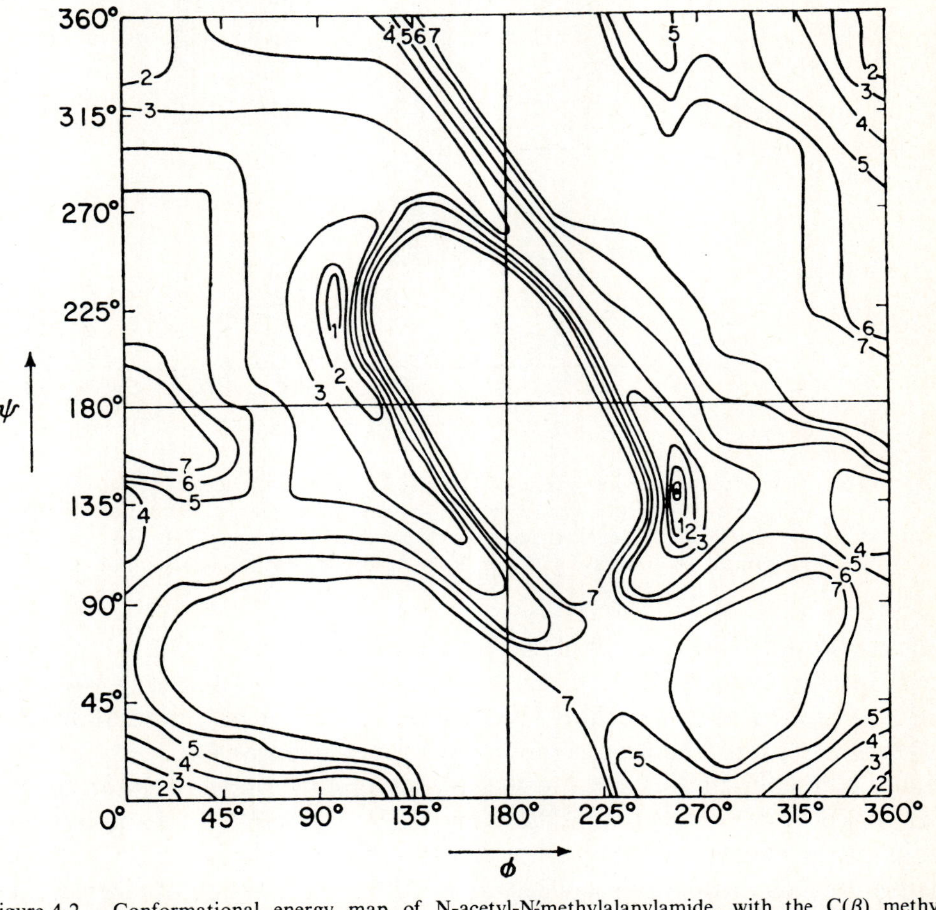

Figure 4-2. Conformational energy map of N-acetyl-N′-methylalanylamide, with the C(β) methyl staggered with respect to the C(α) bonds. The contours are labelled in kcal/mole (referred to the absolute minimum of the calculation represented by a triangle) [Maigret *et al.* (1970b)]. [Reprinted with permission. Copyright by Academic Press Inc. (London) Ltd.]

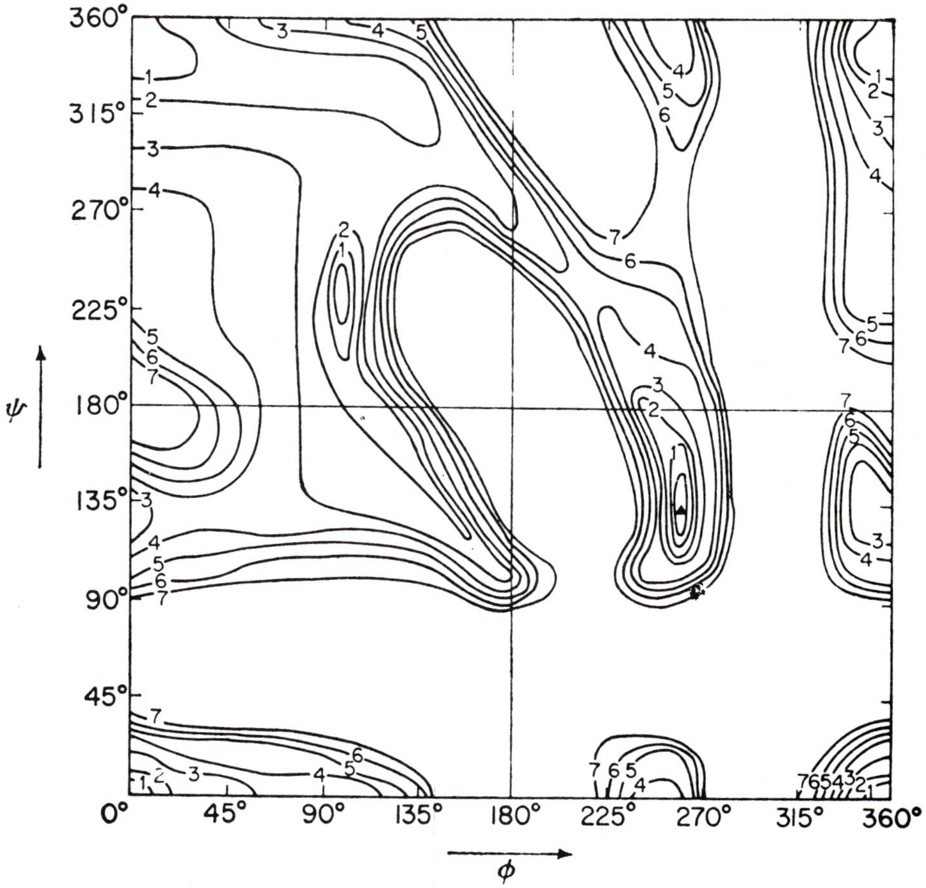

Figure 4-3. Conformational energy map of N-acetyl-N′-methylalanylamide, with the C(β) methyl eclipsed with respect to the C(α) bonds. The contours are labelled in kcal/mole (referred to the absolute minimum of the calculation represented by a triangle) [Maigret *et al.* (1970b)]. [Reprinted with permission. Copyright by Academic Press Inc. (London) Ltd.]

the peptide bond taken in the planar *trans* conformation ($\omega = 0°$), with the side-chain methyl group either staggered (Fig. 4-2) or eclipsed (Fig. 4-3) with respect to the backbone chain. The staggered form is always more stable than the eclipsed form. The deepest minimum is found in both cases at $\phi = 260°$, $\psi = 140°$, corresponding to a conformation resembling a seven-membered ring, stabilized by a hydrogen-bond between H(i + 1) and O(i − 1) and two local minima (at 1 kcal/mole above the deepest minimum) appear at $\phi = 0°$, $\psi = 0°$ (fully extended form) and $\phi = 100°$, $\psi = 220°$. The rotation of the side-chain methyl group leads to a stable conformation when the C(α)-C′, C(α)-N, and C(α)-H bonds are staggered with respect to the C(β)-H bonds; the rotational barrier (at the lowest minimum) is, in both cases, 2.0 kcal/mole. Additional

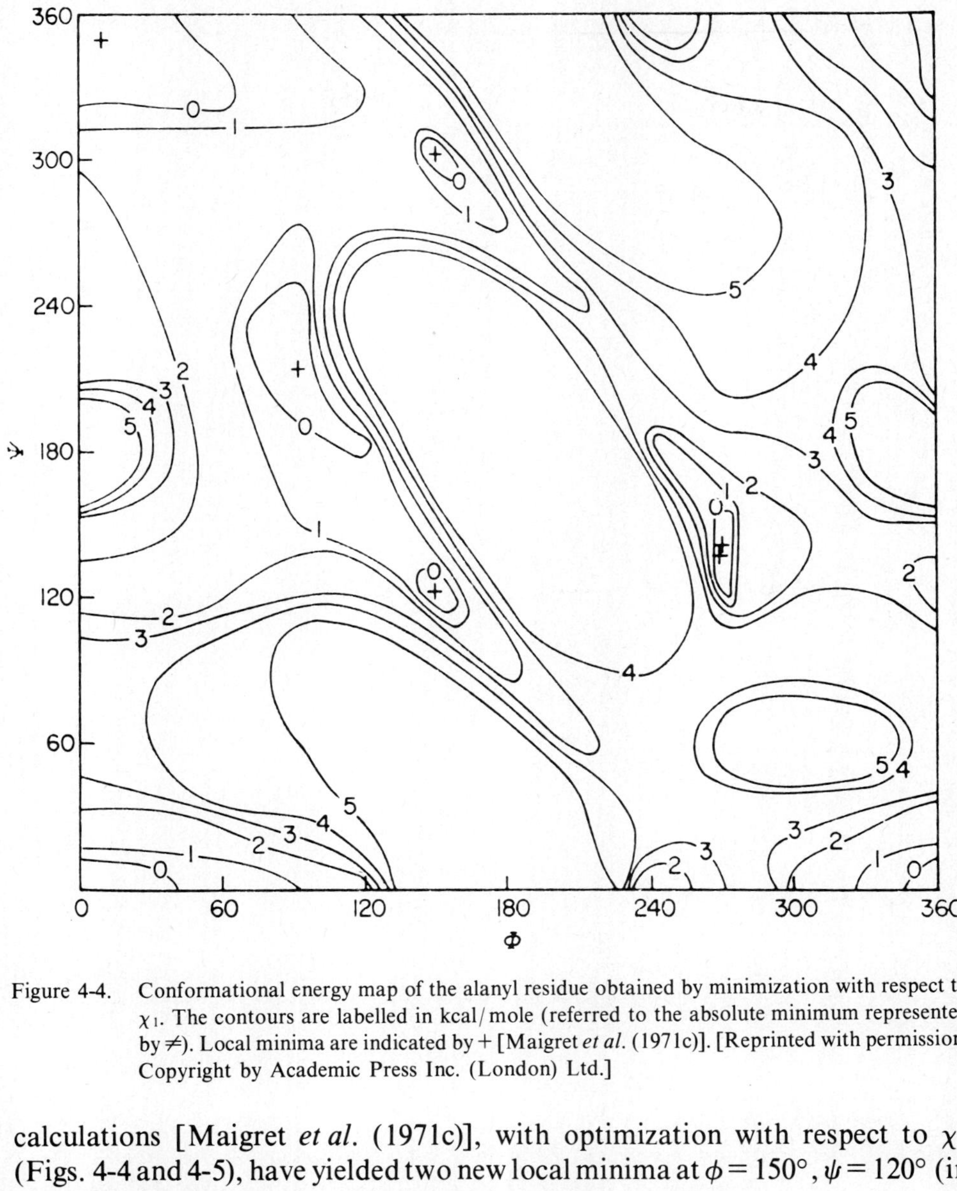

Figure 4-4. Conformational energy map of the alanyl residue obtained by minimization with respect to χ_1. The contours are labelled in kcal/mole (referred to the absolute minimum represented by ≠). Local minima are indicated by + [Maigret *et al.* (1971c)]. [Reprinted with permission. Copyright by Academic Press Inc. (London) Ltd.]

calculations [Maigret *et al.* (1971c)], with optimization with respect to χ_1 (Figs. 4-4 and 4-5), have yielded two new local minima at $\phi = 150°$, $\psi = 120°$ (in the right-handed α-helix region) and $\phi = 150°$, $\psi = 300°$.

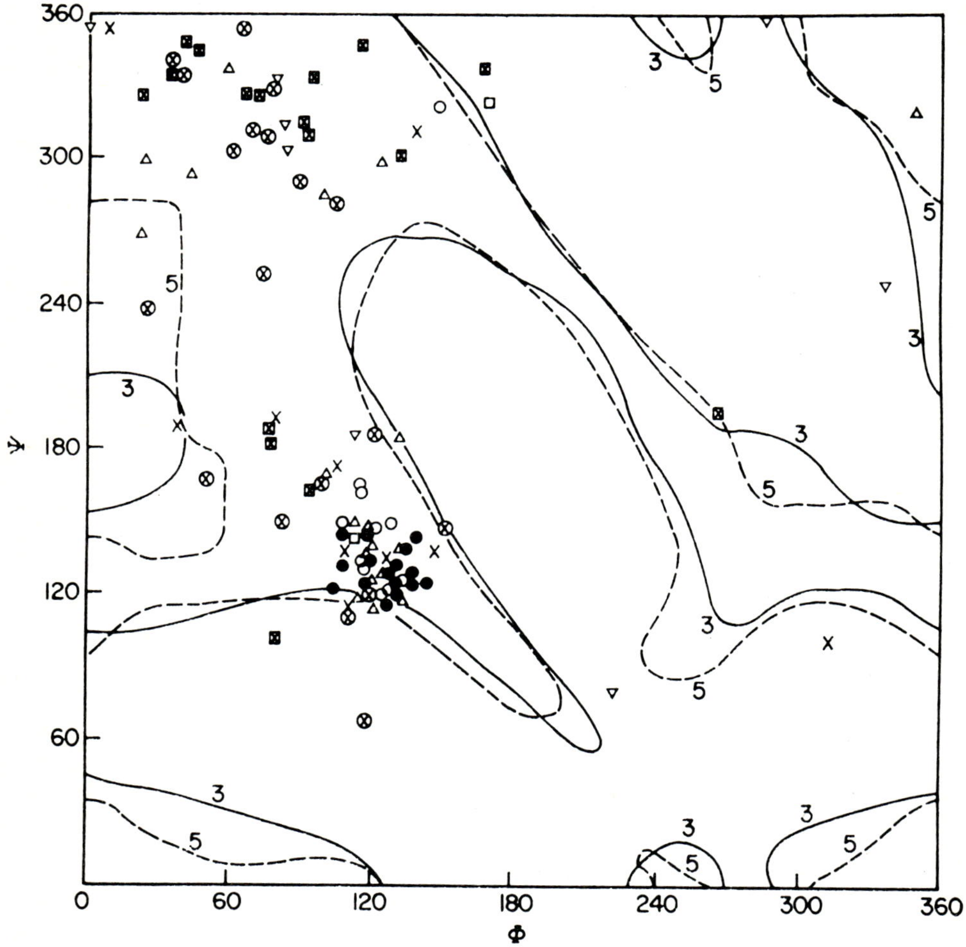

Figure 4-5. Conformational energy map of the alanyl residue. The dashed lines represent the contours (within 5 kcal/mole above the global minimum) of the original calculation [Maigret *et al.* (1970a,b)] and the full lines represent the contours (within 3 kcal/mole above the minimum) of the refined calculation [Maigret *et al.* (1971c)]. Observed conformations of alanyl residues in eight globular proteins: circle, lysozyme; shaded circle, myoglobin; square, α-lactalbumin; inverted triangle, ribonuclease-S; triangle, carboxypeptidase A; squared ×, α-chymotrypsin; circled ×, subtilisin-BPN′; ×, erythrocruorin [Maigret *et al.* (1971c)]. [Reprinted with permission. Copyright by Academic Press Inc. (London) Ltd.]

2) arginyl residue [Pullman *et al.* (1971)]

The calculations were carried out for N-acetyl-N′-methylarginylamide, with the peptide bond taken in the planar *trans* conformation ($\omega = 0°$), the two backbone terminal methyl groups fixed so that the N-H and C = O terminal bonds are each eclipsed by one of the C-H bonds of these groups, each methylene group of the side-chain staggered with respect to the preceding one, and the guanidyl group considered planar. Submaps for ϕ, ψ were obtained for $\chi_1 = 60°$, $\chi_2 = 180°$, $\chi_3 = 180°$, $\chi_4 = 180°$ (Fig. 4-6), $\chi_1 = 180°$, $\chi_2 = 180°$, $\chi_3 = 180°$, $\chi_4 = 180°$ (Fig. 4-7), and $\chi_1 = 300°$, $\chi_2 = 180°$, $\chi_3 = 180°$, $\chi_4 = 180°$ (Fig. 4-8), and a composite map (Fig. 4-9) was constructed by superposition (i.e., taking the lowest energy points) of the preceding submaps. The global minimum occurs at $\phi = 90°$, $\psi = 120°$ and a local minimum is found at $\phi = 0°$-$60°$, $\psi = 330°$-$360°$.

Table 4-3. Geometry and effective charges of arginine*

Atom	*Class*	*x*	*y*	*z*	*Charge*
O	10	7.091	9.353	21.244	−0.41
O	9	7.398	13.432	20.046	−0.55
N	11	11.568	7.995	19.198	−0.58
N	11	4.744	8.355	11.515	−0.45
N	11	2.102	11.819	9.426	−0.61
N	12	1.290	7.388	8.700	−0.50
C	5	8.182	11.079	20.096	0.51
C	8	10.564	10.521	18.587	−0.12
C	7	9.997	10.757	15.761	−0.37
C	7	7.939	8.995	14.792	−0.41
C	7	6.570	10.108	12.596	−0.28
C	5	2.701	9.143	9.823	0.34
H	2	11.926	12.000	10.090	0.18
H	1	11.454	7.718	21.081	0.27
H	1	13.402	7.978	18.810	0.26
H	2	9.539	12.727	15.397	0.19
H	2	11.743	10.425	14.772	0.20
H	2	8.762	7.230	14.212	0.21
H	2	6.654	8.506	16.208	0.22
H	2	5.567	11.728	13.305	0.19
H	2	7.808	10.841	11.226	0.19
H	1	5.215	6.481	11.429	0.31
H	1	3.111	13.239	10.196	0.28
H	1	0.608	12.290	8.297	0.28
H	1	−0.023	7.950	7.423	0.22
H	4	5.896	13.529	21.024	0.42

* The coordinates (in bohrs) have been taken from the work of Lehman *et al.* (1973) [as given by Clementi *et al.* (1977)] and the effective charges from the work of Clementi *et al.* (1977). The notation for the atoms is: O(1), O(2), N, N(ϵ), N(η_1), N(η_2), C′, C(α), C(β), C(γ), C(δ), C(ζ), H(α), H(1), H(2), H(β_1), H(β_2), H(γ_1), H(γ_2), H(δ_1), H(δ_2), H(ϵ), H(η_2), H(η_{12}), H(η_{11}), H(O_2).

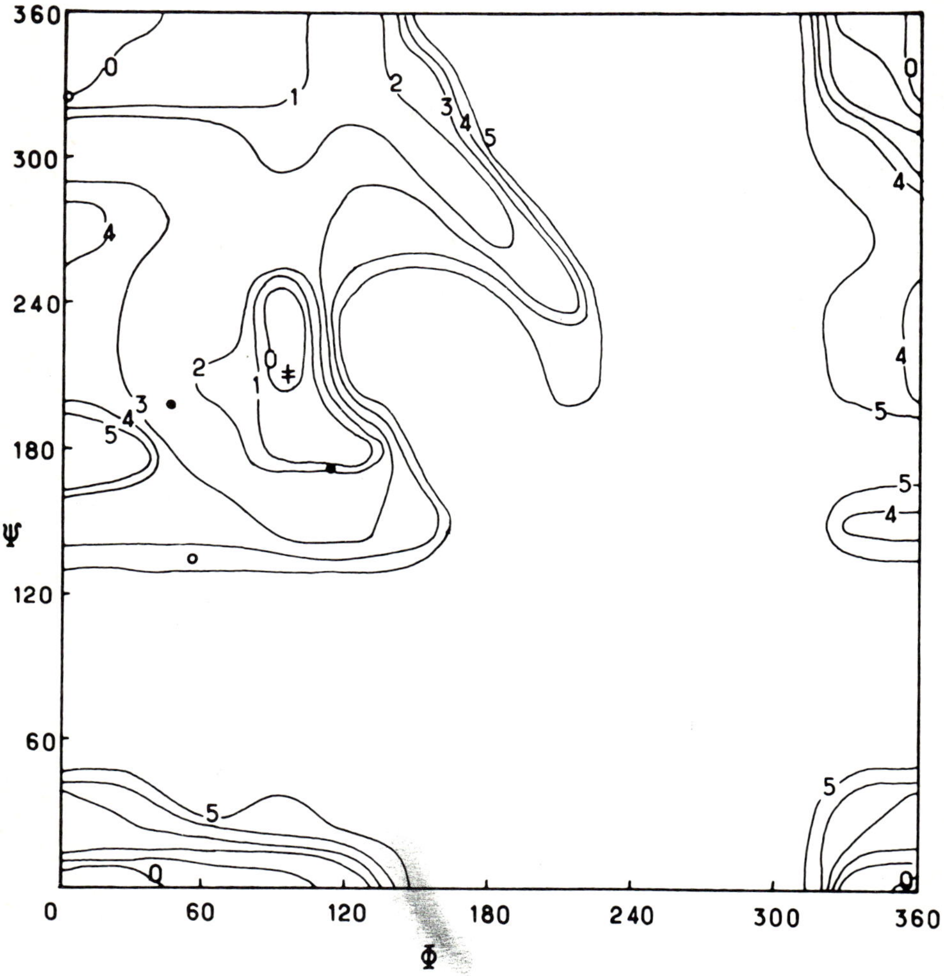

Figure 4-6. Conformational energy submap of the arginyl residue, with $\chi_1 = 60°$, $\chi_2 = 180°$, $\chi_3 = 180°$, $\chi_4 = 180°$. The notation used is: ≠, global minimum; shaded circle, conformations of arginyl residues (with this set of χ) in globular proteins; circle, conformations of other arginyl residues (with $\chi_1 = 60°$) in globular proteins [Pullman *et al.* (1971)].

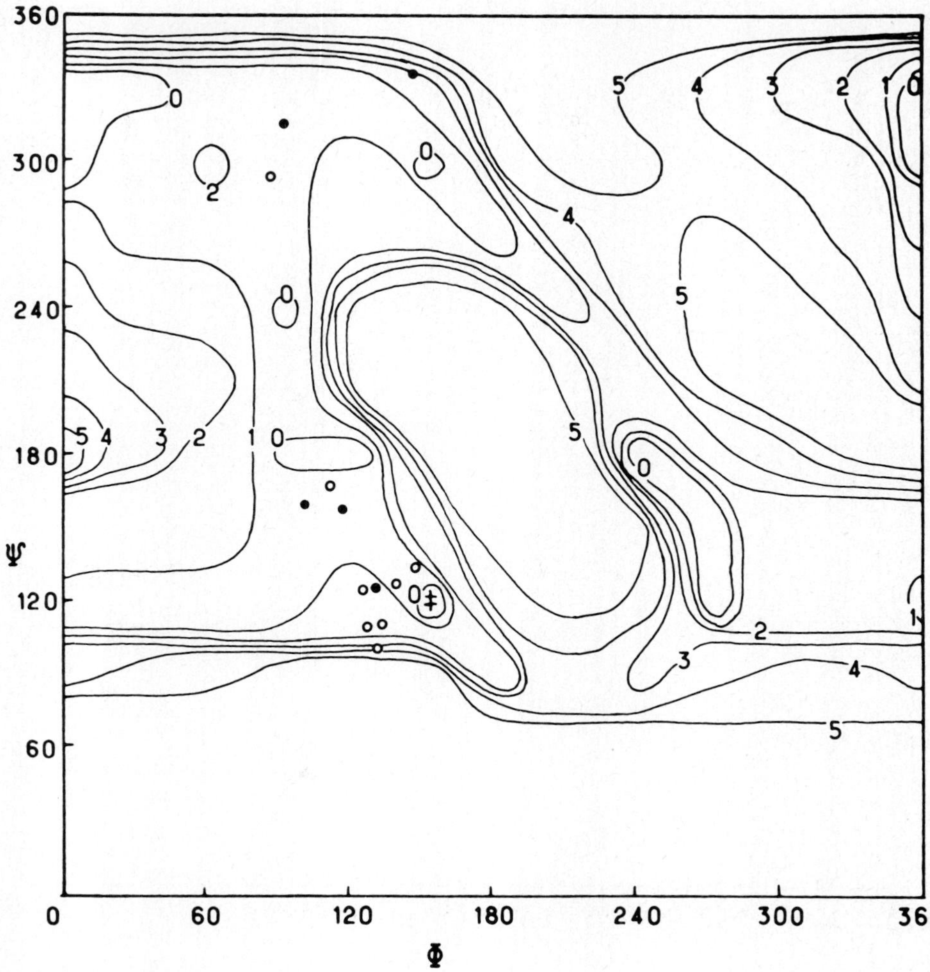

Figure 4-7. Conformational energy submap for the arginyl residue, with $\chi_1 = 180°$, $\chi_2 = 180°$, $\chi_3 = 180°$, $\chi_4 = 180°$. The notation used is: ≠, global minimum; shaded circle, conformations of arginyl residues (with this set of χ) in globular proteins; circle, conformations of other arginyl residues (with $\chi_1 = 180°$) in globular proteins [Pullman *et al.* (1971)].

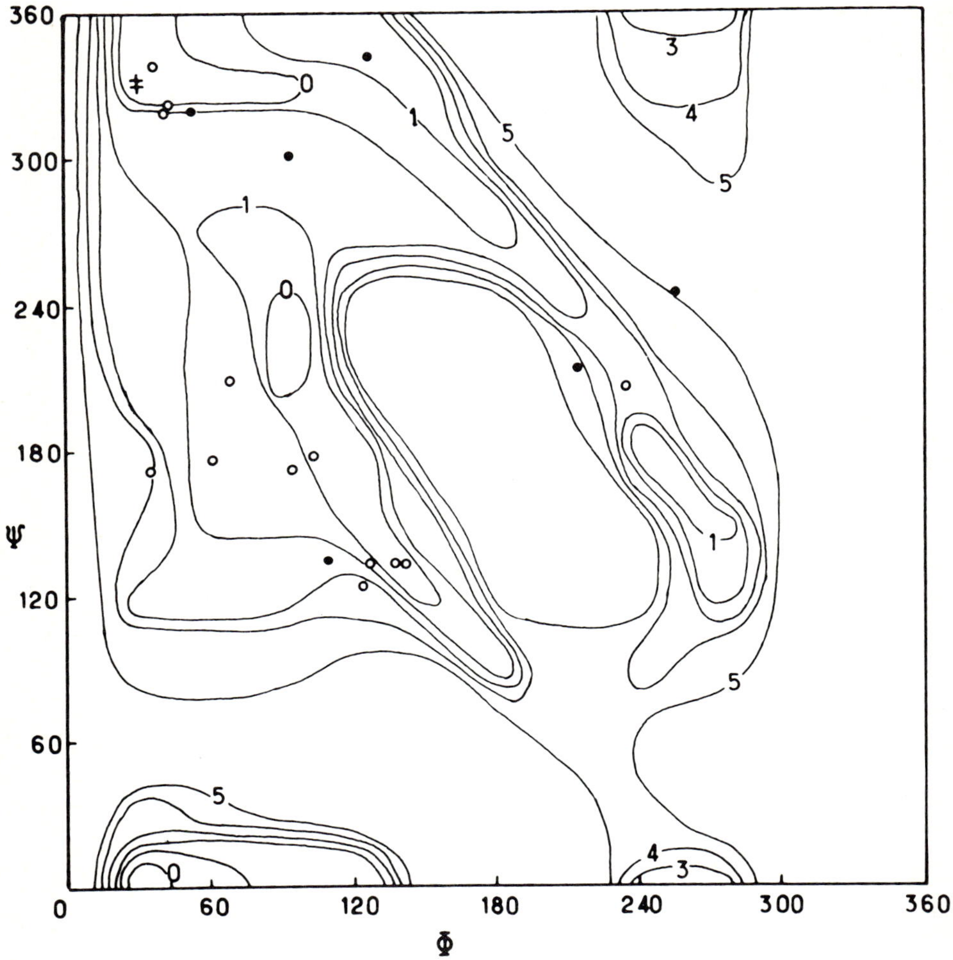

Figure 4-8. Conformational energy submap for the arginyl residue, with $\chi_1 = 300°$, $\chi_2 = 180°$, $\chi_3 = 180°$, $\chi_4 = 180°$. The global minimum is denoted by $\neq$. The notation used is: shaded circle, conformations of arginyl residues (with this set of χ) in globular proteins; circle, conformations of other arginyl residues (with $\chi_1 = 300°$) in globular proteins [Pullman *et al.* (1971)].

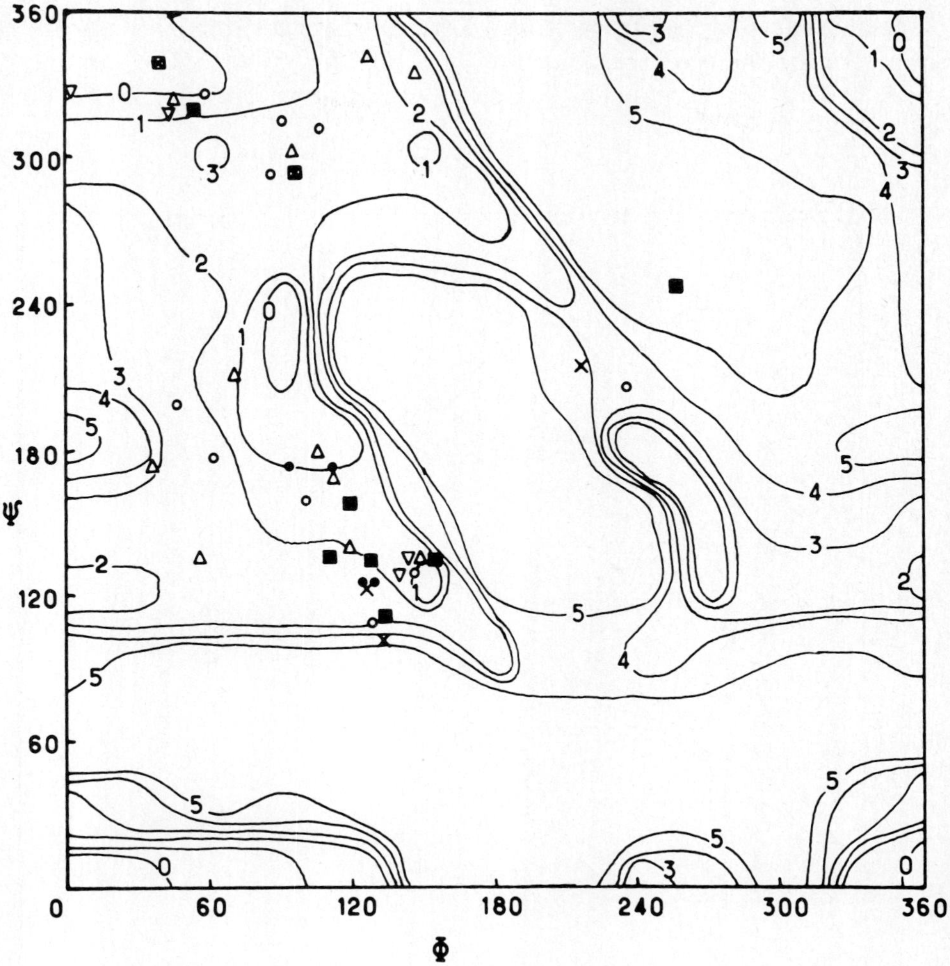

Figure 4-9. Conformational energy map for the arginyl residue. The notation used for the conformations in various systems is: circle, lysozyme; shaded circle, myoglobin; square, α-lactalbumin; inverted triangle, ribonuclease-S; triangle, carboxypeptidase A; squared ×, α-chymotrypsin; ×, subtilisin-BPN′; circled ×, erythrocruorin; shaded square, oxyhemoglobin; shaded triangle, rubredoxin [Pullman *et al.* (1971)].

3) asparaginyl residue [Maigret *et al.* (1971d)]

The calculations were preformed for N-acetyl-N′-methylasparaginylamide, with the peptide group taken in the planar *trans* conformation ($\omega = 0°$), the two backbone terminal methyl groups fixed so that the N-H and C = O terminal bonds are each eclipsed by one of the C-H bonds of these groups, the C(β) methylene group staggered with respect to the C(α)-H bond, and the amino group of the side-chain coplanar with the adjacent carbonyl bond. The complete map (Fig. 4-10) was constructed by superposition (i.e., taking the lowest energy points) of the nine maps for ϕ, ψ with $\chi_1 = 60°, 180°, 300°$ and $\chi_2 = 60°, 180°, 300°$, respectively. Almost equivalent global minima are found at

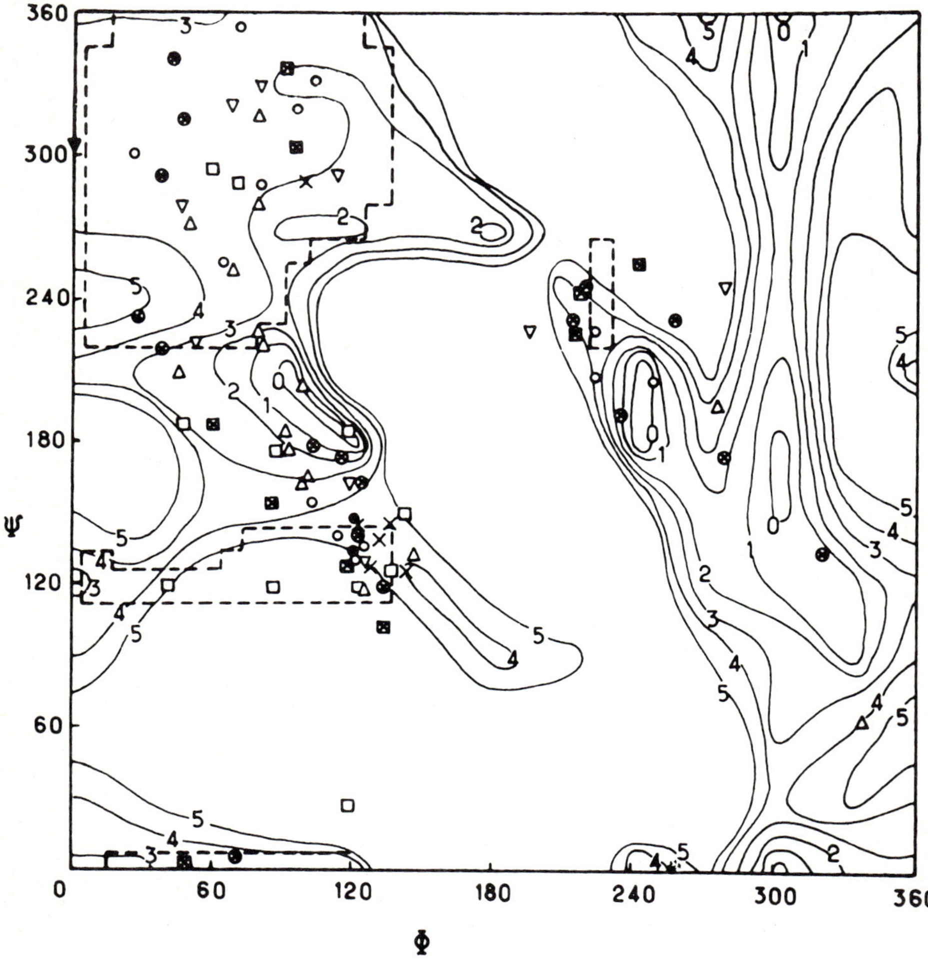

Figure 4-10. Conformational energy map for the asparaginyl residue. The contours are labelled in kcal/mole (referred to the global minimum). The dashed lines indicate the limits of semiempirical calculations. The notation used for the observed conformations of asparaginyl residues in eight globular proteins is: circle, lysozyme; shaded square, myoglobin; square, α-lactalbumin; inverted triangle, ribonuclease-S; triangle, carboxypeptidase A; squared ×, α-chymotrypsin; circled ×, subtilisin-BPN′; ×, erythrocruorin [Maigret *et al.* (1971d)].

$\phi = 90°$-$120°$, $\psi = 180°$-$210°$ (with $\chi_1 = \chi_2 = 60°$, seven-membered hydrogen-bonded ring, stabilized by an additional hydrogen-bond involving the amino group of the side-chain), $\phi = 240°$, $\psi = 180°$-$210°$ (with $\chi_1 = \chi_2 = 60°$, conformation near the left-handed α-helix region, with a seven-membered

Table 4-4. Geometry and effective charges of asparagine*

Atom	*Class*	*x*	*y*	*z*	*Charge*
O	9	−0.027	0.889	0.007	−0.56
O	10	−1.257	−3.223	−0.149	−0.39
O	10	1.208	−1.967	5.312	−0.42
N	13	1.128	1.187	8.217	−0.63
N	11	−4.388	−2.893	3.830	−0.54
C	5	−1.313	−1.075	0.802	0.52
C	8	−3.221	−0.540	2.889	−0.11
C	5	0.174	−0.035	6.190	0.41
C	7	−2.219	1.042	5.071	−0.43
H	1	−3.277	−3.746	5.212	0.26
H	1	−6.172	−2.540	4.576	0.24
H	1	2.643	0.512	9.079	0.33
H	1	0.301	2.785	8.801	0.31
H	2	−4.741	0.643	2.055	0.20
H	2	−1.767	2.959	4.327	0.21
H	2	−3.571	1.391	6.538	0.20
H	4	0.942	0.372	−1.413	0.41

* The coordinates (in bohrs) have been taken from the work of Ramanadham *et al.* (1972) [as given by Clementi *et al.* (1977)] and the effective charges from the work of Clementi *et al.* (1977). The notation for the atoms is: O(2), O(1), O(δ), N(δ), N, C′, C(α), C(γ), C(β), H(1), H(2), H(δ_1), H(δ_2), H(α), H(β_1), H(β_2), H(O_2).

hydrogen-bonded ring), $\phi = 300°$, $\psi = 150°$-$180°$ (with $\chi_1 = 300°$ and $60°$, $\chi_2 = 180°$), and $\phi = 300°$, $\psi = 0°$ (with $\chi_1 = 300°$, $\chi_2 = 180°$, a variation of the fully extended form), and a local minimum appears at $\phi = 180°$, $\psi = 270°$ ($\chi_1 = 180°$, $\chi_2 = 300°$, 2 kcal/mole above the global minima, seven-membered hydrogen-bonded ring involving the amino group of the side-chain).

4) aspartyl residue [Maigret *et al.* (1971d)]

The calculations were performed on N-acetyl-N′-methylaspartylamide, with the peptide bond taken in the planar *trans* conformation ($\omega = 0°$), the two backbone terminal methyl groups fixed so that the N-H and C = O terminal

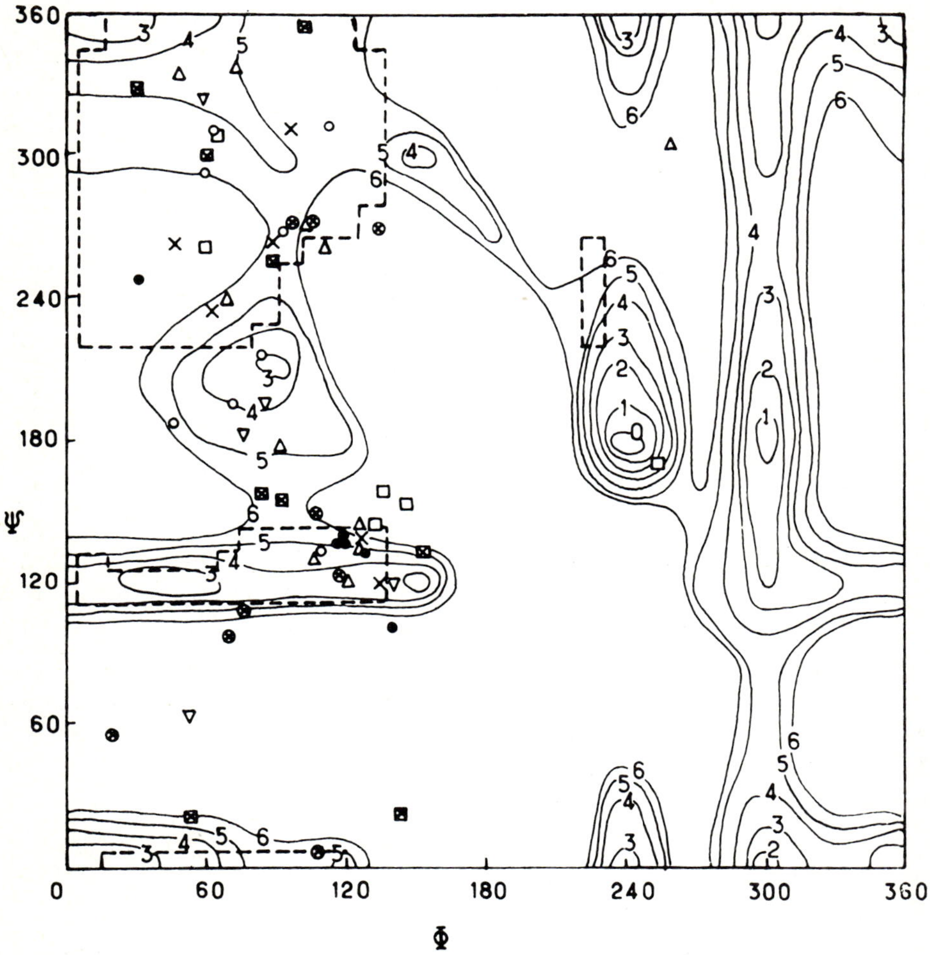

Figure 4-11. Conformational energy map for the aspartyl residue. The contours are labelled in kcal/mole (referred to the global minimum). The dashed lines indicate the limits of semiempirical calculations. The notation used for the conformations of aspartyl residues in eight globular proteins is: circle, lysozyme; shaded circle, myoglobin; square, α-lactalbumin; inverted triangle, ribonuclease-S; triangle, carboxypeptidase A; squared ×, α-chymotrypsin; circled ×, subtilisin-BPN′; ×, erythrocruorin [Maigret *et al.* (1971d)].

bonds are each eclipsed by one of the C-H bonds of these groups, the C(β) methylene group staggered with respect to the C(α)-H bond, the carboxyl group assumed planar ($\chi_3 = 180°$, with the C-OH bond *cis* to the C=O bond), and the hydroxyl group staggered with respect to the adjacent methylene group. The complete map (Fig. 4-11) was constructed by superposition (i.e., taking the lowest energy points) of the nine maps for ϕ, ψ with $\chi_1 = 60°, 180°, 300°$ and $\chi_2 = 60°, 180°, 300°$, respectively. The global minimum is located at $\phi = 240°$, $\psi = 180°$ (with $\chi_1 = 60°$, $\chi_2 = 60°$) and local minima are found at $\phi = 300°$, $\psi = 180°$, (with $\chi_1 = 60°$, $\chi_2 = 180°$, at 1 kcal/mole above the global minimum), $\phi = 30°$, $\psi = 330°$-$360°$ (almost fully extended form), $\phi = 150°$ and $\psi = 120°$, $\phi = 90°$ and $\psi = 210°$, $\phi = 240°$ and $\psi = 0°$, and $\phi = 300°$, $\psi = 0°$.

Table 4-5. Geometry and effective charges of aspartic acid*

Atom	*Class*	*x*	*y*	*z*	*Charge*
O	10	7.577	−0.524	11.055	−0.41
O	9	9.104	−2.528	7.547	−0.54
O	10	−0.522	−2.436	6.969	−0.42
O	9	−1.277	−4.754	10.387	−0.55
N	11	4.615	−2.827	5.277	−0.65
C	5	7.298	−1.616	8.980	0.52
C	8	4.583	−1.854	7.972	−0.16
C	7	2.939	−3.474	9.678	−0.57
C	5	0.214	−3.512	8.866	0.55
H	1	3.004	−2.887	4.708	0.32
H	1	5.350	−4.444	5.207	0.29
H	2	3.903	−0.121	7.868	0.26
H	2	3.013	−2.850	11.361	0.29
H	2	3.531	−5.196	9.713	0.26
H	4	−3.016	−4.687	9.898	0.42
H	4	10.669	−2.212	8.369	0.42

* The coordinates (in bohrs) have been taken from the work of Rao (1973) [as given by Clementi *et al.* (1977)] and the effective charges from the work of Clementi *et al.* (1977). The notation for the atoms is: O(1), O(2), O(δ_1), O(δ_2), N, C′, C(α), C(β), C(γ), H(1), H(2), H(α), H(β_1), H(β_2), H(δ_2), H(O_2).

5) glutamyl residue [Maigret *et al.* (1971d)]

The calculations were carried out for N-acetyl-N′-methylglutamylamide, with the peptide group taken in the planar *trans* conformation ($\omega = 0°$), the two backbone terminal methyl groups fixed so that the N-H and C=O terminal bonds are each eclipsed by one of the C-H bonds of these groups, the C(β)

methylene group staggered with respect to the C(α)-H bond, the C(γ) methylene group staggered with respect to the C(β) group, the carboxyl group

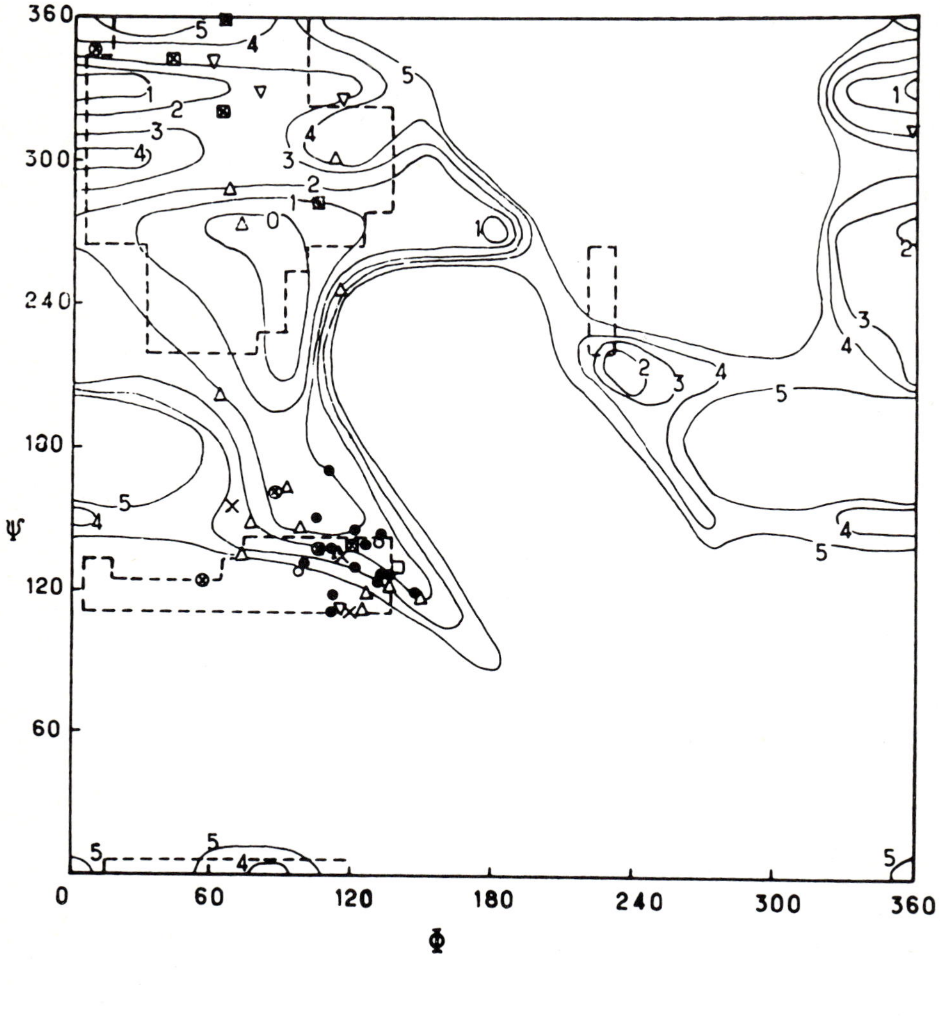

Figure 4-12. Conformational energy map for the glutamyl residue. The contours are labelled in kcal/mole (referred to the global minimum). The dashed lines indicate the limits of semiempirical calculations. The notation used for the conformations of glutamyl residues in eight globular proteins is: circle, lysozyme; shaded circle, myoglobin; square, α-lactalbumin; inverted triangle, ribonuclease-S; triangle, carboxypeptidase A; squared ×, α-chymotrypsin; circled ×, subtilisin-BPN′; ×, erythrocruorin [Maigret *et al.* (1971d)].

assumed planar ($\chi_4 = 180°$, with the C-OH bond *cis* to the C=O bond), and the hydroxyl group staggered with respect to the adjacent methylene group. The complete map (Fig. 4-12) was constructed by superposition (i.e., by taking the lowest energy points) of the twenty-seven maps for ϕ, ψ with $\chi_1 = 60°$, 180°, 300°, $\chi_2 = 60°$, 180°, 300°, and $\chi_3 = 60°$, 180°, 300°, respectively. Two equivalent deepest minima appear at $\phi = 90°$, $\psi = 240°$ (with $\chi_1 = 60°$, $\chi_2 = 60°$, $\chi_3 = 60°$) and $\phi = 90°$, $\psi = 210°$ (with $\chi_1 = 180°$, $\chi_2 = 300°$, $\chi_3 = 300°$, seven-membered hydrogen-bonded ring), and local minima are found at $\phi = 180°$, $\psi = 270°$ (with $\chi_1 = 60°$, $\chi_2 = 60°$, $\chi_3 = 60°$), $\phi = 0°$, $\psi = 330°$ (with $\chi_1 = 60°$, $\chi_2 = 60°$, $\chi_3 = 180°$, resembling the fully extended and the antiparallel sheet forms), and $\phi = 240°$, $\psi = 210°$ (with $\chi_1 = 180°$, $\chi_2 = 300°$, $\chi_3 = 300°$, about 2 kcal/mole over the deepest minima).

Table 4-6. Geometry and effective charges of glutamic acid*

Atom	*Class*	*x*	*y*	*z*	*Charge*
O	9	4.253	10.307	23.389	−0.53
O	10	2.042	6.744	23.932	−0.41
O	10	4.083	9.984	13.747	−0.41
O	9	6.751	7.406	11.764	−0.54
N	11	5.450	3.981	21.250	−0.53
C	5	3.763	7.943	22.970	0.52
C	8	5.539	6.769	21.000	−0.12
C	7	4.679	7.687	18.379	−0.39
C	7	6.443	7.054	16.186	−0.42
C	5	5.630	8.264	13.810	0.52
H	4	2.838	11.089	24.429	0.39
H	4	6.279	8.279	10.197	0.40
H	1	5.651	3.476	23.129	0.25
H	1	6.995	3.195	20.258	0.23
H	2	7.452	7.388	21.399	0.18
H	2	4.453	9.716	18.465	0.22
H	2	2.855	6.914	18.003	0.22
H	2	8.357	7.709	16.668	0.21
H	2	6.633	4.982	15.871	0.22

* The coordinates (in bohrs) have been taken from the work of Sequeira *et al.* (1972) [as given by Clementi *et al.* (1977)] and the effective charges from the work of Clementi *et al.* (1977). The notation for the atoms is: O(2), O(1), O(ϵ_1), O(ϵ_2), N, C′, C(α), C(β), C(γ), C(δ), H(O_2), H(ϵ_2), H(1), H(2), H(α), H(β_1), H(β_2), H(γ_1), H(γ_2).

6) glutaminyl residue [Maigret *et al.* (1971d)]

The calculations were carried out for N-acetyl-N′-methylglutaminylamide, with the peptide group taken in the planar *trans* conformation ($\omega = 0°$), the two backbone terminal methyl groups fixed so that the N-H and C = O terminal

Table 4-7. Geometry and effective charges of glutamine*

Atom	*Class*	*x*	*y*	*z*	*Charge*
C	5	14.907	8.505	7.153	0.49
C	8	12.424	9.768	6.085	−0.12
C	7	10.348	7.771	5.953	−0.38
C	7	11.005	5.518	4.306	−0.42
C	5	9.016	3.475	4.365	0.42
N	11	12.809	11.001	3.571	−0.55
N	13	9.246	1.704	2.593	−0.63
O	10	16.648	8.273	5.480	−0.40
O	9	14.897	7.866	9.548	−0.54
O	10	7.323	3.450	5.952	−0.42
H	1	13.853	9.851	2.366	0.25
H	1	11.120	11.353	2.704	0.24
H	2	11.870	11.262	7.398	0.21
H	2	9.987	7.129	7.899	0.21
H	2	8.607	8.660	5.266	0.19
H	2	11.335	6.070	2.346	0.19
H	2	12.739	4.648	4.966	0.20
H	1	10.726	1.702	1.391	0.30
H	1	8.080	0.216	2.553	0.32
H	4	16.494	7.400	9.929	0.42

* The coordinates (in bohrs) have been taken from the work of Koetzle *et al.* (1973b) [as given by Clementi *et al.* (1977)] and the effective charges from the work of Clementi *et al.* (1977). The notation for the atoms is: C′, C(α), C(β), C(γ), C(δ), N, N(ϵ), O(1), O(2), O(ϵ), H(1), H(2), H(α), H(β_1), H(β_2), H(γ_1), H(γ_2), H(ϵ_2), H(ϵ_1), H(O_2).

bonds are each eclipsed by one of the C-H bonds of these groups, the C(β) methylene group staggered with respect to the C(α)-H bond, the C(γ) methylene group staggered with respect to the C(β) group, and the amino group of the side-chain coplanar with the adjacent carbonyl bond. The complete map (Fig. 4-13) was constructed by superposition (i.e., by taking the lowest energy points) of the twenty-seven maps for ϕ, ψ with $\chi_1 = 60°$, 180°, 300°, $\chi_2 = 60°$, 180°, 300° and $\chi_3 = 60°$, 180°, 300°, respectively. Three practically equivalent minima are found at $\phi = 90°$, $\psi = 210°$ (with $\chi_1 = 60°$, $\chi_2 = 300°$, $\chi_3 = 300°$, with two hydrogen-bonds), $\phi = 30°$, $\psi = 330°$-360° (with $\chi_1 = 300°$, $\chi_2 = 60°$, $\chi_3 = 180°$, slightly distorted fully extended form), and $\phi =$ 150°-210°, $\psi = 240°$-300° (with $\chi_1 = 300°$, $\chi_2 = 60°$, $\chi_3 = 60°$ and also with $\chi_1 =$ 60°, $\chi_2 = 300°$, $\chi_3 = 180°$, seven-membered hydrogen-bonded ring, involving the carbonyl of the side-chain).

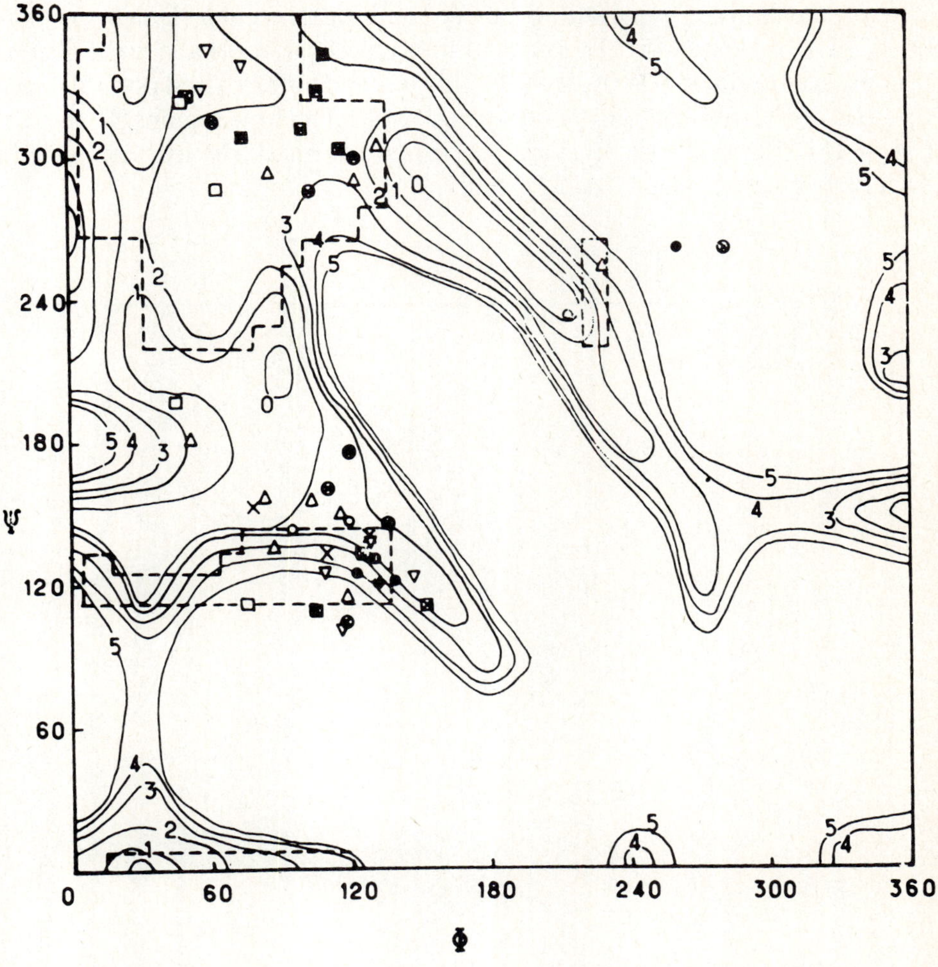

Figure 4-13. Conformational energy map for the glutaminyl residue. The contours are labelled in kcal/mole (referred to the global minimum). The dashed lines indicate the limits of semiempirical calculations. The notation used for the observed conformations of glutaminyl residues in eight globular proteins is: circle, lysozyme; shaded square, myoglobin; square, α-lactalbumin; inverted triangle, ribonuclease-S; triangle, carboxypeptidase A; squared ×, α-chymotrypsin; circled ×, subtilisin-BPN′; ×, erythrocruorin [Maigret *et al.* (1971d)].

7) glycyl residue [Maigret *et al.* (1970b), Pullman *et al.* (1970)]

The calculations were carried for N-methyl-N′-acetylglycylamide, with the peptide bond taken in the planar *trans* conformation ($\omega = 0°$). The deepest minima are found at $\phi = 260°$, $\psi = 140°$ and $\phi = 100°$, $\psi = 220°$, corresponding to seven-membered rings stabilized by an intramolecular hydrogen-bond between H(i + 1) and O(i − 1). Local minima (at 2 kcal/mole above the deepest minima) are found at $\phi = 0°$, $\psi = 0°$ (fully extended form) and $\phi = 180°$, $\psi = 90°$ (and at the symmetrical points) (Fig. 4-14). The four regions A, B, C, D (Fig. 4-15) communicate by large corridors of relatively low energy, the corridor between B and D being deeper.

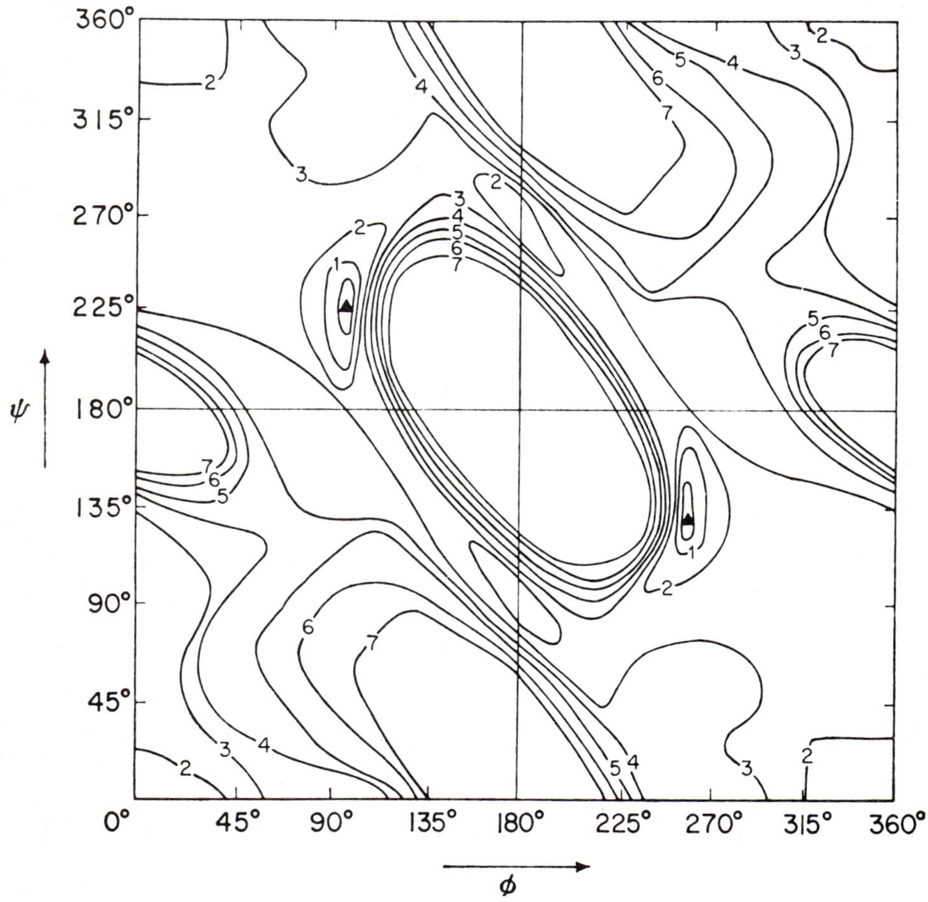

Figure 4-14. Conformational energy map of N-acetyl-N′-methylglycylamide. The contours are labelled in kcal/mole (referred to the global minimum, represented by a shaded triangle) [Maigret *et al.* (1970b)]. [Reprinted with permission. Copyright by Academic Press Inc. (London) Ltd.]

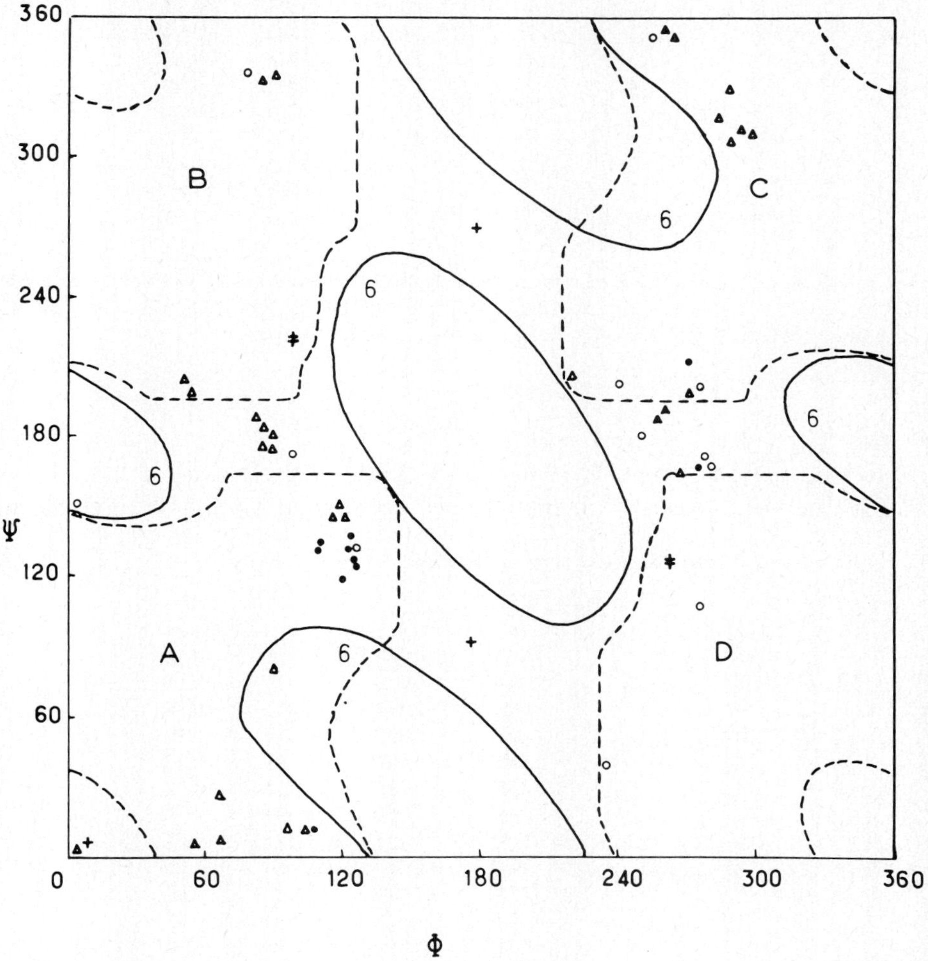

Figure 4-15. Conformational energy map of the glycyl residue. The dashed lines represent the limits of semiempirical calculations and the full lines represent the limits of the theoretical calculations (within 6 kcal/mole of the global minimum, indicated by ≠). The local minima are indicated by +. The notation used for the conformations of glycyl residues in various systems is: circle, lysozyme; shaded circle, myoglobin; triangle, small compounds [Pullman *et al.* (1970)].

Table 4-8. Geometry and effective charges of glycine*

Atom	*Class*	*x*	*y*	*z*	*Charge*
O	9	2.835	2.101	1.389	−0.50
O	10	−1.320	3.200	1.631	−0.38
N	11	2.700	2.032	−3.749	−0.55
C	5	0.673	2.824	0.416	0.46
C	8	0.580	3.276	−2.431	−0.33
H	1	2.597	2.270	−5.723	0.24
H	1	4.433	2.698	−3.125	0.25
H	2	0.689	5.303	−2.786	0.18
H	2	−1.194	2.587	−3.213	0.22
H	4	2.607	1.920	3.160	0.40

* The coordinates (in bohrs) have been taken from the work of Jonsson and Kvick (1972) [as given by Clementi *et al.* (1977)] and the effective charges from the work of Clementi *et al.* (1977). The notation for the atoms is: O(2), O(1), N, C′, C(α), H(1), H(2), H(α_1), H(α_2), H(O_2).

8) histidyl residue [Maigret *et al.* (1971a)]

The calculations were carried out for N-acetyl-N′-methylhistidylamide, with the peptide bond taken in the planar *trans* conformation ($\omega = 0°$) and the two

Table 4-9. Geometry and effective charges of histidine*

Atom	*Class*	*x*	*y*	*z*	*Charge*
O	10	2.200	0.065	7.068	−0.41
O	9	−1.836	0.142	8.618	−0.55
N	11	4.031	−2.474	11.080	−0.70
N	12	5.215	−0.837	15.940	−0.29
N	15	3.662	−0.007	19.734	−0.46
C	5	0.609	−0.235	8.783	0.52
C	8	1.611	−1.059	11.400	−0.15
C	7	2.044	1.255	13.101	−0.52
C	14	3.015	0.551	15.680	0.07
C	14	5.507	−1.129	18.411	−0.03
C	14	2.047	1.071	18.008	−0.15
H	1	4.764	−2.574	12.588	0.34
H	1	5.126	−1.702	10.020	0.31
H	16	3.414	0.000	21.503	0.39
H	2	0.401	−2.242	12.196	0.24
H	2	0.430	2.103	13.230	0.25
H	2	3.131	2.546	12.303	0.24
H	16	6.887	−2.020	19.114	0.25
H	16	0.519	2.020	18.436	0.23
H	4	−2.196	0.691	6.947	0.42

* The coordinates (in bohrs) have been taken from the work of Eddington and Harding (1974) [as given by Clementi *et al.* (1977)] and the effective charges from the work of Clementi *et al.* (1977). The notation for the atoms is O(1), O(2), N, N(δ), N(ϵ), C′, C(α), C(β), C(γ), C(ϵ), C(δ), H(1), H(2), H(ϵ_2), H(α), H(β_1), H(β_2), H(ϵ_1), H(δ), H(O_2).

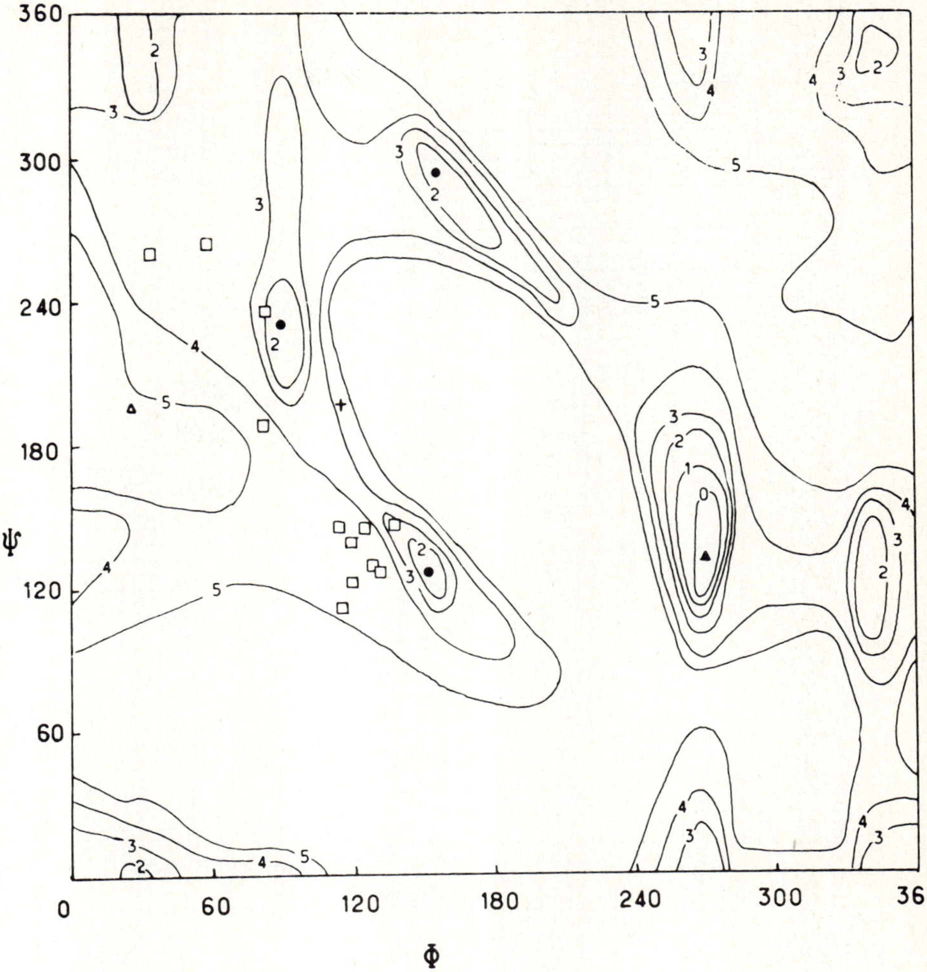

Figure 4-16. Conformational energy map for the histidyl residue. The contours are labelled in kcal/mole (referred to the global minimum represented by a shaded triangle). The local minima are represented by shaded circles. The notation used for the conformations of histidyl residues in various systems is: +, lysozyme; square, myoglobin; triangle, small compounds [Maigret *et al.* (1971a)].

terminal methyl groups fixed so that the N-H and C = O terminal bonds are each eclipsed by one of the C-H bonds of these groups. The resulting conformational map (Figs. 4-16 and 4-17) was constructed by superposition (i.e., taking the lowest energy points) of the six maps for ϕ, ψ with $\chi_1 = 60°$, 180°, 300°, and $\chi_2 = 60°$, 300°, respectively. The global minimum is located at $\phi = 260°$, $\psi = 140°$ (seven-membered hydrogen-bonded ring) and local minima (less than 2 kcal/mole above the global minimum) are found at $\phi = 150°$, $\psi =$

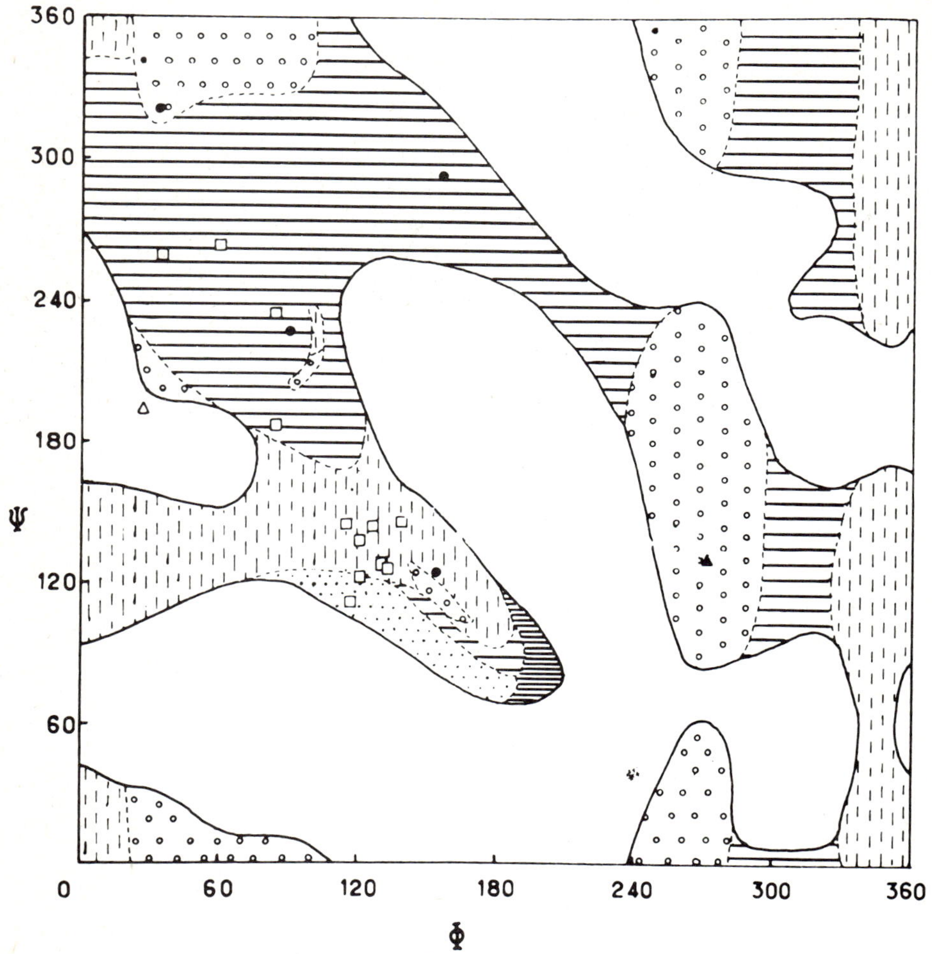

Figure 4-17. Conformational energy map for the histidyl residue, showing the preferred values for the angles χ_1 and χ_2. The outside contour corresponds to a level of 5 kcal/mole above the global minimum, denoted by a shaded triangle. The local minima are represented by shaded circles. The notation used for the histidyl residues in various systems is: +, lysozyme; square, myoglobin; triangle, small compounds. The various regions correspond to the following values of χ_1, χ_2 : vertical dashed lines, $\chi_1 = 60°$, $\chi_2 = 300°$; horizontal lines (wide), $\chi_1 = 180°$, $\chi_2 = 60°$; horizontal lines (narrow), $\chi_1 = 180°$, $\chi_2 = 300°$; circles, $\chi_1 = 300°$, $\chi_2 = 60°$; dots, $\chi_1 = 300°$, $\chi_2 = 300°$ [Maigret *et al.* (1971a)].

120° (in the right-handed α-helix region), $\phi = 40°$, $\psi = 320°$ (in the antiparallel pleated-sheet region), $\phi = 100°$, $\psi = 240°$ (another seven-membered hydrogen-bonded ring), and $\phi = 150°$, $\psi = 280°$.

9) isoleucyl residue [Maigret *et al.* (1971b)]

The calculations were carried out for N-acetyl-N′-methylisoleucylamide, with the peptide bond taken in the planar *trans* conformation ($\omega = 0°$), the two backbone terminal groups fixed so that the N-H and C=O terminal bonds are each eclipsed by one of the C-H bonds of these groups, and the two terminal C(γ_2) and C(δ) methyl groups staggered with respect to the C(β)-H and C(γ_1)-

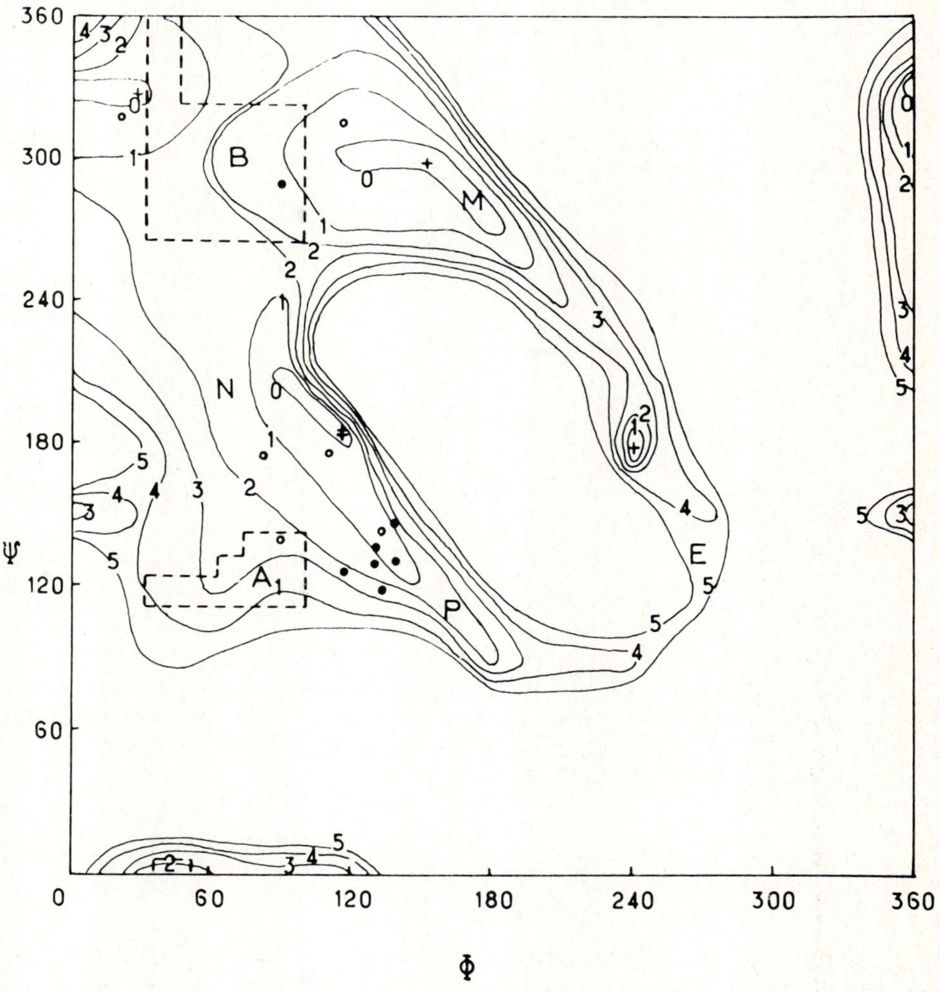

Figure 4-18. Conformational energy map for the isoleucyl residue. The contours are labelled in kcal/mole (referred to the global minimum indicated by a double cross). The local minima are represented by +. The dashed lines represent the limits of semiempirical calculations. The notation used for the conformations of isoleucyl residues in various systems is: circle, lysozyme; shaded circle, myoglobin [Maigret *et al.* (1971b)].

H_2 groups, respectively. The global minimum appears at $\phi = 120°$, $\psi = 180°$ and local minima are found at $\phi = 150°$ and $\psi = 300°$, $\phi = 30°$ and $\psi = 330°$ (antiparallel pleated-sheet region), and $\phi = 240°$, $\psi = 180°$, respectively (Fig. 4-

Table 4-10. Geometry and effective charges of isoleucine*

Atom	*Class*	*x*	*y*	*z*	*Charge*
O	9	11.414	5.430	9.769	−0.54
O	10	12.923	9.392	8.907	−0.41
N	11	16.254	3.561	8.916	−0.50
C	5	13.103	6.999	8.857	0.50
C	8	15.369	5.839	7.445	−0.12
C	7	14.634	5.076	4.741	−0.18
C	7	16.932	3.888	3.294	−0.37
C	6	16.344	3.200	0.539	−0.58
C	6	13.594	7.367	3.315	−0.58
H	1	16.764	4.129	10.809	0.23
H	1	17.880	2.744	7.990	0.23
H	3	18.006	2.397	−0.332	0.20
H	3	15.788	4.880	−0.478	0.19
H	3	14.824	1.840	0.487	0.20
H	3	13.083	6.813	1.419	0.19
H	3	15.020	8.825	3.237	0.19
H	3	11.944	8.081	4.284	0.20
H	2	17.511	2.195	4.276	0.18
H	2	18.472	5.227	3.314	0.18
H	2	13.156	3.672	4.854	0.20
H	2	16.879	7.210	7.363	0.18
H	4	10.071	6.405	10.453	0.42

* The coordinates (in bohrs) have been taken from the work of Torii and Iitaka (1971) [as given by Clementi *et al.* (1977)] and the effective charges from the work of Clementi *et al.* (1977). The notation for the atoms is: O(2), O(1), N, C′, C(α), C(β), C(γ_1), C(δ), C(γ_2), H(1), H(2), H(δ_1), H(δ_2), H(δ_3), H(γ_{21},), H(γ_{22}), H(γ_{23}), H(γ_{11}), H(γ_{12}), H(β), H(α), H(O_2).

18). The map for the preferred conformation of the side-chain (Fig. 4-19) was constructed by superposition (i.e., taking the lowest energy points) of the nine maps for ϕ, ψ with $\chi_1 = 60°$, 180°, 300° and $\chi_2 = 60°$, 180°, 300°, respectively.

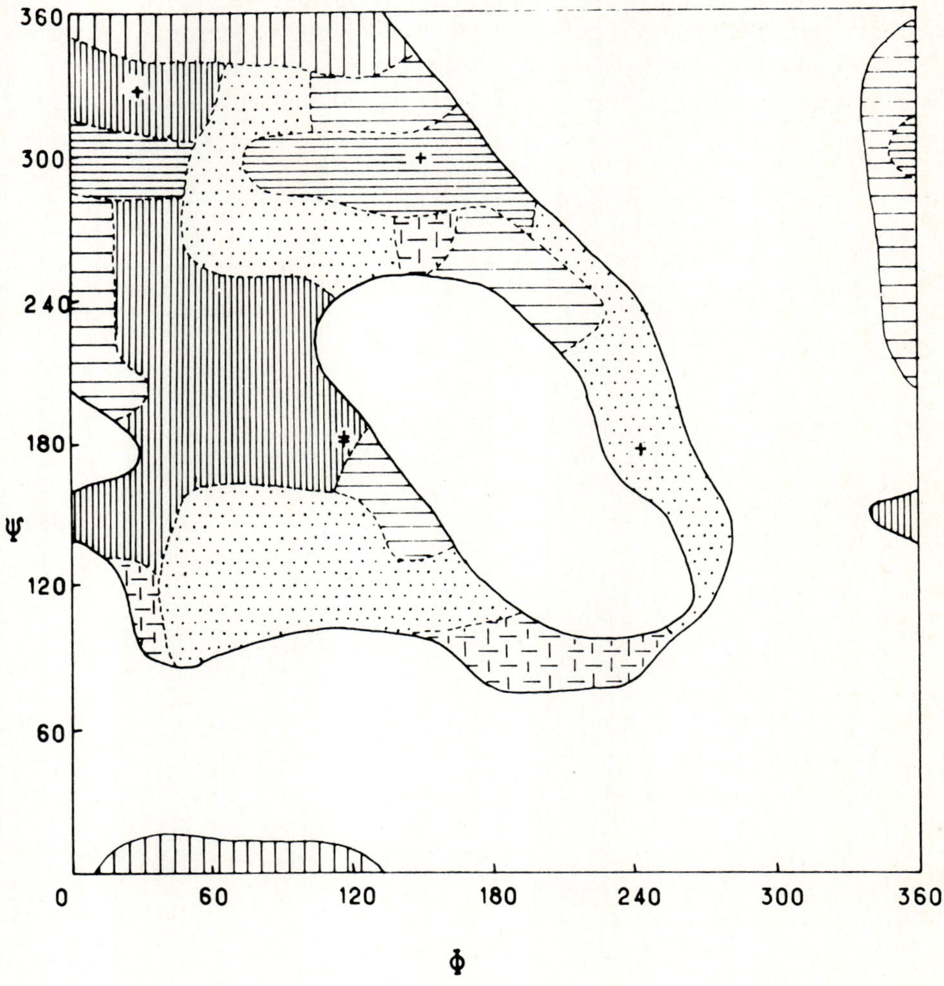

Figure 4-19. Conformational energy map for the isoleucyl residue, showing the preferred values for the angles χ_1 and χ_2. The outside contour corresponds to a level of 5 kcal/mole above the global minimum, indicated by a double cross. The local minima are represented by +. The various regions correspond to the following values of χ_1, χ_2 : blank, $\chi_1 = 60°$, $\chi_2 = 180°$; horizontal lines (wide), $\chi_1 = 180°$, $\chi_2 = 60°$; vertical lines, $\chi_1 = 180°$, $\chi_2 = 180°$; horizontal lines (narrow), $\chi_1 = 180°$, $\chi_2 = 300°$; vertical and horizontal dashed lines, $\chi_1 = 300°$, $\chi_2 = 180°$; dots, $\chi_1 = 300°$, $\chi_2 = 300°$ [Maigret *et al.* (1971b)].

10) leucyl residue [Maigret *et al.* (1971b)]

The calculations were carried out for N-acetyl-N′-methylleucylamide, with the peptide bond taken in the planar *trans* conformation ($\omega = 0°$), the two backbone terminal groups fixed so that the N-H and C=O terminal bonds are each eclipsed by one of the C-H bonds of these groups, and the two terminal C(δ_1) and C(δ_2) methyl groups fixed in a staggered configuration with respect

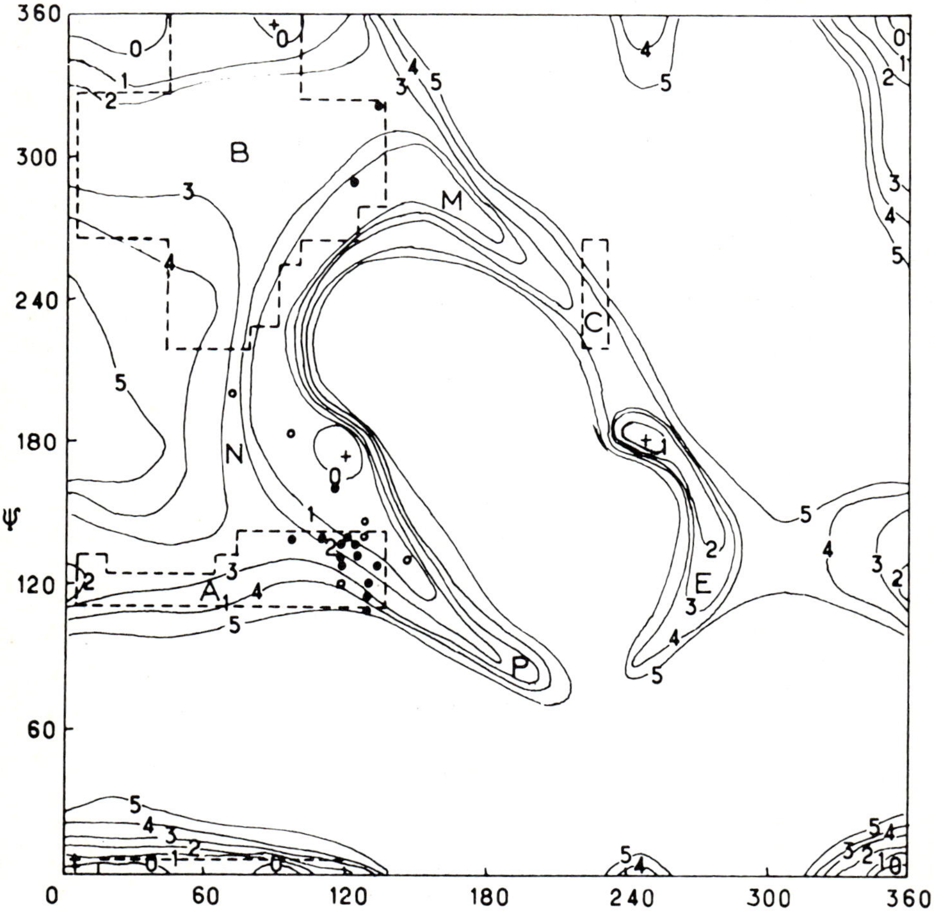

Figure 4-20. Conformational energy map for the leucyl residue. The contours are labelled in kcal/mole (referred to the global minimum, indicated by a double cross). The local minima are represented by +. The dashed lines represent the limits of semiempirical calculations. The notation used for the leucyl residues in various systems is: circle, lysozyme; shaded circle, myoglobin [Maigret *et al.* (1971b)].

to the C(γ)-H group. The global minimum appears at $\phi = 0°$, $\psi = 0°$ (fully extended form) and local minima (at less than 1 kcal/mole above the global minimum) are found near $\phi = 120°$ and $\psi = 180°$, $\phi = 240°$ and $\psi = 180°$, and ϕ

Table 4-11. Geometry and effective charges of leucine*

Atom	*Class*	*x*	*y*	*z*	*Charge*
O	10	10.418	5.759	−1.406	−0.41
O	9	13.771	7.422	0.510	−0.53
N	11	7.538	9.774	−0.638	−0.57
C	5	11.430	7.392	−0.101	0.53
C	8	9.911	9.583	0.872	−0.17
C	7	9.175	9.376	3.666	−0.47
C	7	11.308	9.443	5.589	−0.21
C	6	10.187	9.039	8.217	−0.69
C	6	12.710	11.956	5.490	−0.68
H	2	10.905	11.090	0.648	0.21
H	2	8.044	10.761	4.042	0.23
H	2	8.291	7.782	3.896	0.25
H	2	12.497	8.106	5.222	0.23
H	3	9.240	7.417	8.277	0.23
H	3	11.526	9.118	9.524	0.23
H	3	8.911	10.427	8.654	0.22
H	3	13.397	12.269	3.766	0.23
H	3	14.059	12.019	6.794	0.23
H	3	11.524	13.445	5.870	0.22
H	1	6.493	11.213	−0.041	0.26
H	1	7.929	9.913	−2.473	0.27
H	4	14.950	5.820	0.255	0.39

* The coordinates (in bohrs) have been taken from the work of Golic and Hamilton (1972) [as given by Clementi *et al.* (1977)] and the effective charges from the work of Clementi *et al.* (1977). The notation for the atoms is: O(1), O(2), N, C′, C(α), C(β), C(γ), C(δ_1), C(δ_2), H(α), H(β_1), H(β_2), H(γ), H(δ_{11}), H(δ_{12}), H(δ_{13}), H(δ_{21}), H(δ_{22}), H(δ_{23}), H(1), H(2), H(O_2).

$= 100°$, $\psi = 350°$ (Fig. 4-20). The map for the preferred conformation of the side chain (Fig. 4-21) was constructed by superposition (i.e., taking the lowest energy points) of the nine maps for ϕ, ψ with $\chi_1 = 60°, 180°, 300°$ and $\chi_2 = 60°, 180°, 300°$, respectively.

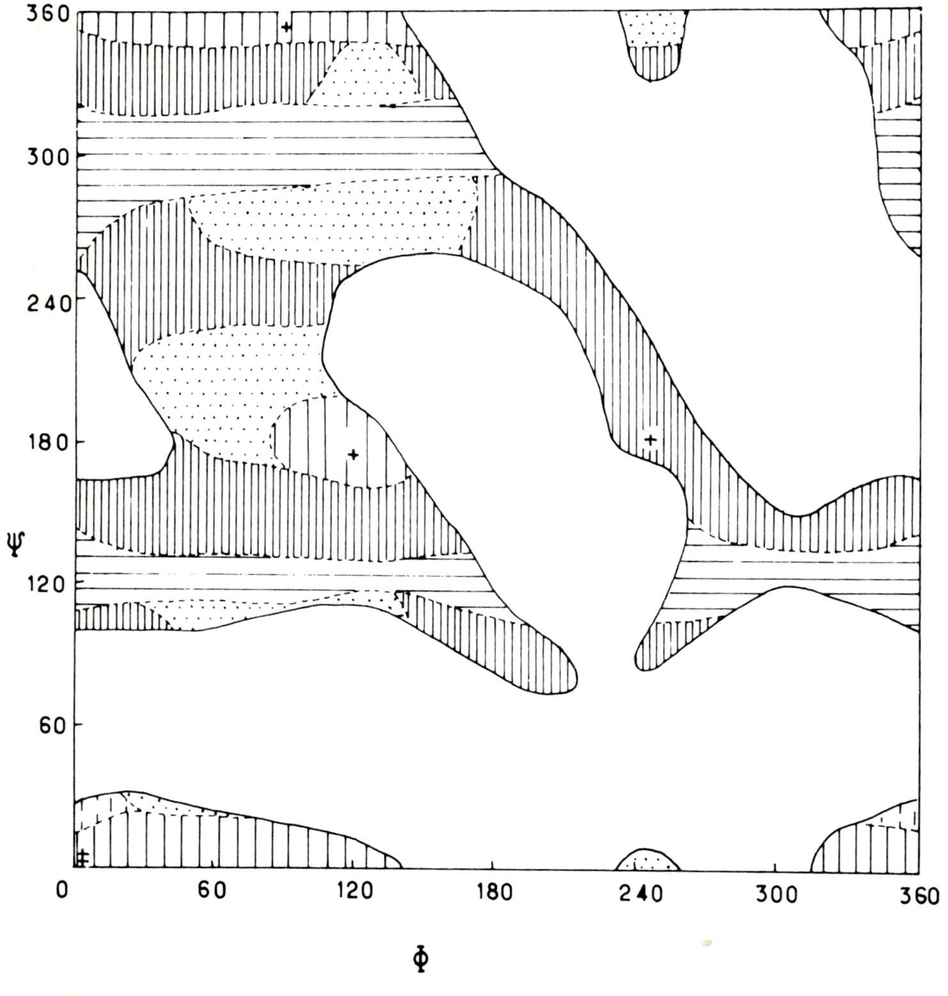

Figure 4-21. Conformational energy map for the leucyl residue, showing the preferred values for the angles χ_1 and χ_2. The outside contour corresponds to a level of 5 kcal/mole above the global minimum, indicated by a double cross. The local minima are represented by +. The various regions correspond to the following values of χ_1, χ_2 : vertical full lines (wide), $\chi_1 = 60°$, $\chi_2 = 180°$; vertical dashed lines, $\chi_1 = 60°$, $\chi_2 = 300°$; horizontal lines (wide), $\chi_1 = 180°$, $\chi_2 = 60°$; vertical full lines (narrow), $\chi_1 = 180°$, $\chi_2 = 180°$; dots, $\chi_1 = 300°$, $\chi_2 = 300°$ [Maigret *et al.* (1971b)].

11) lysyl residue [Pullman *et al.* (1971)]

The calculations were carried out for N-acetyl-N′-methyllysylamide, with the peptide bond taken in the planar *trans* conformation ($\omega = 0°$), the two backbone terminal methyl groups fixed so that the N-H and C = O terminal bonds are each eclipsed by one of the C-H bonds of these groups, each methylene group of the side-chain staggered with respect to the preceding one, and the amino group considered planar. Submaps for ϕ, ψ were obtained for χ_1

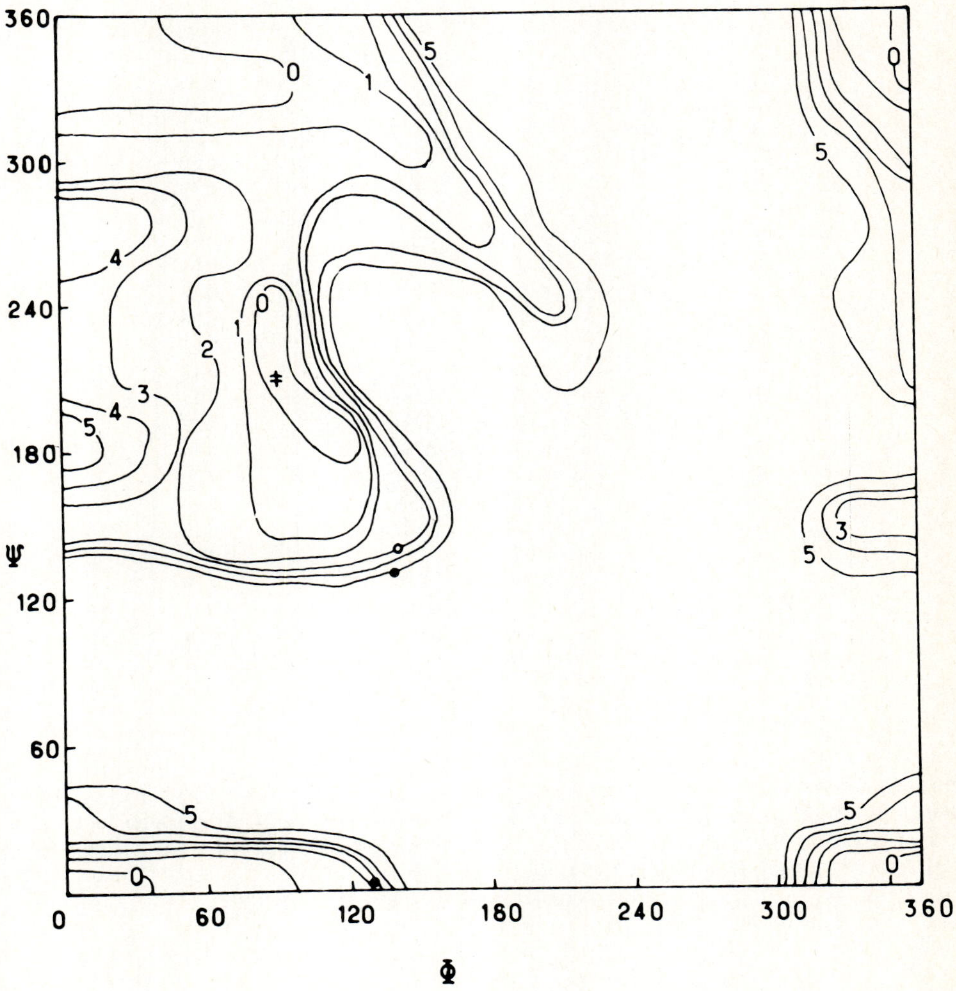

Figure 4-22. Conformational energy submap for the lysyl residue, with $\chi_1 = 60°$, $\chi_2 = 180°$, $\chi_3 = 180°$, $\chi_4 = 180°$. The global minimum is denoted by ≠. The notation used is: shaded circle, conformations of lysyl residues (with this set of χ) in globular proteins; circle, conformations of other lysyl residues (with $\chi_1 = 60°$) in globular proteins [Pullman *et al.* (1971)].

$= 60°$, $\chi_2 = 180°$, $\chi_3 = 180°$, $\chi_4 = 180°$ (Fig. 4-22), $\chi_1 = 180°$, $\chi_2 = 60°$, $\chi_3 = 180°$, $\chi_4 = 180°$ (Fig. 4-23), $\chi_1 = 180°$, $\chi_2 = 180°$, $\chi_3 = 180°$, $\chi_4 = 180°$ (Fig. 4-24), and $\chi_1 = 300°$, $\chi_2 = 180°$, $\chi_3 = 180°$, $\chi_4 = 180°$ (Fig. 4-25), and a composite map (Fig. 4-26) was constructed by superposition (i.e., taking the lowest

Table 4-12. Geometry and effective charges of lysine*

Atom	*Class*	*x*	*y*	*z*	*Charge*
O	10	1.457	3.059	−3.412	−0.39
O	9	5.084	1.476	−5.108	−0.54
N	11	2.778	2.150	1.345	−0.55
N	11	3.923	−10.137	4.074	−0.56
C	5	3.507	1.925	−3.248	0.50
C	8	4.265	0.858	−0.659	−0.15
C	7	3.826	−1.990	−0.638	−0.37
C	7	4.400	−3.225	1.897	−0.36
C	7	4.100	−6.110	1.736	−0.38
C	8	4.260	−7.360	4.305	−0.28
H	1	2.880	4.083	1.069	0.26
H	1	0.918	1.591	1.315	0.26
H	2	6.256	1.259	−0.325	0.18
H	2	1.863	−2.388	−1.136	0.19
H	2	5.070	−2.791	−2.089	0.17
H	2	6.333	−2.788	2.514	0.18
H	2	3.145	−2.557	3.369	0.20
H	2	2.273	−6.566	0.911	0.18
H	2	5.567	−6.889	0.506	0.17
H	2	6.067	−7.012	5.198	0.19
H	2	2.739	−6.685	5.519	0.19
H	1	2.335	−10.525	3.068	0.25
H	1	5.499	−10.913	3.262	0.24
H	4	6.486	0.580	−4.434	0.41

* The coordinates (in bohrs) have been taken from the work of Koetzle *et al.* (1972) [as given by Clementi *et al.* (1977)] and the effective charges from the work of Clementi *et al.* (1977). The notation for the atoms is: O(1), O(2), N, N(ζ), C′, C(α), C(β), C(γ), C(δ), C(ϵ), H(1), H(2), H(α), H(β_1), H(β_2), H(γ_1), H(γ_2), H(δ_1), H(δ_2), H(ϵ_1), H(ϵ_2), H(ζ_1), H(ζ_2), H(O_2).

energy points) of the preceding submaps. The global minimum occurs at $\phi = 90°$, $\psi = 210°$ (seven-membered hydrogen-bonded ring) and local minima are found at $\phi = 0°$, $\psi = 0°$ (fully extended form), $\phi = 270°$, $\psi = 140°$ (seven-membered hydrogen-bonded ring), and $\phi = 110°$-$120°$, $\psi = 120°$-$130°$.

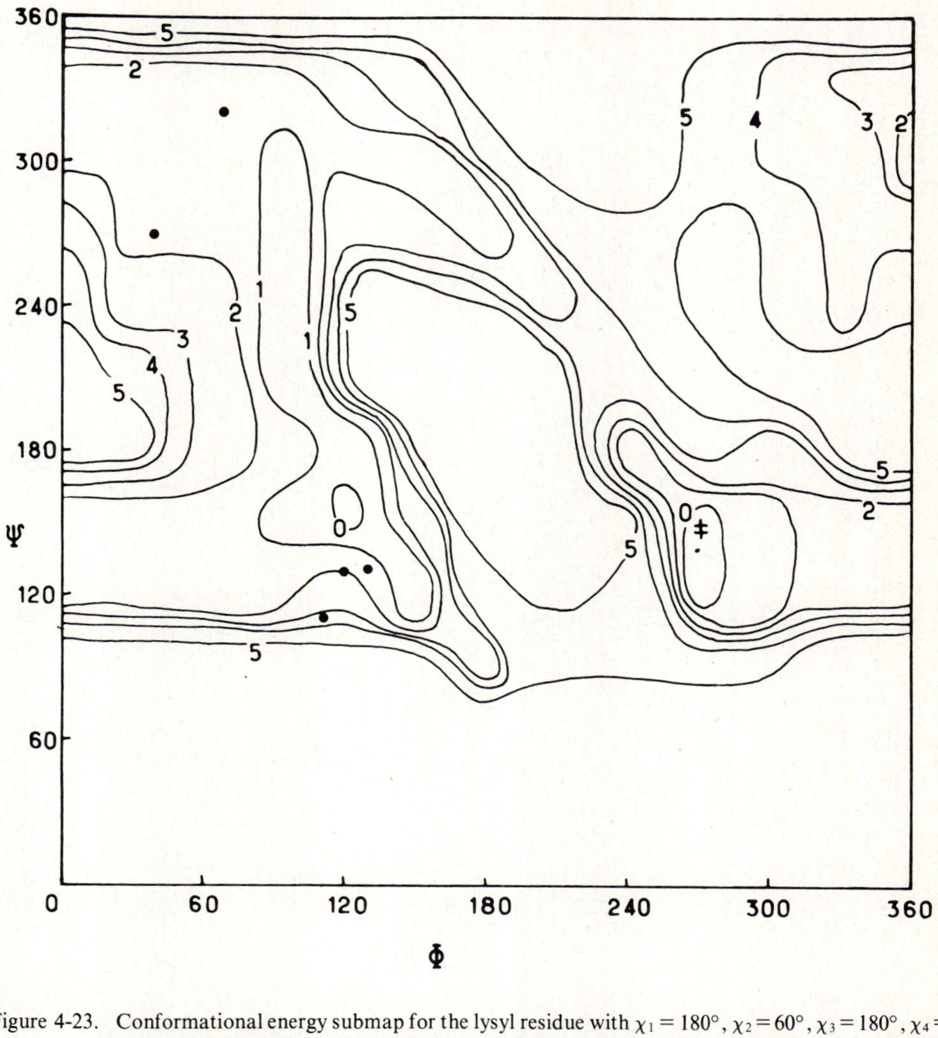

Figure 4-23. Conformational energy submap for the lysyl residue with $\chi_1 = 180°$, $\chi_2 = 60°$, $\chi_3 = 180°$, $\chi_4 = 180°$. The global minimum is denoted by ≠. The notation used is: shaded circle, conformations of lysyl residues (with this set of χ) in globular proteins; circle, conformations of other lysyl residues (with $\chi_1 = 180°$) in globular proteins [Pullman *et al* (1971)].

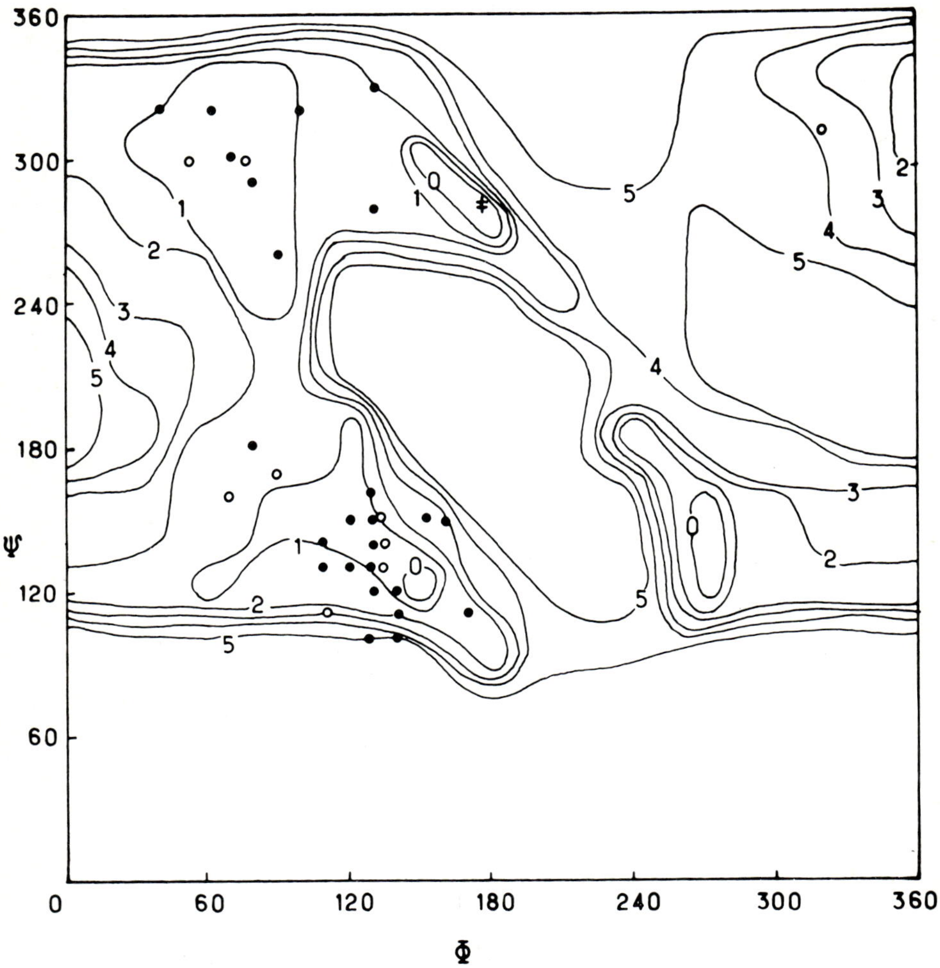

Figure 4-24. Conformational energy submap for the lysyl residue, with $\chi_1 = 180°$, $\chi_2 = 180°$, $\chi_3 = 180°$, $\chi_4 = 180°$. The global minimum is denoted by ≠. The notation used is: shaded circle, conformations of the lysyl residues (with this set of χ) in globular proteins; circle, conformations of other lysyl residues (with $\chi_1 = 180°$) in globular proteins [Pullman *et al.* (1971)].

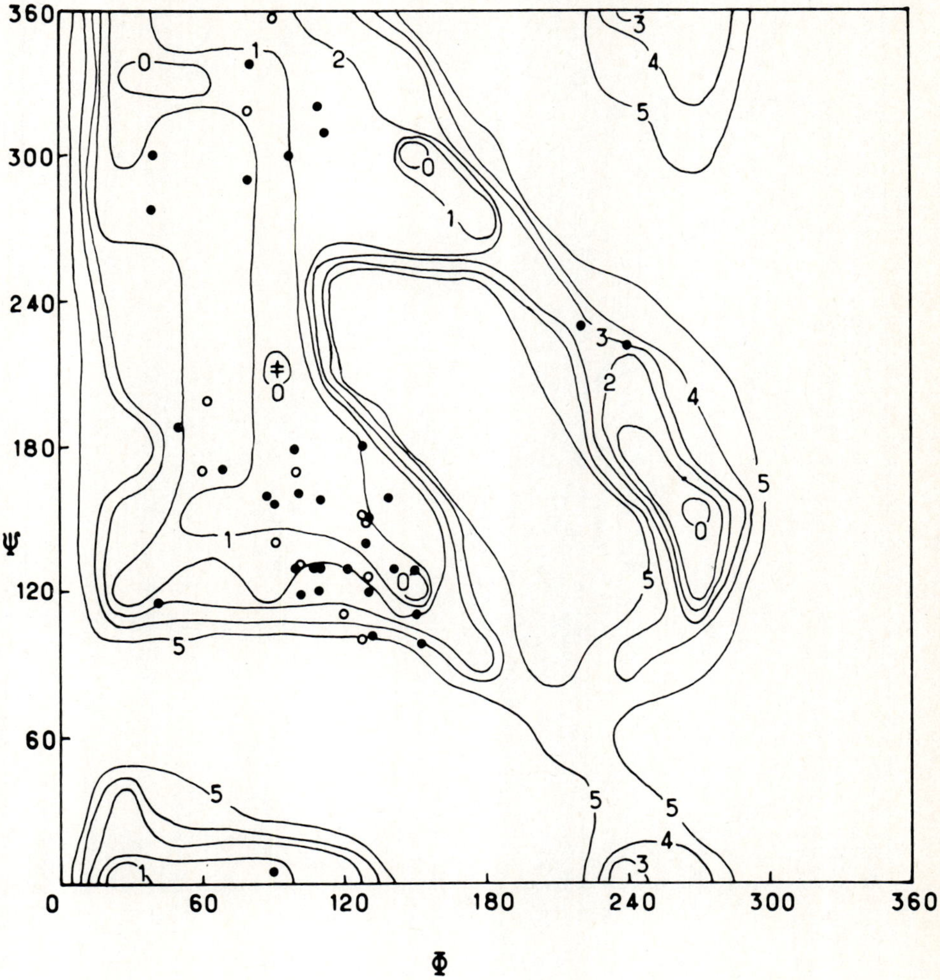

Figure 4-25. Conformational energy submap for the lysyl residue, with $\chi_1 = 300°$, $\chi_2 = 180°$, $\chi_3 = 180°$, $\chi_4 = 180°$. The global minimun is denoted by $\neq$. The notation used is: shaded circle, conformations of the lysyl residues (with this set of χ) in globular proteins; circle, conformations of other lysyl residues (with $\chi_1 = 300°$) in globular proteins [Pullman *et al.* (1971)].

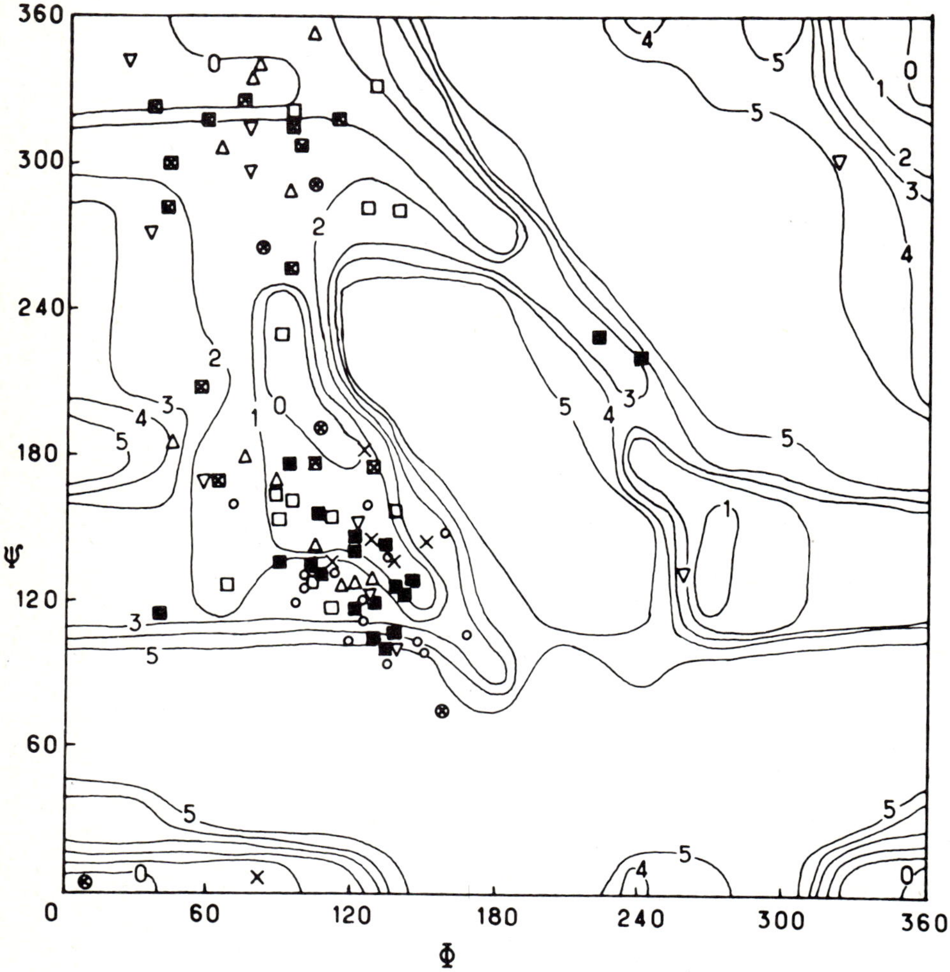

Figure 4-26. Conformational energy map for the lysyl residue. The notation used for the conformations of lysyl residues in various systems is: circle, lysozyme; shaded circle, myoglobin; square, α-lactalbumin; inverted triangle, ribonuclease-S; triangle, carboxypeptidase A; squared ×, α-chymotrypsin; circled ×, subtilisin-BPN′; ×, erythrocruorin; shaded square, oxyhemoglobin; shaded triangle, rubredoxin [Pullman *et al.* (1971)].

12) phenylalanyl residue [Maigret *et al.* (1971a)]

The calculations were carried out for N-acetyl-N′-methylphenylalanylamide, with the peptide bond taken in the planar *trans* conformation ($\omega = 0°$) and the two terminal methyl groups fixed so that the N-H and C = O terminal bonds are each eclipsed by one of the C-H bonds of these groups. The resulting conformational map (Figs. 4-27 and 4-28) was constructed by superposition (i.e., taking the lowest energy points) of the six

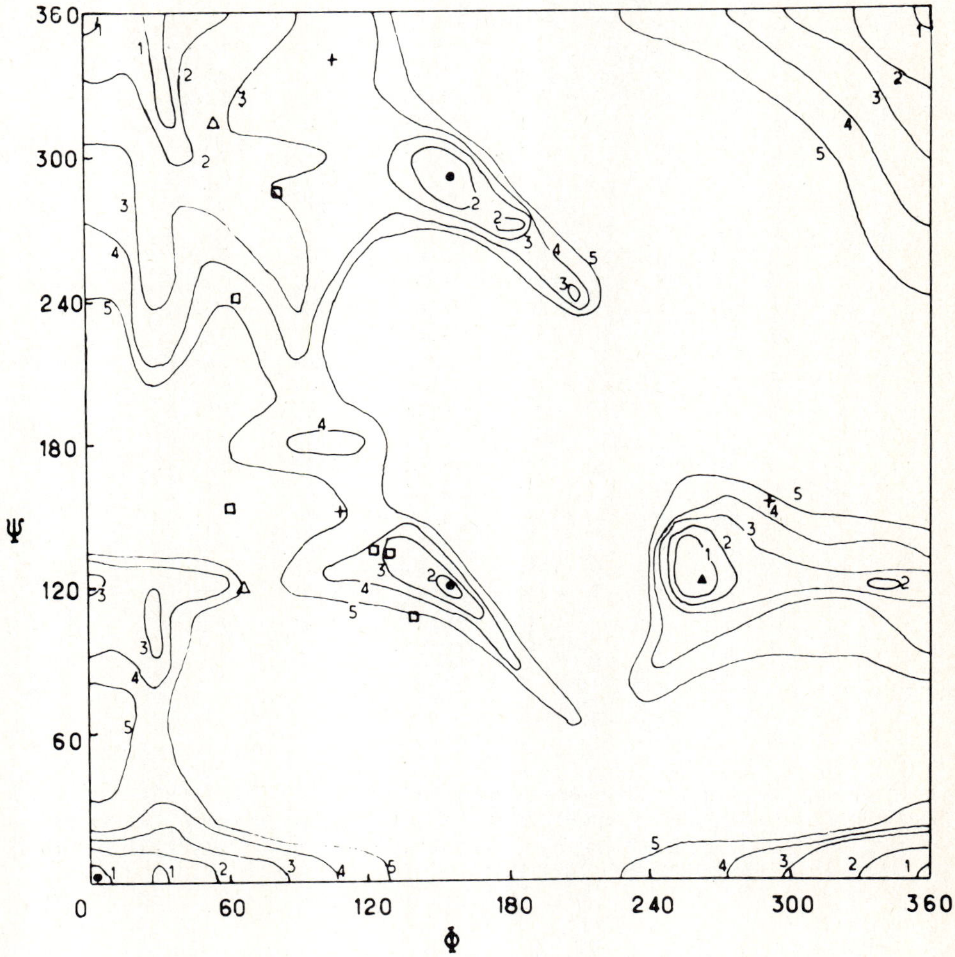

Figure 4-27. Conformational energy map for the phenylalanyl residue. The contours are labelled in kcal/mole (referred to the global minimum indicated by a shaded triangle). The local minima are represented by shaded circles. The notation used for the conformations of phenylalanyl residues in various systems is: +, lysozyme; square, myoglobin; triangle, small compounds [Maigret *et al.* (1971a)].

maps for ϕ, ψ with $\chi_1 = 60°$, 180°, 300° and $\chi_2 = 60°$, 300°, respectively. The global minimum occurs at $\phi = 260°$, $\psi = 130°$ (seven-membered hydrogen-bonded ring) and local minima (about 1-2 kcal/mole above the global

Table 4-13. Geometry and effective charges of phenylalanine*

Atom	*Class*	*x*	*y*	*z*	*Charge*
O	9	4.499	11.490	3.621	−0.51
O	10	2.391	8.694	1.328	−0.38
N	11	2.218	5.312	5.001	−0.52
C	7	6.308	7.188	6.524	−0.38
C	19	8.024	5.801	4.547	−0.01
C	18	9.253	7.151	2.671	−0.22
C	18	10.763	5.902	0.936	−0.21
C	18	11.045	3.303	1.078	−0.21
C	18	9.816	1.954	2.954	−0.21
C	18	8.306	3.203	4.688	−0.21
C	5	3.374	9.249	3.217	0.48
C	8	3.588	7.676	5.559	−0.15
H	2	7.166	8.980	6.987	0.21
H	2	6.204	6.044	8.211	0.21
H	16	9.034	9.177	2.561	0.20
H	16	11.722	6.954	−0.527	0.21
H	16	12.225	2.329	−0.274	0.21
H	16	10.033	−0.073	3.064	0.21
H	16	7.349	2.150	6.152	0.21
H	1	3.081	4.386	3.401	0.24
H	1	0.272	5.741	4.568	0.23
H	2	2.741	8.637	7.149	0.22
H	4	4.303	12.412	2.114	0.37

* The coordinates (in bohrs) have been taken from the work of Gurskaya (1965) [as given by Bolis and Clementi (1977)] and the effective charges from the work of Bolis and Clementi (1977). The notation for the atoms is: O(2), O(1), N, C(β), C(γ), C(δ_1), C(ϵ_1), C(2), C(ϵ_2), C(δ_2), C′, C(α), H(β_1), H(β_2), H(δ_1), H(ϵ_1), H(ζ), H(ϵ_2), H(δ_2), H(1), H(2), H(α), H(O_2).

minimum) are found at $\phi = 150°$, $\psi = 120°$ (in the right-handed α-helix region), $\phi = 40°$, $\psi = 320°$ (in the antiparallel pleated-sheet region), $\phi = 0°$, $\psi = 0°$ (fully extended form), and $\phi = 150°$, $\psi = 300°$.

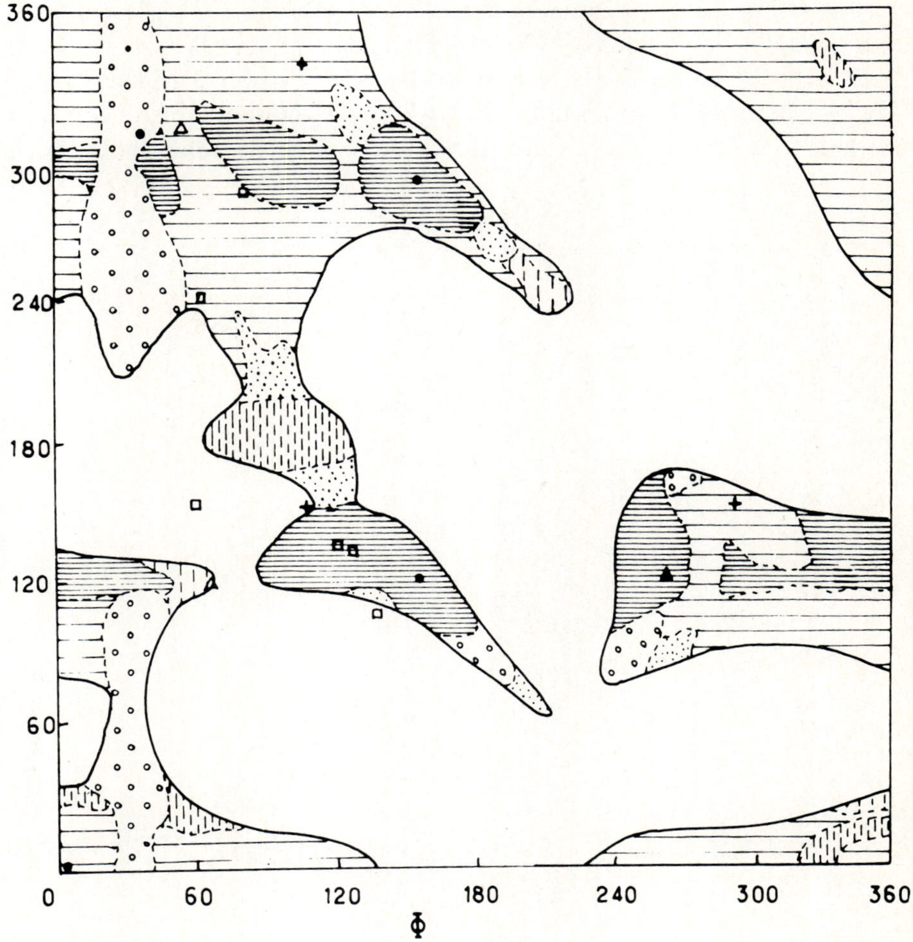

Figure 4-28. Conformational energy map for the phenylalanyl residue, showing the preferred values for the angles χ_1 and χ_2. The outside contour corresponds to a level of 5 kcal/mole above the global minimum, indicated by a shaded triangle. The local minima are represented by shaded circles. The notation used for the conformations of the phenylalanyl residues in various compounds is: +, lysozyme; square, myoglobin; triangle, small compounds. The various regions correspond to the following values of χ_1, χ_2: vertical dashed lines (narrow), $\chi_1 = 60°$, $\chi_2 = 60°$; vertical dashed lines (wide), $\chi_1 = 60°$, $\chi_2 = 300°$; horizontal lines (wide), $\chi_1 = 180°$, $\chi_2 = 60°$; horizontal lines (narrow), $\chi_1 = 180°$, $\chi_2 = 300°$; circles, $\chi_1 = 300°$, $\chi_2 = 60°$; dots, $\chi_1 = 300°$, $\chi_2 = 300°$ [Maigret *et al.* (1971a)].

13) prolyl residue [Maigret *et al.* (1970c)]

The calculations were carried out for N-acetyl-N′-methylprolylamide, with the peptide bond taken both in the planar *trans* ($\omega = 0°$) and *cis* ($\omega = 180°$) conformations, $\phi = 122°$, and the terminal methyl groups fixed so that the N-H and C = O bonds are eclipsed by one of the C-H bonds of these groups. Two minima, at $\psi = 150°$, and $\psi = 340°$, respectively, are found for *trans*-N-acetyl-N′-methylprolylamide, with the barrier at $\psi = 230°$ lower than the barrier at $\psi = 60°$. The global minimum for *cis*-N-acetyl-N′-methylprolylamide corresponds to $\psi = 140°$ and a second minimum (about 1.6 kcal/mole above the

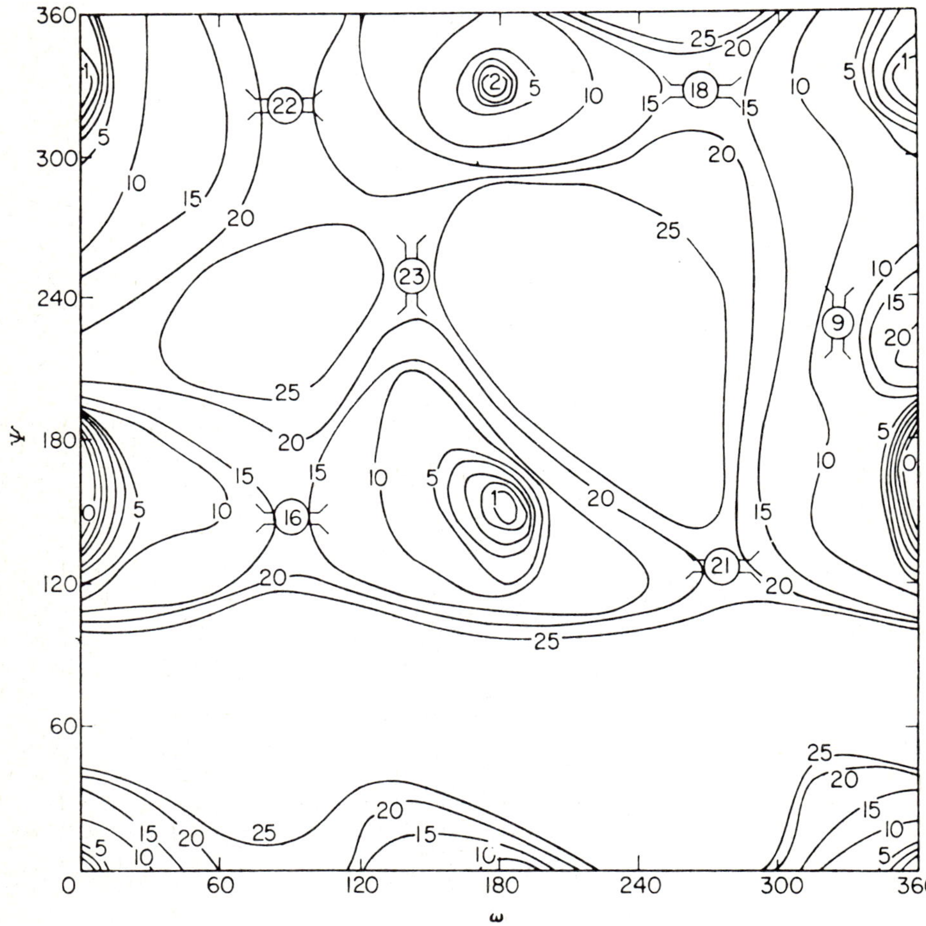

Figure 4-29. Conformational energy map for the *cis-trans* interconversion of N-acetyl-N′-methylprolylamide. The circled values correspond to the energy barriers [Maigret *et al.* (1970c)].

other) appears at $\psi = 330°$. The conformational map (ϕ, ψ) for the *cis-trans* interconversion (Fig. 4-29) indicates that at $\psi = 160°$, the rotation of ω between 180° and 0° and between 180° and 360° involves barriers of about 16 and 26 kcal/mole, respectively; for $\psi = 340°$, the corresponding barriers are about 21 and 18 kcal/mole, respectively.

Table 4-14. Geometry and effective charges of proline*

Atom	*Class*	*x*	*y*	*z*	*Charge*
C	5	10.362	12.748	2.204	0.52
C	8	9.759	10.011	2.835	−0.13
C	7	9.521	9.528	5.700	−0.38
C	7	7.942	7.161	5.710	−0.40
C	7	6.493	6.989	3.272	−0.30
N	11	7.353	9.214	1.683	−0.39
O	10	9.175	13.964	0.594	−0.41
O	9	12.209	13.670	3.578	−0.55
H	1	7.631	8.652	−0.259	0.25
H	2	4.486	7.100	3.625	0.20
H	2	6.906	5.232	2.320	0.18
H	2	6.640	7.211	7.281	0.21
H	2	9.153	5.530	5.904	0.20
H	2	11.342	9.225	6.569	0.21
H	2	8.579	11.069	6.651	0.20
H	2	11.253	8.831	2.099	0.18
H	4	12.439	15.386	3.106	0.41

* The coordinates (in bohrs) have been taken from the work of Mitsui *et al.* (1969) [as given by Clementi *et al.* (1977)] and the effective charges from the work of Clementi *et al.* (1977). The notation for the atoms is: C′, C(α), C(β), C(γ), C(δ), N, O(1), O(2), H(1), H(δ_1), H(δ_2), H(γ_1), H(γ_2), H(β_1), H(β_2), H(α), H(O_2).

14) hydroxyprolyl residue [Maigret *et al.* (1970c)]

The calculations were carried out for N-acetyl-N′-methylhydroxyprolylamide, with the peptide bond taken both in the planar *trans* ($\omega = 0°$) and *cis* ($\omega = 180°$) conformations, $\phi = 122°$, and the terminal methyl groups fixed so that the N-H and C = O bonds are eclipsed by one of the C-H bonds of these groups. Two minima, at ψ = 160° and ψ = 340°, respectively, are found for *trans*-N-acetyl-N′-methylhydroxyprolylamide, with the barrier at ψ = 230° lower than the one at ψ = 60°. The global

Table 4-15. Geometry and effective charges of hydroxyproline*

Atom	*Class*	*x*	*y*	*z*	*Charge*
O	9	9.460	8.278	21.723	−0.54
O	10	6.828	4.966	22.454	−0.42
O	9	3.818	7.431	13.469	−0.53
N	11	6.848	9.892	17.686	−0.40
C	5	7.483	6.864	21.236	0.51
C	8	5.878	7.540	18.928	−0.13
C	7	6.037	5.501	16.894	−0.41
C	17	6.246	6.883	14.384	0.00
C	7	7.689	9.274	15.078	−0.31
H	1	5.393	11.229	17.695	0.26
H	2	3.952	7.895	19.525	0.21
H	2	7.657	4.349	17.190	0.20
H	2	4.417	4.293	16.927	0.23
H	2	7.300	5.788	12.983	0.17
H	4	3.966	7.703	11.661	0.36
H	2	9.700	8.918	15.147	0.20
H	2	7.310	10.812	13.841	0.20
H	4	10.286	7.590	23.160	0.41

* The coordinates (in bohrs) have been taken from the work of Koetzle *et al.* (1973a) [as given by Clementi *et al.* (1977)] and the effective charges from the work of Clementi *et al.* (1977). The notation for the atoms is: O(2), O(1), O(δ), N, C′, C(α), C(β), C(γ), C(δ), H(1), H(α), H(β_1), H(β_2), H(γ), H(δ_1), H(β_1), H(β_2), H(O_2).

minimum for the *cis*-N-acetyl-N′-methylhydroxyprolylamide appears at ψ = 330°, while a second minimum (about 10 kcal/mole above the global minimum) is found at ψ = 150°.

15) seryl residue [Perahia *et al.* (1970)]

The calculations were carried out for N-acetyl-N′-methylserylamide, with the peptide bond taken in the planar *trans* conformation ($\omega = 0°$), the two backbone terminal methyl groups fixed so that the N-H and C = O terminal bonds are each eclipsed by one of the C-H bonds of these groups, and the OH group staggered with respect to the C(β)-H group. The global minimum occurs at $\phi = 90°$, $\psi = 240°$ (with $\chi_1 = \chi_2 = 60°$, corresponding to a seven-membered hydrogen-bonded ring, stabilized by two hydrogen-bonds) and local minima

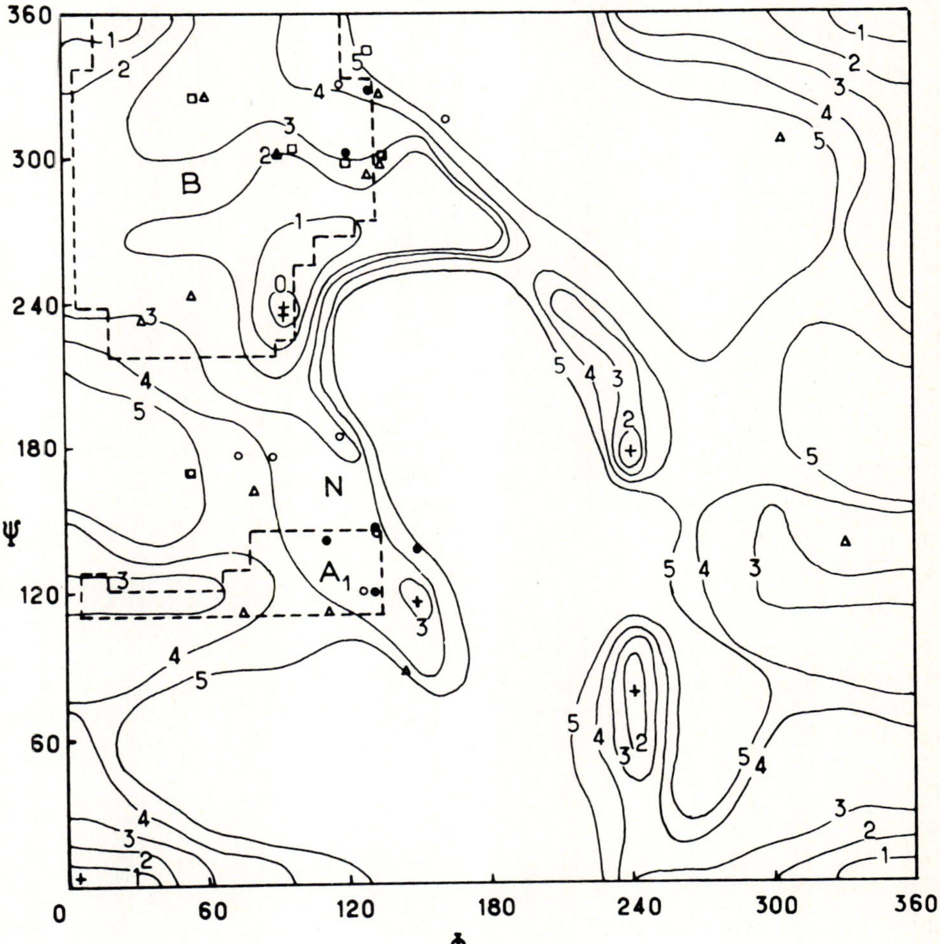

Figure 4-30. Conformational energy map for the seryl residue. The contours are labelled in kcal/mole (referred to the global minimum, denoted by ≠). The local minima are represented by +. The notation used for the conformations of seryl residues in various systems is: circle, lysozyme; shaded circle, myoglobin; square, α-lactalbumin; triangle, ribonuclease-S [Perahia *et al.* (1970)].

are found at $\phi = 0°$, $\psi = 0°$ (with $\chi_1 = \chi_2 = 180°$, fully extended form, about 1 kcal/mole above the global minimum), $\phi = 150°$, $\psi = 120°$ (with $\chi_1 = 60°$, $\chi_2 = 180°$, corresponding to the right-handed α-helix region, about 3 kcal/mole above the global minimum), $\phi = 240°$, $\psi = 180°$ (with $\chi_1 = 60°$, $\chi_2 = 300°$, about 2 kcal/mole above the global minimum, possibly stabilized by a second hydrogen-bond between the side-chain hydroxyl group and the carbonyl group), and $\phi = 240°$, $\psi = 80°$ (with $\chi_1 = 60°$, $\chi_2 = 300°$, about 2 kcal/mole

Table 4-16. Geometry and effective charges of serine*

Atom	*Class*	*x*	*y*	*z*	*Charge*
O	10	3.189	6.944	−1.571	−0.41
O	9	6.521	8.809	0.419	−0.55
O	9	8.385	2.913	−0.373	−0.52
N	11	2.961	2.935	1.358	−0.54
C	5	4.853	7.003	0.099	0.52
C	8	4.953	4.814	1.988	−0.14
C	17	7.556	3.614	2.068	−0.18
H	2	4.499	5.562	3.875	0.17
H	1	1.245	3.878	1.314	0.26
H	1	2.881	1.591	2.791	0.23
H	2	8.887	5.012	2.811	0.21
H	2	7.558	1.993	3.354	0.17
H	4	7.790	1.198	−0.754	0.36
H	4	6.233	10.053	−0.842	0.42

* The coordinates (in bohrs) have been taken from the work of Frey *et al.* (1973b) [as given by Clementi *et al.* (1977)] and the effective charges from the work of Clementi *et al.* (1977). The notation for the atoms is: O(1), O(2), O(γ), N, C′, C(α), C(β), H(α), H(2), H(1), H(β_1), H(β_2), H(γ), H(O_2).

above the global minimum, possibly stabilized by a hydrogen bond between the side-chain hydroxyl group and the carbonyl group) (Fig. 4-30). The map for the preferred conformation of the side chain (Fig. 4-31) was constructed by superposition (i.e., taking the lowest energy points) of the nine maps for ϕ, ψ with $\chi_1 = 60°, 180°, 300°$ and $\chi_2 = 60°, 180°, 300°$, respectively.

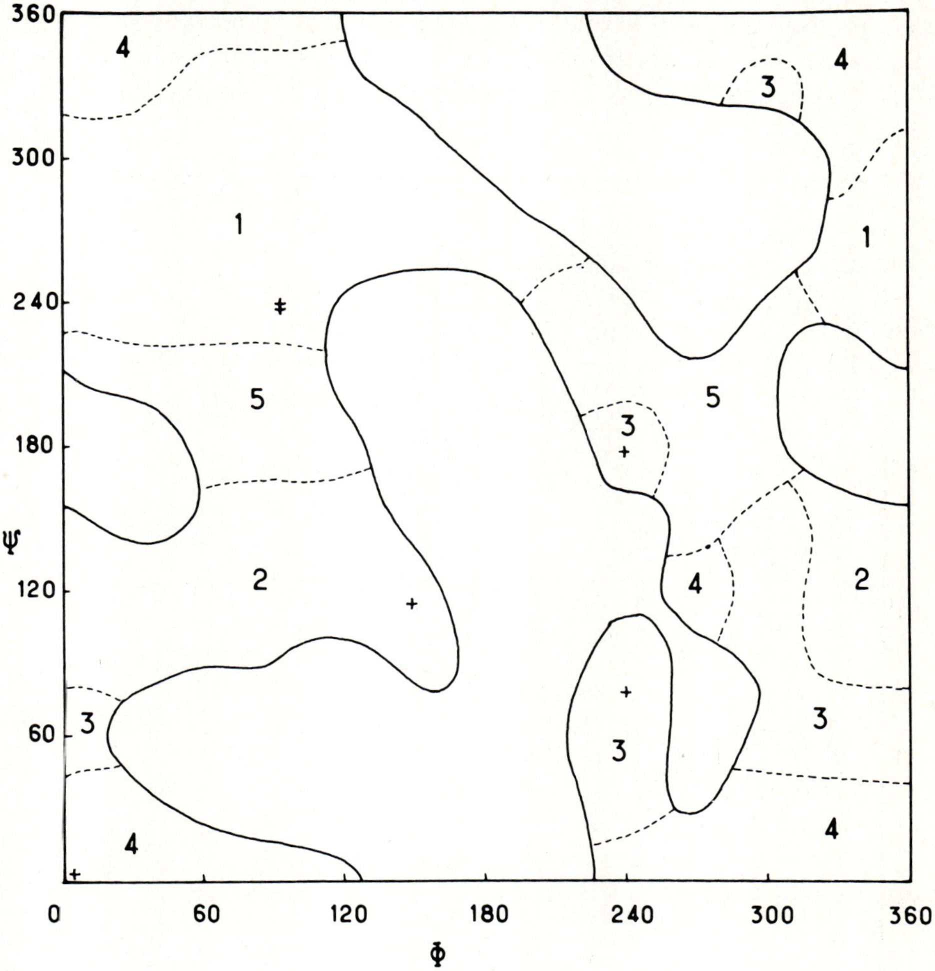

Figure 4-31. Conformational energy map for the seryl residue showing the preferred values for the angles χ_1 and χ_2. The contour corresponds to a level of 5 kcal/mole above the global minimum (denoted by ≠). The local minima are represented by +. The various regions correspond to the following values of χ_1, χ_2 : 1, $\chi_1 = 60°$, $\chi_2 = 60°$; 2, $\chi_1 = 60°$, $\chi_2 = 180°$; 3, $\chi_1 = 60°$, $\chi_2 = 300°$; 4, $\chi_1 = 180°$, $\chi_2 = 180°$; 5, $\chi_1 = 180°$, $\chi_2 = 300°$ [Perahia *et al.* (1970)].

16) threonyl residue [Perahia *et al.* (1970)]

The calculations were carried out for N-acetyl-N′-methylthreonylamide, with the peptide bond taken in the planar *trans* conformation ($\omega = 0°$), the two backbone terminal methyl groups fixed so that the N-H and C = O terminal bonds are each eclipsed by one of the C-H bonds of these groups, the C(β) methyl group staggered with respect to the C(α)-H group, and the hydroxyl group staggered with respect to the C(β)-H group. Two equivalent deepest

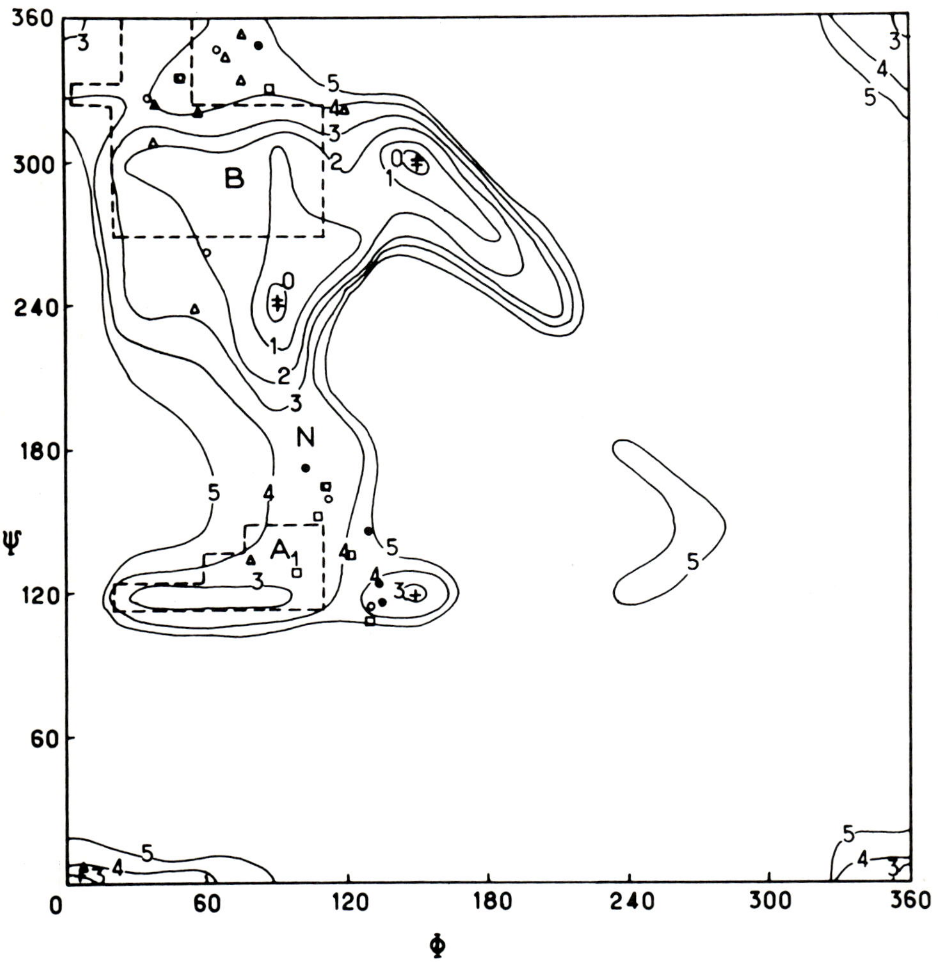

Figure 4-32. Conformational energy map for the threonyl residue. The contours are labelled in kcal/mole (referred to the global minimum, denoted by ≠). The local minima are represented by +. The notation used for the conformations of the threonyl residues in various systems is: circle, lysozyme; shaded square, myoglobin; square, α-lactalbumin; triangle, ribonuclease-S [Perahia *et al.* (1970)].

minima appear at $\phi = 90°$, $\psi = 240°$ (with $\chi_1 = \chi_2 = 60°$) and $\phi = 150°$, $\psi = 300°$ (with $\chi_1 = \chi_2 = 60°$) and local minima (about 3 kcal/mole above the deepest minima) are found at $\phi = 150°$, $\psi = 120°$ and $\phi = 0°$, $\psi = 0°$ (fully extended

Table 4-17. Geometry and effective charges of threonine*

Atom	*Class*	*x*	*y*	*z*	*Charge*
O	10	14.342	3.170	4.536	−0.42
O	9	12.936	3.109	0.464	−0.55
O	9	6.086	2.147	5.125	−0.41
N	11	10.626	0.203	6.234	−0.49
C	5	12.748	2.685	2.900	0.50
C	8	10.263	1.571	3.797	−0.15
C	17	8.174	3.562	4.188	0.00
C	6	7.472	4.893	1.773	−0.60
H	2	9.954	0.102	2.526	0.23
H	2	8.540	5.162	5.393	0.18
H	3	6.006	6.263	2.152	0.18
H	3	6.793	3.527	0.417	0.21
H	3	9.113	5.843	1.018	0.20
H	1	11.320	1.502	7.645	0.24
H	1	11.976	−1.304	5.973	0.23
H	4	3.987	3.305	5.199	0.24
H	4	14.535	3.867	0.161	0.41

* The coordinates (in bohrs) have been taken from the work of Shoemaker *et al.* (1950) [as given by Clementi *et al.* (1977)] and the effective charges from the work of Clementi *et al.* (1977). The notation for the atoms is: O(1), O(2), O(γ), N, C′, C(α), C(β), C(γ), H(α), H(β), H(γ_1), H(γ_2), H(γ_3), H(1), H(2), H(γ), H(O_2).

conformation) (Fig. 4-32). The map for the preferred conformation of the side chain (Fig. 4-33) was constructed by superposition (i.e., taking the lowest energy points) of the nine maps for ϕ, ψ with $\chi_1 = 60°, 180°, 300°$ and $\chi_2 = 60°, 180°, 300°$, respectively.

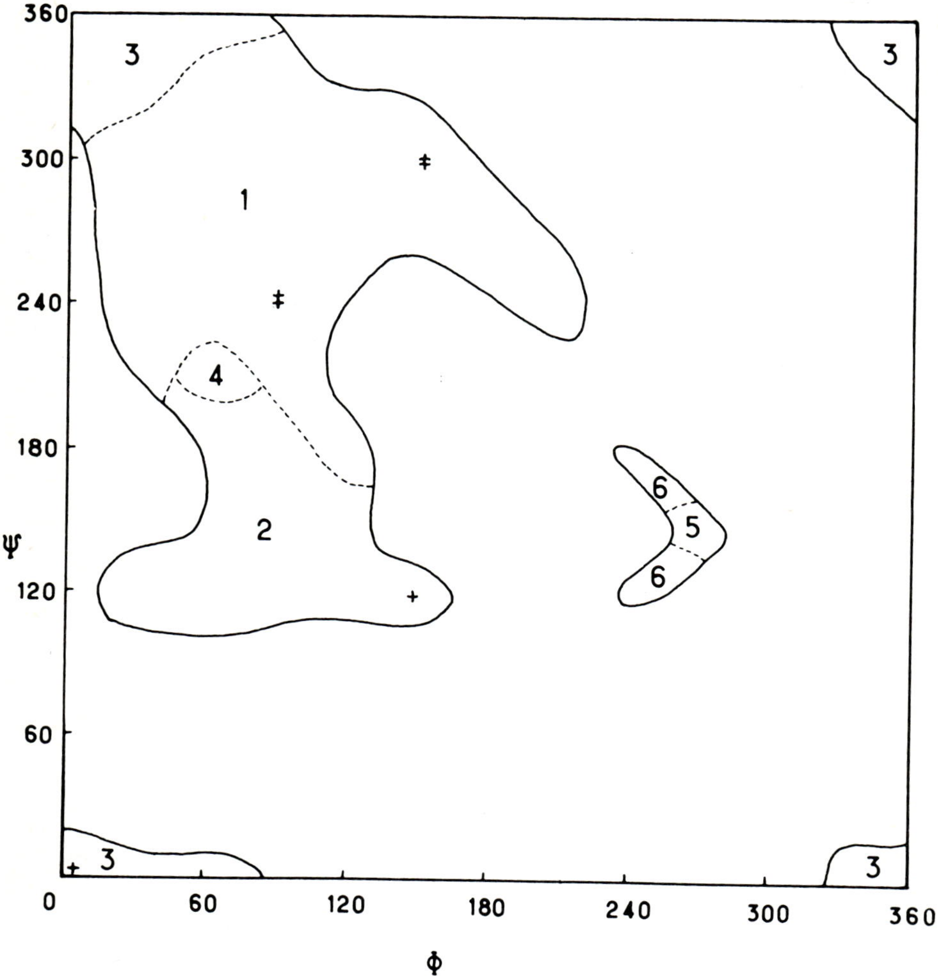

Figure 4-33. Conformation energy map for the threonyl residue, showing the preferred values for the angles χ_1 and χ_2. The outside contour corresponds to a level of 5 kcal/mole above the global minimum (denoted by ≠). The local minima are represented by +. The various regions correspond to the following values of χ_1, χ_2 : 1, $\chi_1 = 60°$, $\chi_2 = 60°$; 2, $\chi_1 = 60°$, $\chi_2 = 180°$; 3, $\chi_1 = 180°$, $\chi_2 = 180°$; 4, $\chi_1 = 180°$, $\chi_2 = 300°$; 5, $\chi_1 = 300°$, $\chi_2 = 60°$; 6, $\chi_1 = 300°$, $\chi_2 = 180°$ [Perahia *et al.* (1970)].

17) tryptophanyl residue [Maigret *et al.* (1971a)]

The calculations were carried out for N-acetyl-N′-methyltryptophanylamide, with the peptide bond taken in the planar *trans* conformation ($\omega = 0°$) and the two terminal methyl groups fixed so that the N-H and C = O terminal bonds are each eclipsed by one of the C-H bonds of these groups. The resulting conformational map (Figs. 4-34 and 4-35) was constructed by superposition (i.e., taking the lowest energy points) of the maps

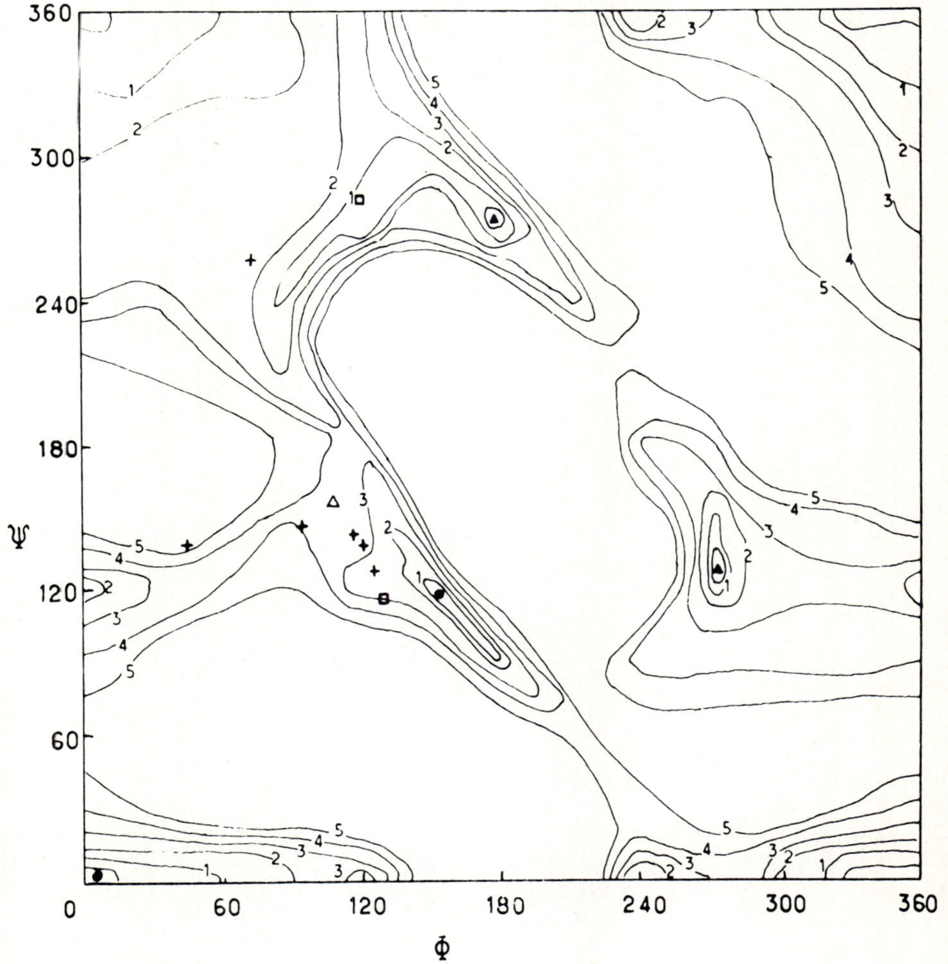

Figure 4-34. Conformational energy map for the tryptophanyl residue. The contours are labelled in kcal/mole (referred to the global minimum, denoted by a shaded triangle). The local minima are represented by shaded circles. The notation used for the conformations of tryptophanyl residues in various systems is: +, lysozyme; square, myoglobin; triangle, small compounds [Maigret *et al.* (1971a)].

for ϕ, ψ with $\chi_1 = 60°$, $180°$, $300°$ and $\chi_2 = 60°$, $300°$, respectively. Two equivalent deepest minima are located at $\phi = 270°$, $\psi = 130°$ (seven-membered

Table 4-18. Geometry and effective charges of tryptophan*

Atom	*Class*	*x*	*y*	*z*	*Charge*
O	9	19.563	0.205	8.010	−0.55
O	10	6.401	−0.673	10.644	−0.41
N	15	−0.066	3.270	0.042	−0.46
N	11	3.129	2.855	8.055	−0.54
C	18	5.463	−0.005	−1.321	−0.20
C	18	5.445	0.308	−3.968	−0.25
C	18	3.536	1.511	−5.302	−0.22
C	18	1.555	2.650	−4.114	−0.23
C	20	1.569	2.376	−1.487	0.14
C	19	3.456	1.079	−0.070	−0.05
C	14	2.819	1.199	2.492	−0.04
C	14	0.622	2.457	2.569	−0.09
C	7	4.365	−0.025	4.719	−0.43
C	8	5.281	1.765	6.792	−0.15
C	5	7.188	0.448	8.682	0.51
H	16	6.859	−0.952	−0.311	0.21
H	16	6.837	−0.437	−5.084	0.21
H	16	3.474	1.639	−7.262	0.21
H	16	0.000	3.606	−5.139	0.19
H	16	−0.443	3.184	4.036	0.21
H	16	−1.615	4.168	−0.292	0.37
H	2	5.974	−0.827	4.081	0.23
H	2	3.164	−1.342	5.669	0.20
H	2	6.084	3.387	6.042	0.26
H	1	1.482	1.936	7.796	0.24
H	1	2.002	1.369	8.884	0.23
H	4	10.366	−0.862	9.210	0.42

* The coordinates (in bohrs) have been taken from the work of Pasternak (1950) [as given by Clementi *et al.* (1977)] and the effective charges from the work of Clementi *et al.* (1977). The notation for the atoms is: O(2), O(1), N(ϵ), N, C(ζ_2), C(η), C(ζ_3), C(ζ_1), C(ϵ), C(δ_2), C(γ), C(δ_1), C(β), C(α), C′, H(ζ_2), H(η), H(ζ_3), H(ζ_1), H(δ_1), H(ϵ), H(β_1), H(β_2), H(α), H(1), H(2), H(O_2).

hydrogen-bonded ring) and $\phi = 180°$, $\psi = 280°$, and local minima (less than 1 kcal/mole above the deepest minima) are found at $\phi = 150°$, $\psi = 120°$ (in the right-handed α-helix region) and $\phi = 0°$, $\psi = 0°$ (fully extended form).

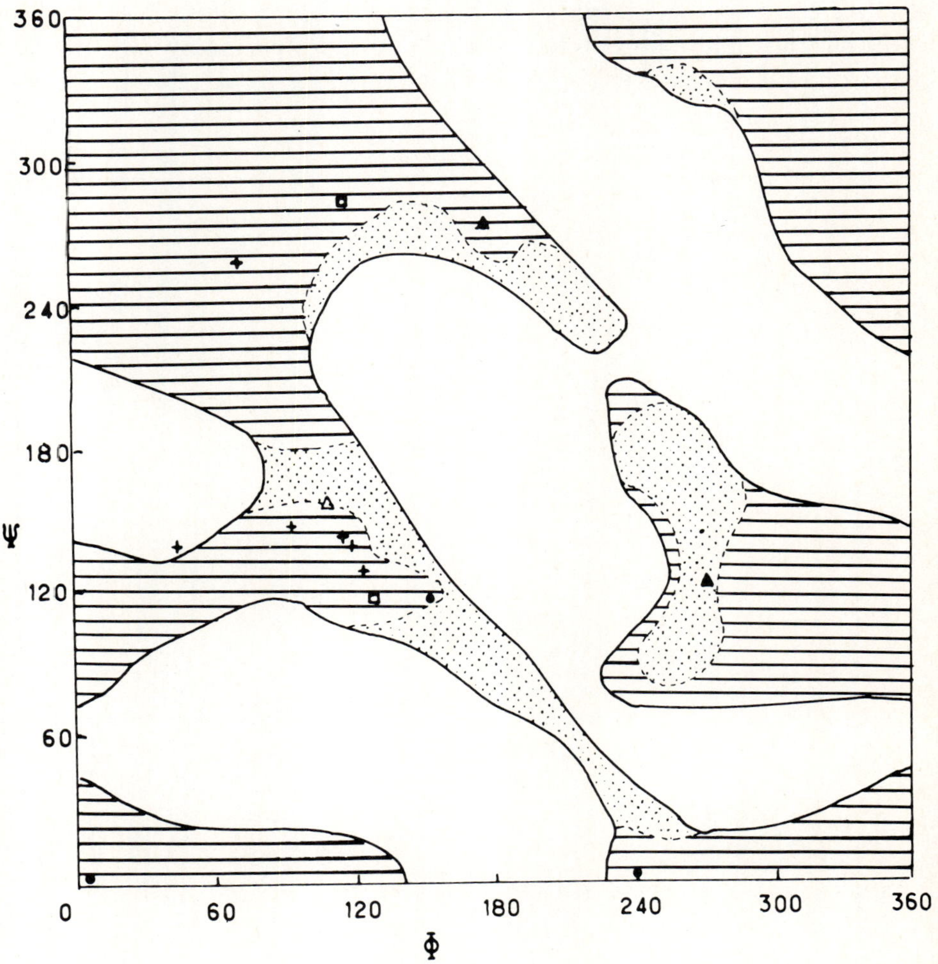

Figure 4-35. Conformational energy map for the tryptophanyl residue, showing the preferred values for the angles χ_1 and χ_2. The outside contour corresponds to a level of 5 kcal/mole above the global minimum (denoted by a shaded triangle). The local minima are represented by shaded circles. The notation used for the conformations of tryptophanyl residues in various systems is: +, lysozyme; square, myoglobin; triangle, small compounds. The various regions correspond to the following values of χ_1, χ_2 : horizontal lines, $\chi_1 = 180°, \chi_2 = 240°$; dots, $\chi_1 = 300°, \chi_2 = 120°$ [Maigret *et al.* (1971a)].

18) tyrosyl residue [Maigret *et al.* (1971a)]

The calculations were carried out for N-acetyl-N′-methyltyrosylamide, with the peptide bond taken in the planar *trans* conformation ($\omega = 0°$) and the two terminal methyl groups fixed so that the N-H and C = O terminal bonds are each eclipsed by one of the C-H bonds of these groups. The resulting conformational map (Figs. 4-36 and 4-37) was constructed by superposition (i.e., taking the lowest energy points) of the six maps for ϕ, ψ with $\chi_1 = 60°$,

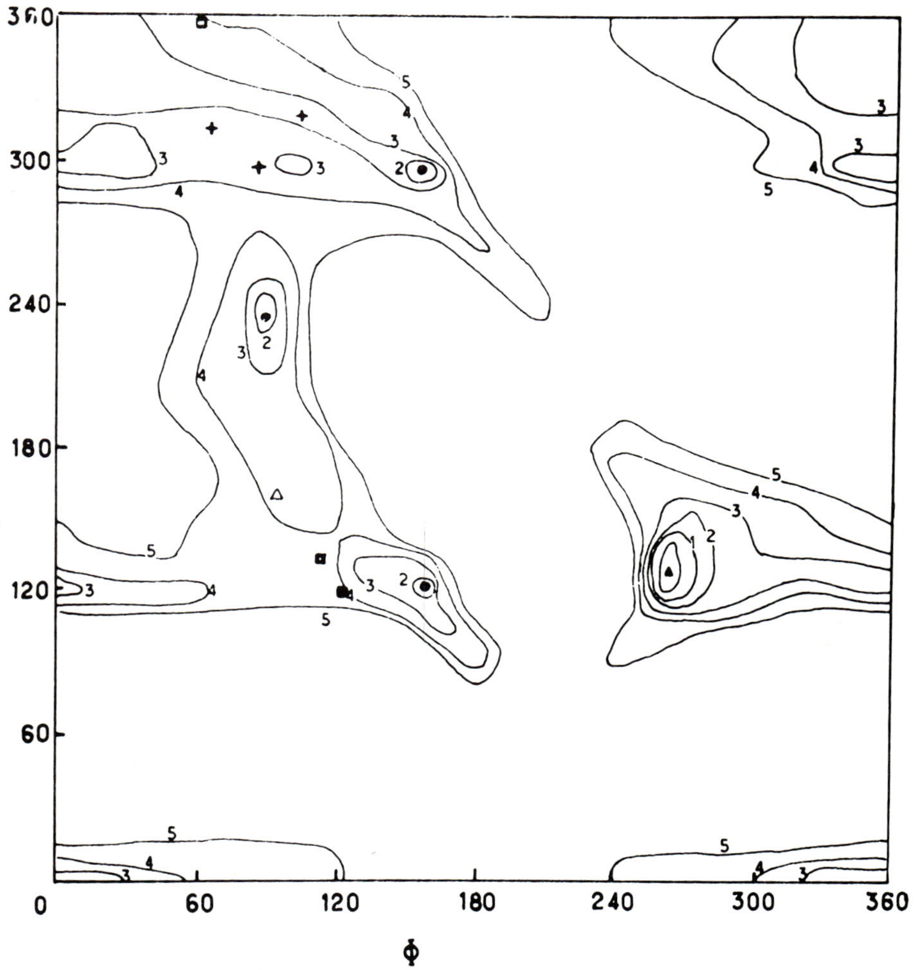

Figure 4-36. Conformational energy map for the tyrosyl residue. The contours are labelled in kcal/mole (referred to the global minimum, denoted by a shaded triangle). The local minima are represented by shaded circles. The notation used for the conformations of tyrosyl residues in various systems is: +, lysozyme; square, myoglobin; triangle, small compounds [Maigret *et al.* (1971a)].

180°, 300° and $\chi_2 = 60°, 300°$, respectively. The global minimum appears at $\phi = 260°$, $\psi = 130°$ (seven-membered hydrogen-bonded ring) and local minima

Table 4-19. Geometry and effective charges of tyrosine*

Atom	*Class*	*x*	*y*	*z*	*Charge*
O	10	2.183	13.293	6.896	−0.42
O	9	5.659	11.372	8.565	−0.55
O	9	6.280	23.914	5.176	−0.51
C	5	4.237	12.175	6.700	0.50
C	8	5.222	11.701	4.024	−0.12
C	7	7.511	13.396	3.438	−0.38
C	19	7.070	16.180	3.899	−0.04
C	18	5.994	17.731	2.050	−0.19
C	18	7.799	17.289	6.175	−0.19
C	18	5.665	20.311	2.454	−0.26
C	18	7.500	19.867	6.602	−0.24
C	17	6.460	21.377	4.726	0.21
N	11	3.158	12.135	2.164	−0.56
H	1	2.425	13.901	2.304	0.27
H	1	1.684	10.879	2.452	0.25
H	2	5.796	9.722	3.851	0.21
H	2	9.066	12.752	4.621	0.20
H	2	8.082	13.070	1.477	0.21
H	16	5.458	16.934	0.237	0.21
H	16	8.666	16.127	7.635	0.20
H	16	4.837	21.484	0.994	0.18
H	16	8.120	20.721	8.362	0.22
H	4	5.583	24.805	3.707	0.39
H	4	4.837	11.798	10.104	0.41

* The coordinates (in bohrs) have been taken from the work of Frey *et al.* (1973a) [as given by Clementi *et al.* (1977)] and the effective charges from the work of Clementi *et al.* (1977). The notation for the atoms is: O(1), O(2), O(ϵ), C′, C(α), C(β), C(γ), C(δ_1), C(δ_2), C(ϵ_1), C(ϵ_2), C(ζ), N, H(1), H(2), H(α), H(β_1), H(β_2), H(δ_1), H(δ_2), H(ϵ_1), H(ϵ_2), H(ϵ), H(O_2).

(about 2 kcal/mole above this global minimum) are found at $\phi = 150°$, $\psi = 120°$ (in the right-handed α-helix region), $\phi = 150°$, $\psi = 300°$, and $\phi = 90°$, $\psi = 240°$ (another seven-membered hydrogen-bonded ring).

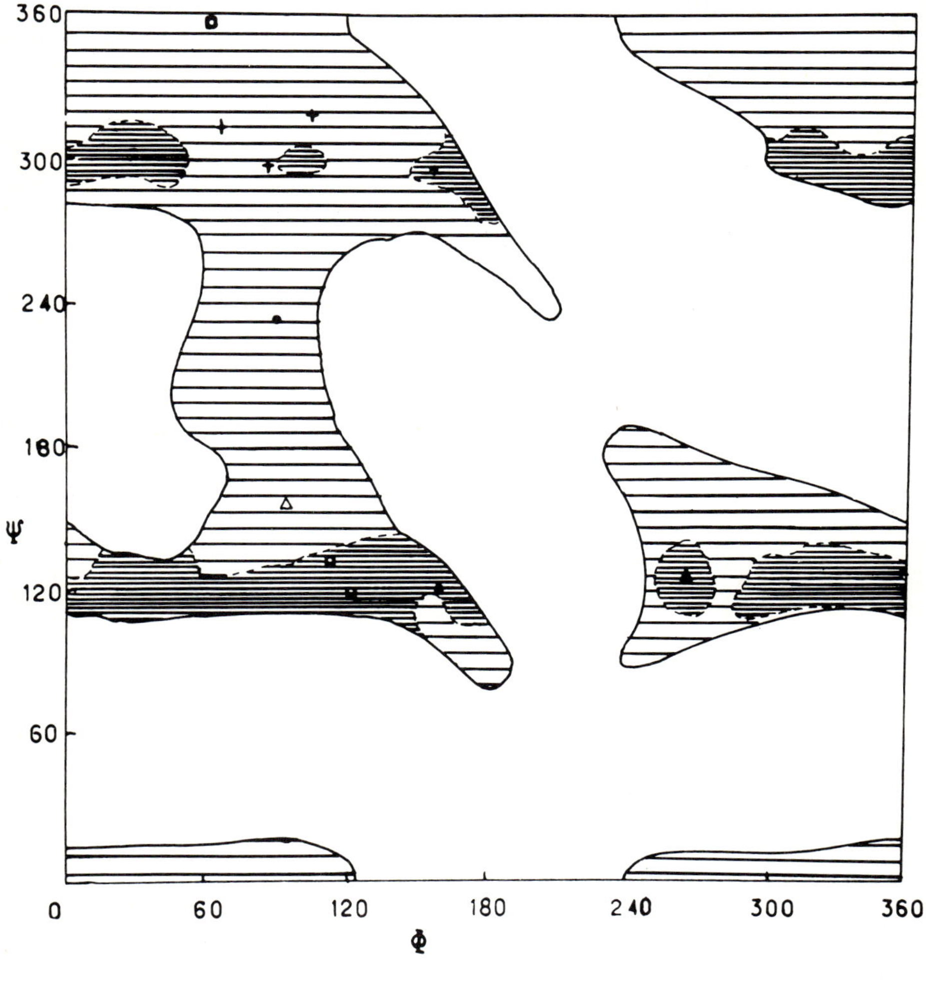

Figure 4-37. Conformation energy map for the tyrosyl residue, showing the preferred values for the angles χ_1 and χ_2. The outside contour corresponds to a level of 5 kcal/mole (referred to the global minimum, denoted by a shaded triangle). The local minima are represented by shaded circles. The notation used for the conformations of tyrosyl residues in various systems is: +, lysozyme; square, myoglobin; triangle, small compounds. The various regions correspond to the following values of χ_1, χ_2 : horizontal lines (wide), χ_1 = 180, χ_2 = 60°; horizontal lines (narrow), χ_1 = 180°, χ_2 = 300° [Maigret *et al.* (1971a)].

19) valyl residue [Maigret *et al.* (1971b)]

The calculations were carried out for N-acetyl-N′-methylvalylamide, with the peptide bond taken in the planar *trans* conformation ($\omega = 0°$), the two backbone terminal methyl groups fixed so that the N-H and C = O terminal bonds are each eclipsed by one of the C-H bonds of these groups, and the C(γ_1) and C(γ_2) methyl groups are both assumed to be staggered with respect to the C(β)-H_2 group. The global minimum appears at $\phi = 120°$, $\psi = 180°$

Table 4-20. Geometry and effective charges of valine*

Atom	*Class*	*x*	*y*	*z*	*Charge*
O	10	15.972	10.125	8.884	−0.40
O	9	14.534	6.355	7.448	−0.55
N	11	11.462	12.281	9.221	−0.50
C	5	14.249	8.734	8.085	0.50
C	8	11.653	9.897	7.716	−0.13
C	7	10.963	10.369	4.916	−0.17
C	6	9.956	7.969	3.640	−0.60
C	6	13.289	11.478	3.454	−0.57
H	1	11.952	11.900	11.165	0.24
H	1	9.552	12.994	9.129	0.23
H	3	12.779	11.796	1.503	0.19
H	3	14.842	10.155	3.544	0.18
H	3	13.851	13.245	4.308	0.21
H	3	10.225	6.690	2.072	0.21
H	3	8.091	8.795	3.531	0.19
H	3	10.144	6.943	5.394	0.19
H	2	9.477	11.768	4.870	0.19
H	2	10.282	8.587	8.471	0.17
H	4	16.257	5.920	7.708	0.42

* The coordinates (in bohrs) have been taken from the work of Torii and Iitaka (1970) [as given by Clementi *et al.* (1977)] and the effective charges from the work of Clementi *et al.* (1977). The notation for the atoms is: O(1), O(2), N, C′, C(α), C(β), C(γ_2), C(γ_1), H(1), H(2), H(γ_{11}), H(γ_{12}), H(γ_{13}), H(γ_{21}), H(γ_{22}), H(γ_{23}), H(β), H(α), H(O_2).

(corresponding to a seven-membered hydrogen-bonded conformation) and local minima (within 1.0-1.5 kcal/mole above the global minimum) are found at $\phi = 150°$, $\psi = 130°$ [region R(α), of the α and 3_{10} helices], $\phi = 30°$, $\psi = 330°$ (region of the antiparallel pleated-sheet), $\phi = 150°$, $\psi = 300°$ (M region), and $\phi = 240°$, $\psi = 180°$ (corresponding to another seven-membered hydrogen-bonded ring) (Fig. 4-38). The map for the preferred conformation of the side chain (Fig. 4-39) was constructed by superposition (i.e., taking the lowest energy points) of the three maps for ϕ, ψ with $\chi_1 = 60°, 180°, 300°$, respectively.

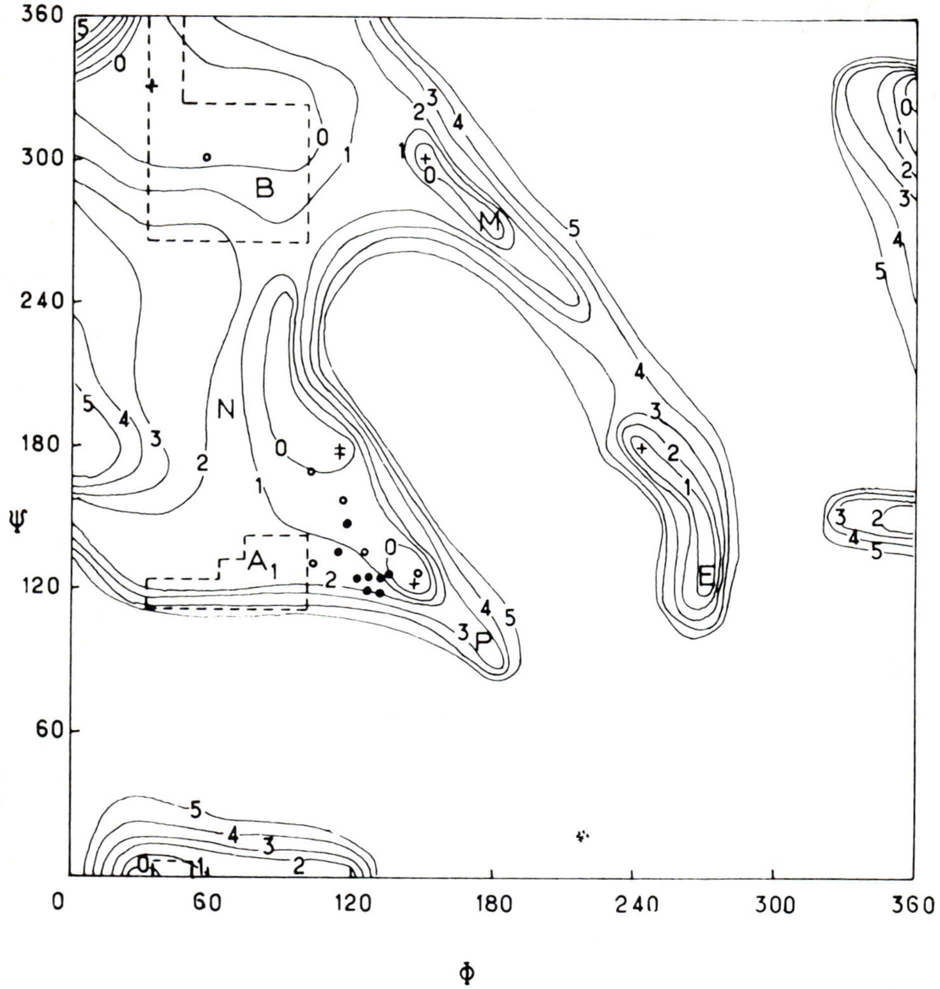

Figure 4-38. Conformational energy map for the valyl residue. The contours are labelled in kcal/mole (referred to the global maximum, denoted by a double cross). The dashed lines represent the limits of semiempirical calculations. The notation used for the conformations of the valyl residues in various systems is: circle, lysozyme; shaded circle, myoglobin [Maigret *et al.* (1971b)].

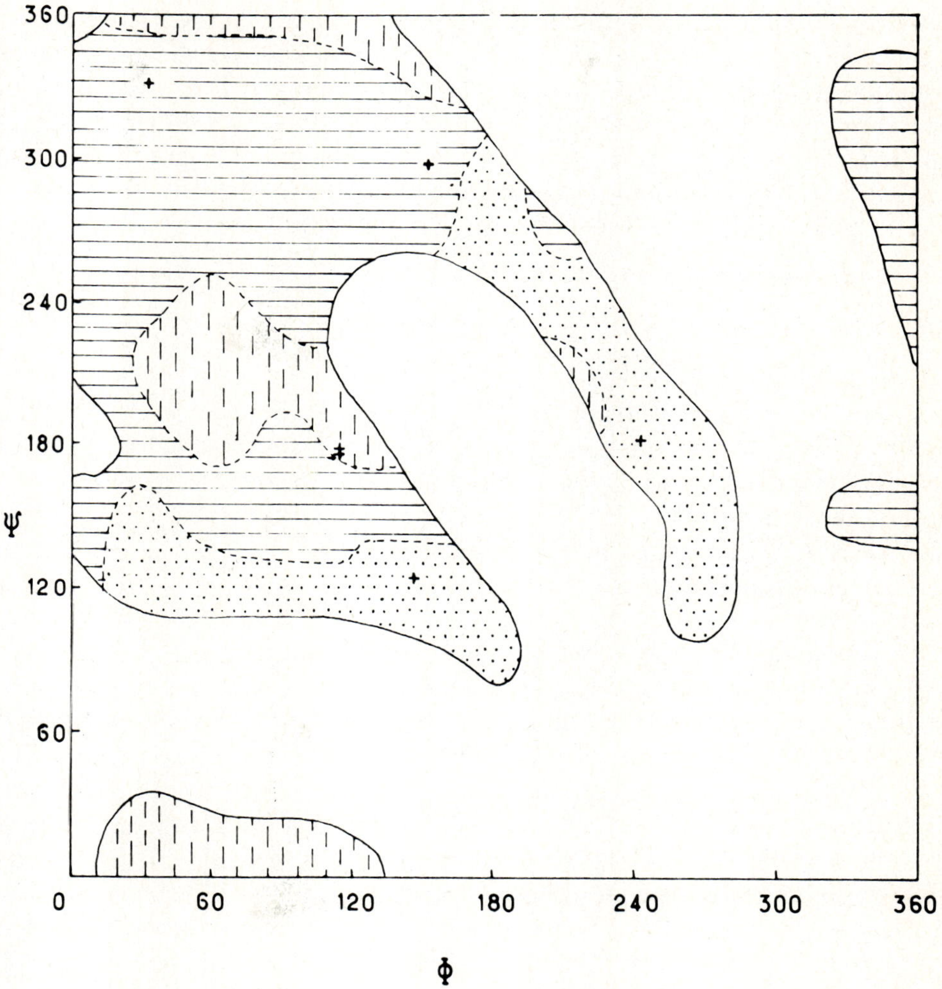

Figure 4-39. Conformation energy map for the valyl residue, showing the preferred values for the angle χ_1. The outside contour corresponds to a level of 5 kcal/mole above the global minimum (denoted by a double cross). The local minima are represented by +. The various regions correspond to the following values of χ_1: vertical dashed lines, $\chi_1 = 60°$; horizontal lines, $\chi_1 = 180°$; dots, $\chi_1 = 300°$ [Maigret *et al.* (1971b)].

20) other amino acyl residues

For completeness, the geometries and effective charges of methionine and cysteine and cystine are presented in Tables 4-21 through 4-23.

The results summarized above show that the dipeptide model provides satisfactory contours for the stable regions in globular proteins. However, for a more precise comparison with the experimental data, one should take into consideration, in addition to the energy values, the probabilities of occurrence of the preferred conformations [Pullman *et al.* (1971)], as it may happen that important minima of the conformational map may have relatively low probabilities of occurrence [Maigret *et al.* (1971c)].

The probabilities (see Appendix A5) may be approximated as

$$p_i = (1/f)\exp(-E_i(\phi,\psi)/RT)$$

with the partition function

$$f = \Sigma\Sigma \exp(-E_i(\phi,\psi)/RT)$$

evaluated at equal intervals of ϕ and ψ throughout their range; the absolute value of the partition function is subject to the arbitrariness in the choice of the intervals of ϕ and ψ, but the procedure constitutes a satisfactory approximation whenever only the relative probabilities of the different regions of the map are to be determined.

The complete information may be collected in a probability map (i.e., a plot of the probability as a function of ϕ and ψ), analogous to a conformational map but where the isolines correspond to different probability values. For a comparison of the local minima it may be sufficient, however, to determine the probabilities associated with the conformational areas (within, say, a limit of 1 kcal/mol) around each of the minima; for example, for the alanyl residue, the values are 26.9 (seven-membered hydrogen-bonded ring, $\phi = 120°$, $\psi = 220°$), 24.7 (antiparallel pleated sheet, $\phi = 30°$, $\psi = 330°$), 13.6 (fully-extended form, $\phi = 0°$, $\psi = 0°$), 5.9 (seven-membered hydrogen-bonded ring, $\phi = 260°$, $\psi = 140°$), 4.2 (right-handed α-helix, $\phi = 150°$, $\psi = 120°$), and 3.4% ($\phi = 150°$, $\psi = 300°$), which show that the first seven-membered hydroden-bonded ring is the most probable conformation, although the second one corresponds to a slightly deeper minimum [Maigret *et al.* (1971c)].

Preferred conformations of ribose, deoxyribose, and β-nucleosides

Before presenting the corresponding summary, some comments are needed regarding the conformational parameters of these systems. [For completeness of the information, Tables 4-24 through 4-27 present the geometries and effective charges of adenine, cytosine, guanine, and thymine, respectively.]

Table 4-21. Geometry and effective charges of methionine*

Atom	*Class*	x	y	z	*Charge*
S	22	0.306	1.528	3.633	0.21
O	9	−4.937	1.726	13.260	−0.55
O	10	−1.638	−0.693	14.122	−0.43
N	11	1.782	2.718	12.282	−0.51
C	5	−2.556	1.084	13.007	0.51
C	8	−0.882	2.718	11.309	−0.13
C	7	−0.918	1.501	8.542	−0.38
C	21	0.684	2.984	6.676	−0.51
C	21	2.718	3.109	1.904	−0.74
H	2	2.644	2.934	7.240	0.21
H	2	0.063	4.927	6.609	0.20
H	2	−2.844	1.454	7.867	0.20
H	2	−0.205	−0.409	8.652	0.20
H	2	−1.604	4.626	11.246	0.21
H	1	2.502	0.808	12.294	0.24
H	1	1.813	3.470	14.179	0.23
H	3	3.676	4.433	3.127	0.21
H	3	1.885	4.112	0.334	0.21
H	3	4.059	1.740	1.201	0.21
H	4	−5.696	0.578	14.414	0.41

* The coordinates (in bohrs) have been taken from the work of Mathieson (1952) [as given by Clementi *et al.* (1977)] and the effective charges from the work of Clementi *et al.* (1977). The notation for the atoms is: S, O(2), O(1), N, C′, C(α), C(β), C(γ), C(ϵ), H(γ_1), H(γ_2), H(β_1), H(β_2), H(α), H(1), H(2), H(ϵ_1), H(ϵ_2), H(ϵ_3), H(O_2).

Table 4-22. Geometry and effective charges of cysteine*

Atom	*Class*	x	y	z	*Charge*
S	22	0.008	4.954	1.926	0.06
O	10	6.623	2.652	4.243	−0.41
O	9	4.099	4.782	6.862	−0.54
N	11	6.049	5.961	0.467	−0.54
C	5	5.326	4.448	4.744	0.53
C	8	4.946	6.659	2.950	−0.13
C	21	2.191	7.437	2.730	−0.50
H	1	5.698	7.391	−0.849	0.24
H	1	7.970	5.727	0.535	0.26
H	2	6.048	8.245	3.703	0.18
H	2	1.560	8.184	4.567	0.20
H	2	1.993	9.013	1.369	0.19
H	23	0.019	4.954	4.477	0.06
H	4	4.421	3.369	8.062	0.40

* The coordinates (in bohrs) have been taken from the work of Jones *et al.* (1974) [as given by Clementi *et al.* (1977)] and the effective charges from the work of Clementi *et al.* (1977). The notation for the atoms is: S(γ), O(1), O(2), N, C′, C(α), C(β), H(2), H(1), H(α), H(β_1), H(β_2), H(Sγ), H(O_2).

Each conformation of the furanose ring (Fig. 4-40) is unequivocally determined in terms of two parameters, the phase angle of pseudorotation P

Table 4-23. Geometry and effective charges of cystine*

Atom	*Class*	*x*	*y*	*z*	*Charge*
O	10	6.623	2.652	4.243	−0.41
O	9	4.099	4.782	6.862	−0.54
N	11	6.049	5.961	0.467	−0.54
C	5	5.326	4.448	4.744	0.53
C	8	4.946	6.659	2.950	−0.13
C	21	2.191	7.437	2.730	−0.50
S	22	0.008	4.954	1.926	0.11
H	1	5.698	7.391	−0.849	0.24
H	1	7.970	5.727	0.535	0.26
H	2	6.048	8.245	3.703	0.18
H	2	1.560	8.184	4.567	0.22
H	2	1.993	9.013	1.369	0.19
H	4	4.421	3.369	8.062	0.40
O	10	−6.623	2.652	−4.243	−0.41
O	9	−4.099	4.782	−6.862	−0.54
N	11	−6.049	5.961	−0.467	−0.54
C	5	−5.326	4.448	−4.744	0.53
C	8	−4.946	6.659	−2.950	−0.13
C	21	−2.191	7.437	−2.730	−0.50
S	22	−0.008	4.954	−1.926	0.11
H	1	−5.698	7.391	0.849	0.24
H	1	−7.970	5.727	−0.535	0.26
H	2	−6.048	8.245	−3.703	0.18
H	2	−1.560	8.184	−4.567	0.22
H	2	−1.993	9.013	−1.369	0.19
H	4	−4.421	3.369	−8.062	0.40

* The coordinates (in bohrs) have been taken from the work of Jones *et al.* (1974) [as given by Clementi *et al.* (1977)] and the effective charges from the work of Clementi *et al.* (1977). The notation for the atoms is: O(1), O(2), N, C′, C(α), C(β), S(γ), H(2), H(1), H(α), H(β_1), H(β_2), H(O_2), and O(1), O(2), N, C′, C(α), C(β), S(γ), H(2), H(1), H(α), H(β_1), H(β_2), H(O_2).

and the degree of puckering τ [Altona and Sundaralingam (1972)]. The five torsional angles θ_i, $0 \leqslant i \leqslant 4$, of the five-membered ring (Table 4-28 and Fig.

Table 4-24. Geometry and effective charges of adenine*

Atom	*Class*	*x*	*y*	*z*	*Charge*
N	12	4.457	2.596	0.000	−0.33
N	12	2.176	−1.256	0.000	−0.32
N	12	−2.416	3.516	0.000	−0.29
N	15	−2.457	−0.751	0.000	−0.47
N	11	2.280	6.384	0.000	−0.60
C	24	2.280	3.852	0.000	0.25
C	19	0.000	2.588	0.000	−0.01
C	20	0.000	0.000	0.000	0.21
C	25	4.301	0.086	0.000	0.05
C	25	−3.798	1.373	0.000	0.05
H	16	6.091	−0.968	0.000	0.20
H	16	−3.119	−2.521	0.000	0.38
H	16	−5.876	1.373	0.000	0.21
H	1	0.696	7.413	0.000	0.33
H	1	3.865	7.413	0.000	0.32

* The coordinates (in bohrs) have been taken from the work of Spencer (1959) [as given by Scordamaglia *et al.* (1977)] and the effective charges from the work of Scordamaglia *et al.* (1977).

Table 4-25. Geometry and effective charges of cytosine*

Atom	*Class*	*x*	*y*	*z*	*Charge*
O	27	15.682	−0.460	0.000	−0.40
N	15	19.039	2.159	0.000	−0.48
N	12	14.826	3.649	0.000	−0.36
N	11	14.140	7.920	0.000	−0.60
C	26	16.471	1.706	0.000	0.46
C	24	15.768	5.980	0.000	0.26
C	25	18.318	6.522	0.000	−0.31
C	25	19.995	4.524	0.000	−0.01
H	1	12.264	7.689	0.000	0.33
H	1	14.692	9.727	0.000	0.31
H	16	18.995	8.487	0.000	0.19
H	16	22.048	4.849	0.000	0.23
H	16	20.228	0.690	0.000	0.37

* The coordinates (in bohrs) have been taken from the work of Spencer (1959) [as given by Scordamaglia *et al.* (1977)] and the effective charges from the work of Scordamaglia *et al.* (1977).

Table 4-26. Geometry and effective charges of guanine*

Atom	*Class*	*x*	*y*	*z*	*Charge*
O	27	8.535	7.333	0.000	−0.36
N	15	9.288	3.040	0.000	−0.49
N	12	5.899	0.122	0.000	−0.38
N	12	3.158	6.191	0.000	−0.25
N	15	1.748	2.159	0.000	−0.46
N	11	9.997	−1.260	0.000	−0.62
C	24	8.353	0.666	0.000	0.42
C	20	4.334	2.089	0.000	0.21
C	19	5.156	4.544	0.000	−0.05
C	26	7.747	5.166	0.000	0.37
C	25	1.161	4.603	0.000	0.02
H	16	−0.804	5.280	0.000	0.20
H	16	0.559	0.690	0.000	0.38
H	1	9.461	−3.072	0.000	0.35
H	1	11.871	−1.013	0.000	0.31
H	16	11.161	3.286	0.000	0.35

* The coordinates (in bohrs) have been taken from the work of Spencer (1959) [as given by Scordamaglia *et al.* (1977)] and the effective charges from the work of Scordamaglia *et al.* (1977).

Table 4-27. Geometry and effective charges of thymine*

Atom	*Class*	*x*	*y*	*z*	*Charge*
O	27	0.000	7.463	0.000	−0.40
O	27	4.188	0.000	0.000	−0.36
N	15	−2.258	3.854	0.000	−0.47
N	15	2.305	3.939	0.000	−0.49
C	26	0.000	5.158	0.000	0.51
C	25	−2.258	1.304	0.000	−0.05
C	28	0.000	0.000	0.000	−0.09
C	26	2.307	1.332	0.000	0.35
C	6	0.000	−2.910	0.000	−0.58
H	16	3.903	5.082	0.000	0.36
H	16	−3.792	5.082	0.000	0.37
H	16	−3.930	0.134	0.000	0.23
H	3	−1.672	−3.627	−0.965	0.19
H	3	0.000	−3.627	1.930	0.20
H	3	1.672	−3.627	−0.965	0.22

* The coordinates (in bohrs) have been taken from the work of Spencer (1959) [as given by Scordamaglia *et al.* (1977)] and the effective charges from the work of Scordamaglia *et al.* (1977).

4-40) along the pseudorotational pathway are described by

$\theta_i = \tau \cos (P + 144 i)$

which reduces to

$\theta_0 = \tau \cos P$

for i = 0. This angle θ_0 takes the values τ, 0, $-\tau$, 0, and τ when P goes through the full pseudorotational cycle (0, $\pi/2$, π, $3\pi/2$, 2π), so that a pseudorotation by P = π produces the mirror image of the ring.

The equations for θ_i, $0 \leqslant i \leqslant 4$, can be used to find P, which is given by

$\tan P = (\theta_2 + \theta_4 - \theta_1 - \theta_3)/2\,\theta_0\,(\sin 36 + \sin 72)$

so that P can be evaluated from the values for the five torsion angles; for negative values of θ_0, a correction of π must be added to the value of P calculated from the above expression. The standard conformation, with P = 0°, corresponds to the maximum value of the angle τ_2 and represents the C3′-*endo*-C2′-*exo* ($^3T_2$) conformation of the sugar; the symmetrical twist (T) and

Table 4-28. Conventions and notations of the torsion angles in the furanose ring*

Torsion angle	*Defined by the atoms*
$\tau_0 = \theta_3$	C4′-O1′-C1′-C2′
$\tau_1 = \theta_4$	O1′-C1′-C2′-C3′
$\tau_2 = \theta_0$	C1′-C2′-C3′-C4′
$\tau_3 = \theta_1$	C2′-C3′-C4′-O1′
$\tau_4 = \theta_2$	C3′-C4′-O1′-C1′
χ(C-N)	O1′-C1′-N1-C2
χ(C-N)	O1′-C1′-N9-C8
ϕ(C2′-O2′)	C1′-C2′-O2′-H(O2′)
ϕ(C3′-O3′)	C2′-C3′-O3′-H(O3′)
ϕ(C4′-O5′)	C3′-C4′-C5′-O5′
ϕ(C5′-O5′)	C4′-C5′-O5′-H(O5′)

* As given by Saran *et al.* (1973). See Fig. 4-40 for the numbering of the atoms. The two χ(C-N) angles given correspond to pyrimidine and purine nucleosides, respectively. The convention adopted is: the torsion angle ϕ(B-C) of the bonded atoms A-B-C-D is the angle between the planes defined by atoms A, B, C and B, C, D; it is taken as positive for a right-handed rotation (i.e., the far bond C-D rotates in the clockwise direction with respect to the near bond A-B, when looking along the bond B-C) and the zero value corresponds to the cis-planar arrangement of the bonds A-B and C-D.

envelope (E) conformations occur, respectively, at even and odd multiples of 18° in P and the C3′-*endo* and C2′-*endo* conformations correspond to P = 18° and P = 162°, respectively.

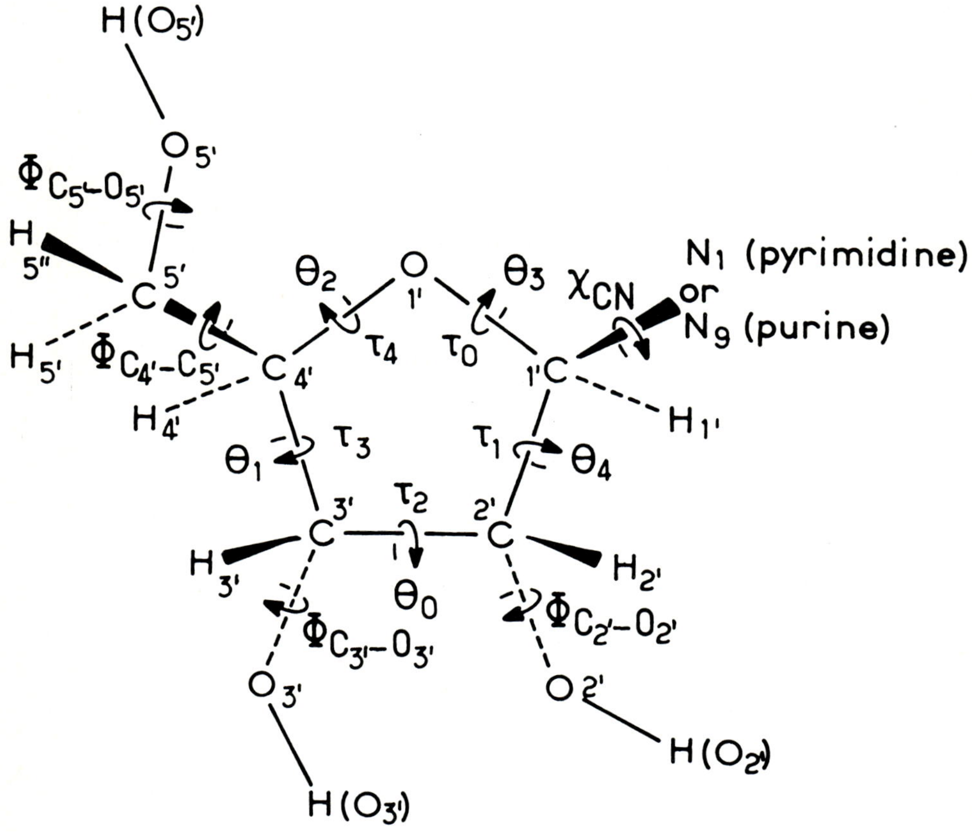

Figure 4-40. Numbering of atoms and designation of the dihedral angles in the furanose ring [Saran *et al.* (1973)].

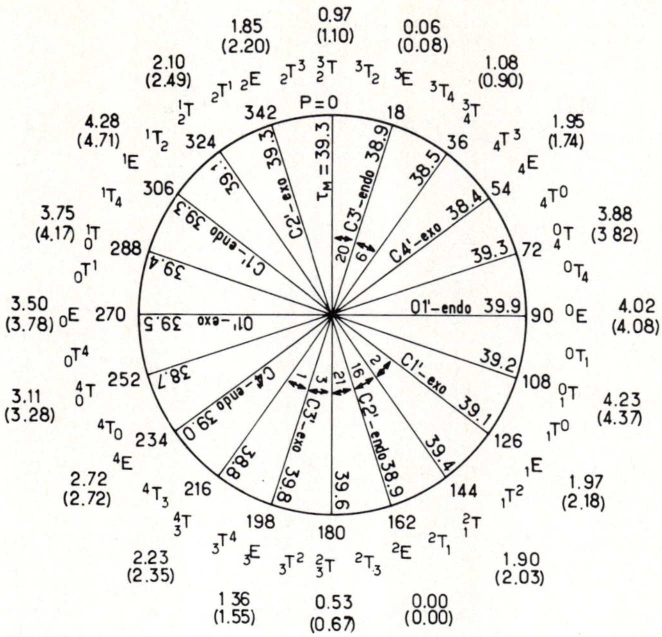

Figure 4-41. Conformational wheel, in the pseudorotational representation, for *D*-ribose and deoxyribose. The outermost numbers are the energies (kcal / mole) referred to the global minimum (with the values for deoxyribose given in parentheses). The values of P are given along the circle and the values of τ along the radii. The distribution of the observed experimental compounds is shown by numbers along the arrows in the populated P values [Saran *et al.* (1973)].

The conformational energies for the ribose and deoxyribose rings are presented in Fig. 4-41 as a function of P, with ϕ(C4′-C5′) = 60° (gg), ϕ(C5′-O5′) = 180°, ϕ(C3′-O3′) = 60°, ϕ(C2′-O2′) = 60°. The results show the existence of two almost identical energy minima at P = 18° and 162°, corresponding to the C3′-*endo* and C2′-*endo* conformations, the latter with a slightly lower energy.

The results for some chosen β-purine and β-pyrimidine ribosides are summarized below:

1) β-purine ribosides

The calculations were carried for adenosine and deoxyadenosine with ϕ(C4′-C5′) = 60°, ϕ(C5′-O5′) = 180°, ϕ(C3′-O3′) = 180° and (for adenosine) ϕ(C2′-O2′) = 180°.

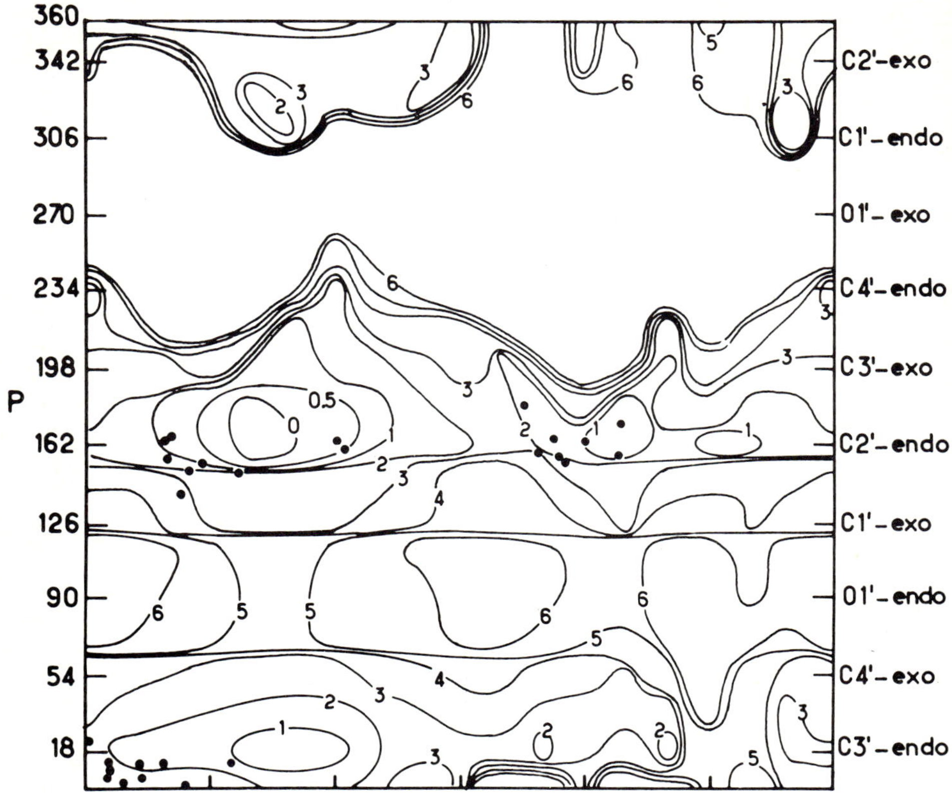

Figure 4-42. Conformational energy map of P vs. χ(C-N) in β-purine ribosides. The contours are labelled in kcal/mole (referred to the global minimum). The experimental X-ray results are represented by dots [Saran *et al.* (1973)].

i) β-adenosine

The conformational energy map (Fig. 4-42) for P vs. χ(C-N) presents a global minimum at P = 162° and χ(C-N) = 70°-100° (C2′-*endo* conformation). Three local minima (about 1 kcal/mole above the global minimum) are located at P = 162° and χ(C-N) = 240-260°, P = 162° and χ(C-N) = 300-320°, and P = 18° and χ(C-N) = 70-125° (C3′-*endo* conformation). The transition between the C3′-*endo* and C2′-*endo* conformations would occur through the C4′-*exo*-C1′-*exo* pathway with an energy barrier of about 5 kcal/mole; the transition through the C3′-*exo*-C2′-*exo* pathway is probably forbidden because of a larger energy barrier.

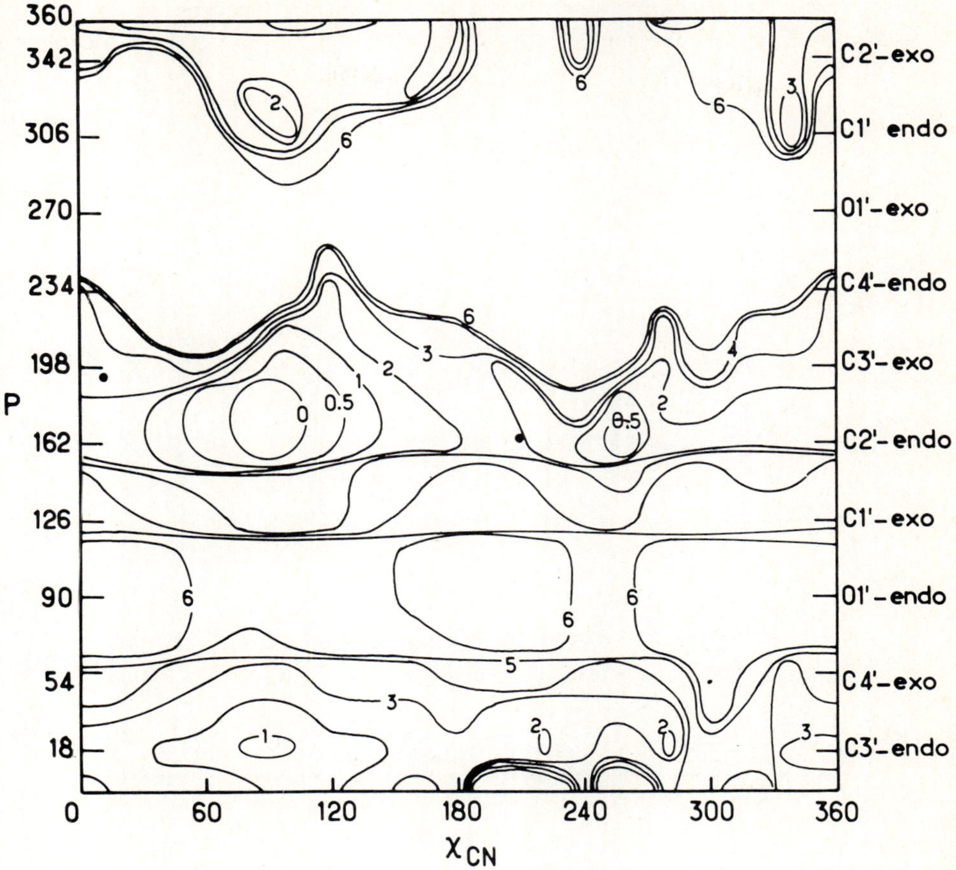

Figure 4-43. Conformational energy map of P vs. χ(C-N) in β-purine deoxyribosides. The contours are labelled in kcal/mole (referred to the global minimum). The experimental X-ray results are represented by dots [Saran *et al.* (1973)].

ii) β-deoxyadenosine

The conformational energy map (Fig. 4-43) for P vs. χ(C-N) is very similar to that of β-adenosine, with only some small differences; the local minimum at P = 162° and χ(C-N) = 240° is only 0.5 kcal/mole above the global minimum and the local minimum at P = 18° and χ(C-N) = 90° occupies a much smaller area.

2) β-pyrimidine ribosides

The calculations were carried out for uridine and deoxyuridine with ϕ(C4'-C5') = 60°, ϕ(C5'-O5') = 180°, ϕ(C3'-O3') = 180°, and (for uridine) ϕ(C2'-O2') = 180°.

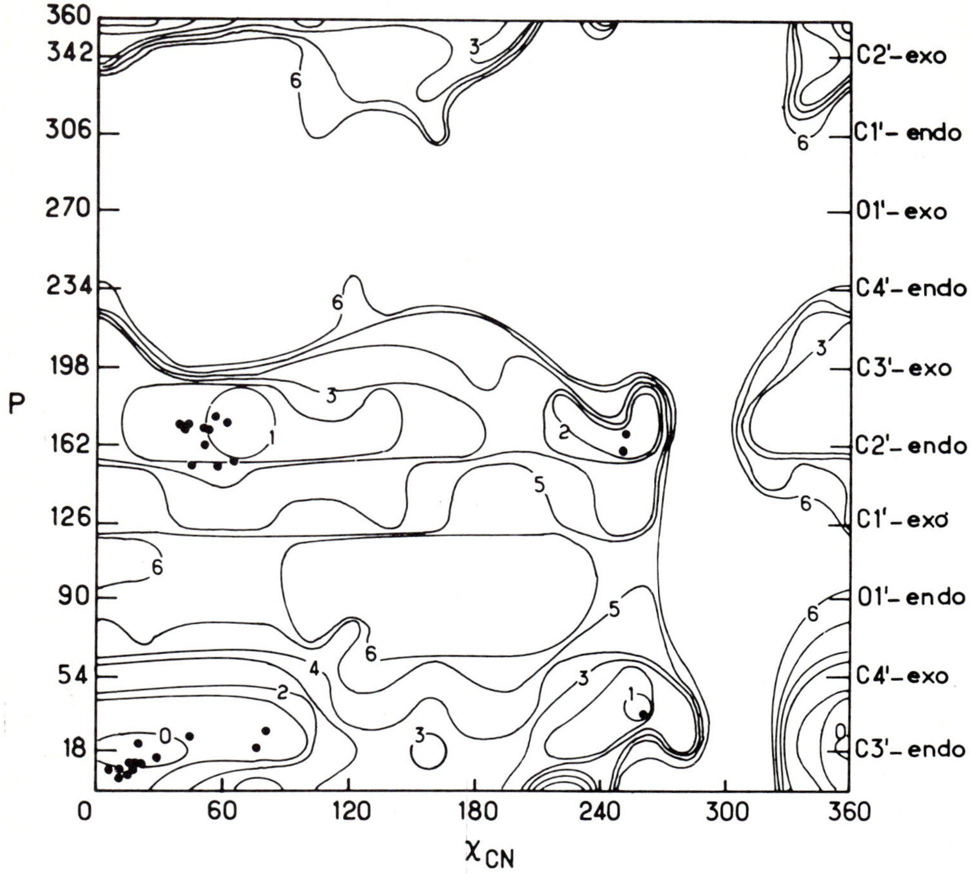

Figure 4-44. Conformational energy map of P vs. χ(C-N) in β-pyrimidine ribosides. The contours are labelled in kcal/mole (referred to the global minimum). The experimental X-ray results are represented by dots [Saran *et al* (1973)].

i) β-uridine

The conformational energy map (Fig. 4-44) for P vs. χ(C-N) presents a global minimum at P = 18° and χ(C-N) = 0-45° (C3'-*endo* conformation), with a large area (within 1 kcal/mole) associated with it. Local minima are found at P = 162° and χ(C-N) = 50-85° (C2'-*endo* conformation, about 1 kcal/mole above the global minimum), P = 18° and χ(C-N) = 260° (also at 1 kcal/mole above the global minimum), and at P = 18° and P = 162° (but 2-3 kcal/mole above the global minimum). The transition between the two minima would occur through the C4'-*exo*-C1'-*exo* pathway, with an energy barrier of about 6 kcal/mole.

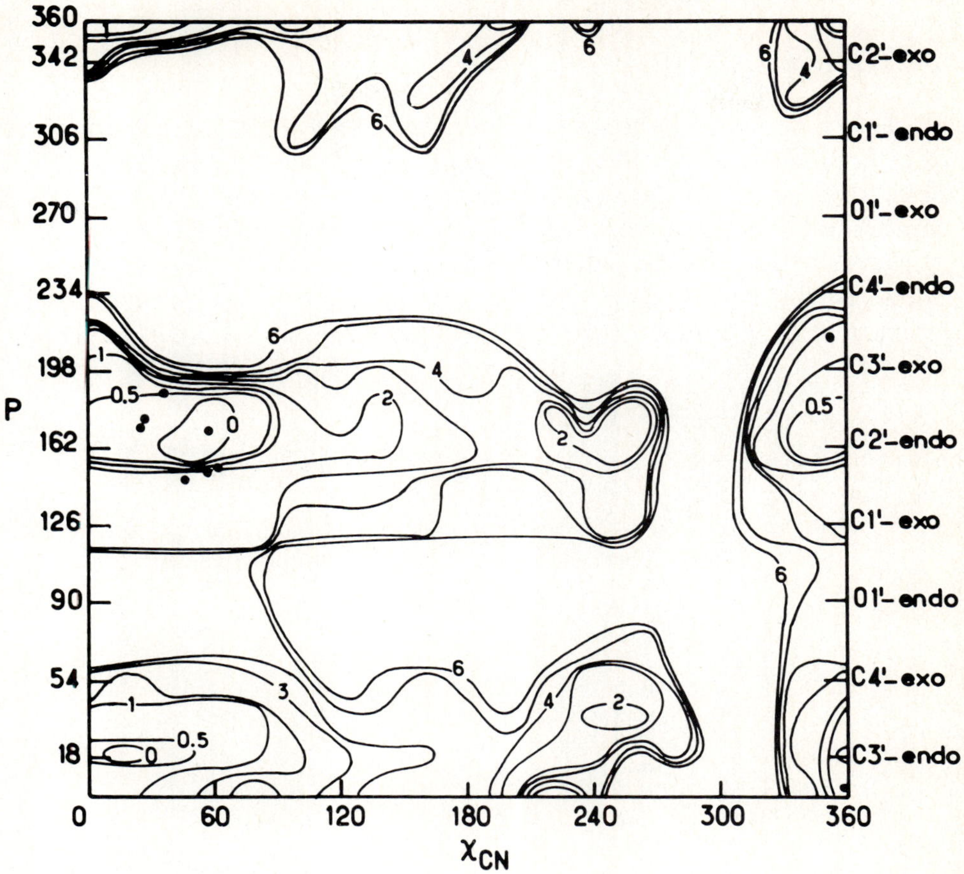

Figure 4-45. Conformational energy map of P vs. χ(C-N) in β-pyrimidine deoxyribosides. The contours are labelled in kcal/mole (referred to the global minimum). The experimental X-ray results are represented by dots [Saran *et al.* (1973)].

ii) β-deoxyuridine

The conformational energy map (Fig. 4-45) for P vs. χ(C-N) shows two almost identical global minima at P = 18° and P = 162°, with χ(C-N) = 0-80°, with a large area (within 0.5 kcal/mole) associated with the minimum at P = 162°. Local minima are found at P = 18° and P = 160° (about 2 kcal/mole above the global minima). The barrier for the transition from the C3'-*endo* to the C2'-*endo* conformation is 5 kcal/mole.

ORGANIZATION OF THE CALCULATIONS

Conformational Analyses

Given a biopolymer built up from N units, the study of its conformation and the determination of its stabilization energy in an 'aqueous' environment will consist of a number of steps as described below or *with suitable modifications* (depending on the data available, existing computing facilities, etc.)

In order to simplify the description of such a procedure, the notation $u_{i\,F}$, $u_{i\,M}$, $u_{i\,P}$ will be used to represent the i-th unit as it exists in free form, in a model system, and as a residue in the polymer. Similarly, $p_{ij\,F}$ will denote the by-

Table 4-29. Classes of atoms in amino acids and puric and pyrimidinic bases*

Class	*Atom*	*Group*	*Remarks*
1	H	amino	
2	H	methyne and methylene	
3	H	methyl	
4	H	carboxyl	
5	C	carboxyl	
6	C	methyl	
7	C	methylene	
8	C	methyne	
9	O	hydroxyl	in carboxyl group
10	O	carbonyl	in carboxyl group
11	N	amino	
12	N		in ring (without H)
13	N	carboxamide	
14	C		in ring, either next to N or fully substituted
15	N		in ring (with H)
16	H		attached to C in ring
17	C		in ring
18	C		phenolic or alcoholic
19	C		ring-junction, geminal to C
20	C		ring-junction, geminal to N
21	C	methylene	α to S
22	S		
23	H	sulfhydrol	
24	C		in ring, α to amino group
25	C		in ring, geminal to N
26	C	carbonyl	in ring
27	O	carbonyl	attached to ring
28	C		in ring, α to methyl

* As given by Clementi *et al.* (1977) and Scordamaglia *et al.* (1977) See Appendix A8 for details on the criterion used for this classificaton.

product liberated in the formation of the dimer, $u_{i\,P}$-$u_{j\,P}$, as described by the chemical reaction

$$u_{i\,F} + u_{j\,F} \leftrightarrow (u_{i\,P}\text{-}u_{j\,P}) + p_{ij\,F}$$

[e.g., $p_{ij\,F}$ will stand for H_2O in the case of reaction of two amino acids].

The suggested steps are:

a) Determine the quantum-mechanical preferred conformation of each $u_{i\,M}$;

a′) Determine the quantum-mechanical preferred conformation of each $u_{i\,F}$;

a″) Determine the quantum-mechanical preferred conformation of each $p_{ii+1\,F}$, that will be formed when the polymer is constructed from the $u_{i\,F}$;

b) Determine the analytical energy function for the interaction of each $u_{i\,P}$ with a water molecule, adopting for $u_{i\,P}$ the conformation determined for the corresponding residue in $u_{i\,M}$;

Table 4-30. Pairwise interaction constants for (molecule-water) complexes: Interacton with a hydrogen atom of water*.

Class	a'_{ab}	c_{ab}	g_{ab}
1	0.100 088 759 + 01	0.207 521 847 + 01	0.173 850 511 + 04
2	0.100 000 000 + 01	0.824 252 534 + 01	0.145 269 011 + 04
3	0.998 704 314 + 00	0.130 720 375 + 01	0.280 374 992 + 03
4	0.100 093 982 + 01	0.159 769 344 + 01	0.463 124 774 + 04
5	0.999 759 927 + 00	0.356 004 411 + 01	0.184 585 125 + 06
6	0.100 027 880 + 01	0.356 004 411 + 01	0.389 186 260 + 05
7	0.100 000 000 + 01	0.190 353 591 + 02	0.618 960 681 + 04
8	0.100 101 370 + 01	0.117 872 994 + 03	0.235 846 649 + 04
9	0.100 005 104 + 01	0.845 311 871 + 02	0.870 600 693 + 03
10	0.999 389 425 + 00	0.163 284 536 + 03	0.762 464 880 + 04
11	0.100 009 498 + 01	0.141 507 664 + 03	0.277 916 666 + 04
12	0.998 375 371 + 00	0.491 524 463 + 02	0.486 883 081 + 04
13	0.100 299 819 + 01	0.347 660 562 + 01	0.682 350 397 + 04
14	0.100 080 614 + 01	0.968 582 392 + 01	0.130 161 777 + 06
15	0.100 051 089 + 01	0.441 930 092 + 03	0.127 116 438 + 05
16	0.998 747 133 + 00	0.356 528 398 + 01	0.104 377 336 + 04
17	0.992 949 978 + 00	0.356 004 411 + 01	0.121 598 151 + 06
18	0.100 236 642 + 01	0.356 004 411 + 01	0.404 495 615 + 05
19	0.874 550 834 + 00	0.356 004 411 + 01	0.157 257 820 + 06
20	0.981 899 209 + 00	0.385 323 176 + 02	0.177 231 436 + 06
21	0.100 149 776 + 01	0.604 173 851 + 01	0.545 685 400 + 05
22	0.101 107 869 + 01	0.259 413 876 + 03	0.918 035 120 + 04
23	0.102 515 377 + 01	0.102 959 207 + 03	0.776 556 488 + 04
24	0.998 235 397 + 00	0.943 062 698 + 03	0.970 811 281 + 06
25	0.995 092 936 + 00	0.649 988 854 + 01	0.173 948 033 + 06
26	0.997 791 417 + 00	0.974 940 002 + 01	0.189 976 014 + 06
27	0.100 600 804 + 01	0.207 783 979 + 03	0.296 326 648 + 04
28	0.856 379 583 + 00	0.649 988 854 + 01	0.649 881 953 + 06

*As given by Bolis and Clementi (1977). The interaction energy is expressed in kcal/mole when the distances are expressed in angstroms. See Table 4-29 for details on the classes of atoms. (Reprinted with permission. Copyright by The American Chemical Society.)

b′) Determine the analytical energy function for the interaction of each $u_{i\ F}$ with a water molecule;

b″) Determine the analytical energy function for the interaction of each $p_{ii+1\ F}$ with a water molecule;

c′) Evaluate the potential energy of each $u_{i\ F}$ in an 'aqueous' environment;

c″) Evaluate the potential energy of each $p_{ii+1\ F}$ in an 'aqueous' environment;

d) Construct the polymer, the conformation of $u_{i\ M}$ being adopted for the corresponding $u_{i\ P}$, and optimize its conformation (see below);

e) Evaluate the potential energy of the polymer in an 'aqueous' environment.

The determination of the conformation of the polymer requires only setps *a* and *d*, while all the steps are needed for the evaluation of its energy of formation. Step *d* might be carried out for the complete polymer [Momany *et al.* (1975)] or step-by-step, as in the following cycle:

Table 4-31. Pairwise interaction constants for (molecule-water) complexes: Interaction with an oxygen atom of water*

Class	a'_{ab}	c_{ab}	g_{ab}
1	0.998 202 369 + 00	0.227 048 409 + 03	0.319 285 464 + 04
2	0.999 966 606 + 00	0.344 879 279 + 01	0.332 071 268 + 04
3	0.998 410 933 + 00	0.344 879 279 + 01	0.704 513 690 + 04
4	0.100 093 916 + 01	0.389 675 373 + 03	0.307 583 102 + 04
5	0.999 941 483 + 00	0.242 698 627 + 03	0.570 163 312 + 04
6	0.999 055 520 + 00	0.102 073 144 + 02	0.294 559 020 + 06
7	0.999 073 327 + 00	0.110 689 075 + 03	0.276 483 570 + 06
8	0.100 345 021 + 01	0.102 073 144 + 02	0.345 947 762 + 06
9	0.999 645 844 + 00	0.112 916 672 + 02	0.305 269 293 + 06
10	0.100 092 629 + 01	0.102 073 144 + 02	0.261 569 141 + 06
11	0.999 096 863 + 00	0.124 076 189 + 02	0.631 822 125 + 06
12	0.996 603 809 + 00	0.220 215 490 + 02	0.200 655 234 + 06
13	0.100 005 582 + 01	0.318 320 038 + 03	0.615 968 625 + 06
14	0.101 622 795 + 01	0.126 181 303 + 03	0.130 406 014 + 06
15	0.100 060 296 + 01	0.161 787 306 + 04	0.626 337 617 + 06
16	0.100 239 015 + 01	0.233 495 670 + 03	0.792 550 488 + 04
17	0.993 570 738 + 00	0.102 073 144 + 02	0.102 467 840 + 06
18	0.999 982 804 + 00	0.102 073 144 + 02	0.520 450 148 + 06
19	0.109 999 990 + 01	0.199 325 260 + 03	0.136 718 561 + 05
20	0.101 060 666 + 01	0.655 663 666 + 03	0.911 956 562 + 06
21	0.100 122 450 + 01	0.229 295 538 + 03	0.155 507 061 + 05
22	0.995 094 240 + 00	0.139 064 225 + 02	0.174 304 908 + 07
23	0.105 506 256 + 01	0.254 667 215 + 03	0.369 866 099 + 05
24	0.100 573 744 + 01	0.167 215 482 + 04	0.576 040 719 + 06
25	0.999 616 608 + 00	0.149 723 616 + 02	0.241 460 602 + 06
26	0.996 846 467 + 00	0.229 376 206 + 02	0.211 941 961 + 06
27	0.994 164 132 + 00	0.233 567 953 + 02	0.341 006 609 + 06
28	0.108 784 823 + 01	0.102 073 144 + 02	0.360 925 835 + 05

* As given by Bolis and Clementi (1977). The interaction energy is expressed in kcal/mol when the distances are expressed in angstroms. See Table 4-29 for details on the classes of atoms. (Reprinted with permission. Copyright by The American Chemical Society.)

a) the (i + 1)-th residue is added to the existing polypeptide; its coordinates (as a part of the polymer) are obtained by appropriate transformation (translation and rotation; see Appendix A2) of the coordinates for the corresponding residue (obtained from those for the monomer in free form);

b) apply the necessary transformation so that it adopts the conformation corresponding to the global minimum (or a local minimum) in the conformational map of this residue (if available);

c) optimize the conformation of the polymer built so far;

d) restart at step *a* for the next residue to be added.

Step *c* may be carried out in a number of ways. For example, if the (global or local) minimum energy conformation of the residue represents a satisfactory approximation to that of u_{iP}, it will suffice to optimize the relative conformations of the residues with respect to each other. In any case, it may be sufficient to consider only the last few units [e.g., for polypeptides, one might include in the optimization the (i − 2)-th through the (i + 1)-th residues, because of the possibility of U-turns]. Finally, in all cases, the preferred conformations should be determined (and this applies to all the systems considered) taking into account the aqueous environment, but economic reasons advise against it.

The actual calculations for polynucleotides involve then four separate sets of data and parameters:

a) the geometries of the amino acids, from crystallographic data;

Table 4-32. Designation of atom types*

Atom	*Type*	*Designation*
H	1	aliphatic hydrogen
	2	primary and secondary amine or amide hydrogen
	3	aromatic hydrogen
	4	hydroxyl or carboxylic hydrogen
C	5	aliphatic carbon
	6	carbonyl, carboxylic, or peptide bond carbon
	7	aromatic carbon
N	8	primary or secondary amide nitrogen
	9	uncharged primary or secondary amine nitrogen (also primary or secondary amide nitrogen as hydrogen acceptor in a hydrogen-bond)
O	10	carbonyl or carboxylic (C = O) oxygen
	11	hydroxyl, carboxylic (C-O-H), or ester oxygen
S	12	sulfur

* Adapted from Momany *et al.* (1974). The sulfhydryl hydrogen, aliphatic carbon with three heavy atoms attached [e.g., C(α) of amino acid], quaternary carbon, carboxylate ion carbon, nitrogen of charged NH_3^+ group, and disulfide sulfur are taken as H(3), C(5), C(7), C(6), N(8), and S(12), respectively. (Reprinted with permission. Copyright by The American Chemical Society.)

b) the optimum conformations of the amino acyl residues, as determined by Pullman and co-workers within the framework of the dipeptide mode;

c) the potential energy function (with the appropriate parametrization) of Scheraga and co-workers for the determination of the preferred conformation;

d) the potential energy function (with the corresponding coefficients) of Clementi and co-workers for the evaluation of the interaction of the polypeptide with an aqueous environment.

Table 4-33. Effective charges for backbone atoms*

Atom	*Type*	*Glycine*	*Proline*	*Aspartate*	*Others*
N	8	−0.344	−0.285	−0.356	−0.356
H(N)	2	0.176		0.176	0.176
C(α)	5	−0.008	0.050	−0.060	0.064
C′	6	0.450	0.455	0.450	0.450
O	10	−0.384	−0.385	−0.384	−0.384
H(α)	1	0.055	0.040	0.024	0.020

* See Table 4-32 for details on the types of atoms. The last column (under the designation 'others') includes all the C(β)-containing residues with charged or uncharged side-chains (excluding aspartate) [Momany *et al.* (1975)]. (Reprinted with permission. Copyright by The American Chemical Society.)

All the data for these calculations have been tabulated or summarized in this work in order to aid in the execution of future calculations. It is unfortunate that, at present, two sets of effective charges are involved in steps *c* and *d*: those in step *c* were obtained in semiempirical calculations (including the valence electrons), while those in step *d* are from *a priori* calculations, including all the electrons. It would be preferable to use the latter, but unfortunately, the potential energy function of step *c* was parametrized using the former; in due time, a new parametrization using the *a priori* charges may be advisable.

In any case, some comments are required regarding the values of the charges in conformational calculations [Momany *et al.* (1975)]:

a) the variation in the charges corresponding to changes in the dihedral angles around the minimum energy conformation are small (for N-acetyl-N′-methylamides);

b) rotation about the peptide bond is accompanied by large changes in the charges of the atoms involved, but the potential barrier is very high so that the planar conformations (*cis* or *trans*) are preferred, the charges for both conformations being very similar.

c) there may be a significant change in the charges when a hydrogen-bond is established, but the effect is simulated by the hydrogen-bond function used.

From the point of view of data, the situation is very similar for polynucleotides but, unfortunately, a parametrized potential energy function for conformational analyses is not available, so that such detailed calculations as those outlined for polypeptides are not possible at present. The actual calculations at this point involve the following sets of data and parameters:

a) the geometries of the purine and pyrimidine bases, ribose and deoxyribose, and the phosphate group from crystallographic data;

Table 4-34. Effective charges for atoms in saturated side chains*

Amino acid	*Atom*	*Charge*	*Atom*	*Charge*	*Atom*	*Charge*
alanine	C(β)	−0.090	H(β)	0.040		
cysteine	C(β)	−0.105	H(β)	0.055	S(γ)	0.015
	H(γ)	0.010				
cystine	C(β)	−0.090	H(β)	0.055	S(γ)	0.010
isoleucine	C(β)	−0.005	H(β)	0.025	$C(\gamma_1)$	−0.020
	$H(\gamma_1)$	0.015	$C(\gamma_2)$	−0.075	$H(\gamma_2)$	0.025
	$C(\delta_1)$	−0.075	H(δ)	0.025		
leucine	C(β)	−0.030	H(β)	0.022	C(γ)	−0.011
	H(γ)	0.025	$C(\delta_1)$	−0.074	$C(\delta_2)$	−0.074
	H(δ)	0.025				
lysine	C(β)	−0.030	H(β)	0.015	C(γ)	−0.025
	H(γ)	0.020	C(δ)	−0.030	H(δ)	0.020
	C(ε)	0.110	H(ε)	−0.005	N(ζ)	−0.395
	H(ζ)	0.150				
lysine (+)	C(β)	−0.030	H(β)	0.015	C(γ)	−0.025
	H(γ)	0.020	C(δ)	−0.115	H(δ)	0.100
	C(ε)	0.050	H(ε)	0.120	N(ζ)	−0.320
	H(ζ)	0.320				
methionine	C(β)	−0.010	H(β)	0.030	C(γ)	−0.120
	H(γ)	0.045	S(δ)	0.035	C(ε)	−0.190
	H(ε)	0.055				
serine	C(β)	0.130	H(β)	0.020	O(γ)	−0.310
	H(γ)	0.170				
threonine	C(β)	0.160	H(β)	0.015	$O(\gamma_1)$	−0.310
	$H(\gamma_1)$	0.170	$C(\gamma_2)$	−0.095	$H(\gamma_2)$	0.030
valine	C(β)	0.008	H(β)	0.016	$C(\gamma_1)$	−0.072
	$C(\gamma_2)$	−0.072	H(γ)	0.025		

* As given by Momany *et al.* (1975). (Reprinted with permission. Copyright by The American Chemical Society.)

b) the optimum conformations of β-adenosine, β-deoxyadenosine, β-uridine, and β-deoxyuridine, as determined by Pullman and co-workers;

Table 4-35. Effective charges for atoms in unsaturated non-aromatic side chains*

Amino acid	*Atom*	*Charge*	*Atom*	*Charge*	*Atom*	*Charge*
arginine	C(β)	−0.030	H(β)	0.015	C(γ)	−0.030
	H(γ)	0.030	C(δ)	0.100	H(δ)	0.010
	N(ϵ)	−0.350	H(ϵ)	0.170	C(ζ)	0.480
	N(η_1)	−0.420	H(η_1)	0.090	N(η_2)	−0.410
	H(η_2)	0.160				
arginine (+)	C(β)	−0.030	H(β)	0.015	C(γ)	−0.030
	H(γ)	0.030	C(δ)	0.120	H(δ)	0.010
	N(ϵ)	−0.300	H(ϵ)	0.230	C(ζ)	0.580
	N(η_1)	−0.390	H(η_1)	0.280	N(η_2)	−0.390
	H(η_2)	0.280				
asparagine	C(β)	−0.120	H(β)	0.055	C(γ)	0.460
	O(δ_1)	−0.375	N(δ_2)	−0.445	H(δ_2)	0.200
aspartic acid	C(β)	−0.100	H(β)	0.060	C(γ)	0.500
	O(δ_1)	−0.360	O(δ_2)	−0.350	H(δ_2)	0.220
aspartate	C(β)	−0.170	H(β)	−0.020	C(γ)	0.500
	O(δ_1)	−0.570	O(δ_2)	−0.570		
glutamic acid	C(β)	−0.030	H(β)	0.030	C(γ)	−0.110
	H(γ)	0.055	C(δ)	0.500	O(ϵ_1)	−0.360
	O(ϵ_2)	−0.350	H(ϵ_2)	0.210		
glutamate	C(β)	−0.120	H(β)	0.020	C(γ)	−0.170
	H(γ)	−0.040	C(δ)	0.500	O(ϵ_1)	−0.570
	O(ϵ_2)	−0.570				
glutamine	C(β)	−0.030	H(β)	0.020	C(γ)	−0.110
	H(γ)	0.050	C(δ)	0.470	O(δ_1)	−0.380
	N(δ_2)	−0.430	H(δ_2)	0.185		

* As given by Momany *et al.* (1975). (Reprinted with permission. Copyright by The American Chemical Society.)

c) the potential energy function (with the corresponding coefficients) of Clementi and co-workers for the evaluation of the interaction of the polynucleotide with an aqueous environment.

A tentative conformational analysis may be carried out without consideration of torsional energies, except for those corresponding to rotation about the β-glycosidic bond and those corresponding to the sugar ring. Considering only the optimum conformations of the ribosides and deoxyribosides as described above, the potential barriers may be taken from the corresponding energy

Table 4-36. Effective charges for atoms in ring-bearing side-chains*

Amino acid	*Atom*	*Charge*	*Atom*	*Charge*	*Atom*	*Charge*
histidine (Hδ₁)	C(β)	−0.040	H(β)	0.015	C(γ)	0.055
	N(δ₁)	−0.250	H(δ₁)	0.150	C(δ₂)	0.080
	H(δ₂)	0.035	C(ε₁)	0.190	H(ε₁)	0.020
	N(ε₂)	−0.240				
histidine (Hε₂)	C(β)	−0.075	H(β)	0.040	C(γ)	0.100
	N(δ₁)	−0.230	C(δ₂)	0.054	H(δ₂)	0.020
	C(ε₁)	0.182	H(ε₁)	0.010	N(ε₂)	−0.255
	H(ε₂)	0.144				
histidine (+)	C(β)	−0.050	H(β)	0.065	C(γ)	0.150
	N(δ₁)	−0.200	H(δ₁)	0.275	C(δ₂)	0.100
	H(δ₂)	0.135	C(ε₁)	0.275	H(ε₁)	0.150
	N(ε₂)	−0.200	H(ε₂)	0.265		
phenylalanine	C(β)	−0.040	H(β)	0.025	C(γ)	0.020
	C(δ)	−0.015	H(δ)	0.010	C(ε)	−0.015
	H(ε)	0.010	C(ζ)	0.015	H(ζ)	0.005
proline	C(β)	−0.025	H(β)	0.015	C(γ)	−0.050
	H(γ)	0.025	C(δ)	0.100	H(δ)	0.010
hydroxyproline	C(β)	−0.050	H(β)	0.020	C(γ)	0.172
	H(γ)	0.006	O(δ₁)	−0.323	H(δ₁)	0.165
	C(δ₂)	0.075	H(δ₂)	0.020		
tryptophan	C(β)	−0.030	H(β)	0.015	C(γ)	−0.025
	C(δ₁)	0.085	H(δ₁)	0.000	C(δ₂)	0.005
	N(ε₁)	−0.280	H(ε₁)	0.130	C(ε₂)	0.150
	C(ε₃)	−0.080	H(ε₃)	0.080	C(ζ₂)	−0.060
	H(ζ₂)	0.030	C(ζ₃)	−0.020	H(ζ₃)	0.010
	C(η₂)	−0.005	H(η₂)	0.010		
tyrosine	C(β)	−0.040	H(β)	0.025	C(γ)	0.020
	C(δ)	−0.010	H(δ)	0.010	C(ε)	−0.060
	H(ε)	0.030	C(ζ)	0.225	O(η)	−0.330
	H(η)	0.165				

* As given by Momany *et al.* (1975). (Reprinted with permission. Copyright by The American Chemical Society.)

Table 4-37. Effective charges for atoms of C-terminal end groups*

End Group	*Atom*	*Type*	*Atomic Charge*	*Group Charge*
−COOH	O	11	−0.380	−0.176
	H	4	0.204	
$-COO^-$	O	10	−0.532	−0.680
$-COCH_3$	C	5	−0.180	0.000
	H	1	0.060	
$-CONH_2$	N	8	−0.390	0.000
	H	2	0.195	
$-CONHCH_3$	N	8	−0.345	0.000
	H	2	0.163	
	C	5	0.050	
	H	1	0.044	
$-CON(CH_3)_2$	N	8	−0.280	0.000
	C	5	0.050	
	H	1	0.030	
$-COOCH_3$	O	11	−0.300	0.000
	C	5	0.135	
	H	1	0.055	
$-COOC_2H_5$	O	11	−0.300	0.000
	C	5	0.155	
	H	1	0.040	
	C	5	−0.100	
	H	1	0.055	

* As given by Momany *et al.* (1975). See Table 4-32 for details on the types of the atoms. (Reprinted with permission. Copyright by The American Chemical Society.)

Table 4-38. Effective charges for atoms in N-terminal groups*

End Group	*Atom*	*Type*	*Atomic Charge*	*Group Charge*
$-NH_2$	H	2	0.176	0.176
$-NH_3^+$	H	2	0.285	0.680
$-NHCH_3$	C	5	−0.060	0.000
	H	1	0.020	
$-NHCOCH_3$	C	6	0.455	0.000
	O	10	−0.386	
	C	5	−0.129	
	H	1	0.020	
−NHCHO	C	6	0.430	0.000
	H	3	−0.050	
	O	10	−0.380	

* As given by Momany *et al.* (1975). The first three end groups are not used with proline-type residues. See Table 4-32 for details on the types of the atoms. (Reprinted with permission. Copyright by The American Chemical Society.)

maps when the systems considered are adenosine, deoxyadenosine, uridine or deoxyuridine; for the remaining glycosides, similar values could be used as an approximation.

Table 4-39. Dipole polarizabilities, $\alpha(A^3)$, and effective number of electrons, n_i*

Atom	*Type*	α	n_i
H	all	0.42	0.85
C	5,7	0.93	5.20
C	6	1.51	5.20
N	8	0.87	6.10
N	9	0.93	6.10
O	10	0.84	7.00
O	11	0.59	7.00
S	12	0.504	14.8

* As given by Dalgarno (1962) and Scott and Scheraga (1966). [Taken from the work of Momany *et al.* (1974, 1975)].

Table 4-40. Parameters for the non-bonded potential interaction terms*

Atom	*Type*	c_{ab} (*kcal* A^6/mol)	g_{ab} (*kcal* A^{12}/mol)	$-\epsilon_{ab}$ (*kcal/mol*	$<r_{ab}>$ (*A*)
H	1	45.5	1.410	0.037	2.92
	2	45.5	0.843	0.062	2.68
	3	45.5	1.439	0.036	2.93
	4	45.5	1.169	0.044	2.83
C	5	370.5	90.603	0.038	4.12
	6	766.5	104.898	0.141	3.74
	7	370.5	47.530	0.073	3.70
N	8	363.1	73.254	0.045	3.99
	9	401.0	37.494	0.107	3.51
O	10	369.0	17.019	0.200	3.12
	11	217.2	12.563	0.094	3.24
S	12	249.0	36.318	0.043	3.78

* As given by Momany *et al.* (1974). (Reprinted with permission. Copyright by The American Chemical Society.)

Table 4-41. Parameters for the hydrogen-bond interaction terms*

Atoms	*Types*	f_{ab} (*kcal A*^{10}/mol)	g'_{ab} (*kcal A*^{12}/mol)	R_{min} (*A*)	U_{min} (*kcal/mol*)
H... O	2... 10	4014	12,040	1.90	−1.11
H... O	4... 10	5783	13,344	1.66	−5.92
H... O	4... 11	4610	13,033	1.84	−1.71
H... N	2... 9	8244	32,897	2.19	−0.55

* As given by Momany *et al.* (1974). (Reprinted with permission. Copyright by The American Chemical Society.)

Evaluation of Interaction Energies

The calculations involve four distinct steps, listed below without any implication of order of execution, as it may be understood from their inspection:

a) evaluation of the interaction energy for given conformations of the interacting systems, placed at given positions;

b) minimization of the interaction energy by optimization of the conformations;

c) minimization of the interaction energy with respect to the separation and relative orientation of the interacting systems;

d) consideration of the aqueous environment.

The main difficulty to be faced in these calculations arises from practical considerations, the optimization of the conformations and the evaluation of the interactions with the aqueous environment being time-consuming and expensive. Therefore, it may be found necessary or convenient to omit the optimization of the conformations and to consider the interaction with the environment only for the end points, that is, the separate systems and the molecular associate. The determination of the minimum with respect to position and orientation does not offer any difficulty, as described in the section on minimization techniques.

Long- through medium-range interactions (without including the $1/R^{12}$ terms) can be calculated whenever the expansion coefficients can be obtained from atomic data, as described above. The evaluation of the complete set of interactions, including hydrogen-bonding, requires the semiempirical coefficients, described and tabulated above; as those coefficients were obtained by a parametrization based on semiempirical charges for the amino acids, calculations involving polynucleotides must be carried out using equivalent charges (available in the literature, but not tabulated here), although approximate results could be obtained using *a priori* charges (with reevaluation of those coefficients directly dependent on the effective charges).

The actual calculations of interactions between polypeptides involve the following sets of data and parameters:

a) effective charges for amino acyl residues, determined by Scheraga and co-workers in semiempirical calculations;

b) the potential energy function (with corresponding coefficients), determined by Scheraga and co-workers;

c) the potential energy function of Clementi and co-workers for the evaluation of the interaction with the aqueous environment.

MINIMIZATION TECHNIQUES

The hypersurface corresponding to the interaction between two molecules (or two distinct units of a polymer) is a function of the variables that describe the relative positions of the two molecules (or components). For example, assuming two systems of coordinates, associated respectively with each molecule (e.g., centered at their centers of mass), those variables are the polar coordinates (R, θ, ϕ), that describe the position of the second cartesian system with respect to the first, and the rotation angles (α, β, γ), around the three axes, that define the position of the second molecule with respect to the coordinate system associated with it. That is, the functional to be considered is, in short,

$$E = E(R,\theta,\phi,\alpha,\beta,\gamma)$$

(which implicitly contains also a dependence on the formulation adopted for the evaluation of the interaction energy) and the determination of its minima represents the problem to be solved. The theoretical and practical aspects of the minimization techniques, that may be used, will be discussed in this section but, first, some general comments are appropriate.

The problem of determining the minima of a functional, already present in the early days of calculus, has now led to the development of a specialized field, denoted as optimization theory. Within this field, the theory of non-linear programming occupies a special position, as a direct consequence of the fact that, in most cases, the functional to be minimized is not linear in the variables.

General Outline

The problem, then, consists of the determination of the minima of the functional

$$E = E(\mathbf{X})$$

where the (column) vector $\mathbf{X}$ has as elements the variables $X_1, X_2, \ldots, X_N$.

It is said that the point $\mathbf{X}(g)$ represents the global minimum of the function if the condition

$$E(\mathbf{X}(g)) \leqslant E(\mathbf{X})$$

is satisfied for any $\mathbf{X}$ belonging to the space considered; on the other hand, the point $\mathbf{X}(l)$ corresponds to a local minimum when the condition

$$E(\mathbf{X}(l)) \leqslant E(\mathbf{X})$$

holds for any $\mathbf{X}$ belonging to the space in the neighbourhood of $\mathbf{X}(l)$.

A necessary condition for a function to have a minimum is that all its first derivatives exist at any point, vanishing at the minimum; a sufficient condition, however, is that all the second derivatives exist at that point and that the hessian matrix (the matrix of the second derivatives, at the minimum) be positive definite.

In order to locate the position of the global minimum, it is necessary to explore the entire space spanned by the independent variables. Many attempts have been made towards the solution of this problem, but only for those cases where the number of variables is not greater than two the existing methods may be considered reasonably useful. [For further details on this problem, see 'Towards Global Optimisation', which collects the proceedings of a workshop held at the University of Cagliari, Italy (1974), edited by L. C. W. Dixon and G. P. Szego, North Holland and Elsevier Scientific Publishing Company, Amsterdam (1975).] Therefore, the discussion is hereafter restricted to those methods dealing with the determination of local minima and the designation 'minimum' will stand for 'local minimum'.

The minimization methods may be grouped into two main categories: search and gradient methods. Search methods are those that find the minimum through a comparison of the values of the function corresponding to a sequence of points $\mathbf{X}(1), \mathbf{X}(2), \ldots$, that may be defined according to different procedures: in tabulation methods, the search is carried out at points of a grid of the variables; the sequential search methods involve the evaluation of the function at the vertices of a given geometrical design in the space of the variables; and the linear search methods use a set of search directions throughout the search. Gradient methods, on the other hand, are based on the fact that given an initial starting point, $\mathbf{X}(0)$, the search should be carried out in that direction for which the directional derivative of the function at $\mathbf{X}(0)$ is negative: the Newton-Raphson type methods require a knowledge of the values of the function and of its first and second derivatives; derivative methods are those which make use only of the values of the function and of its first derivatives; and non-derivative methods, which need only the values of the function.

Only those gradient methods (and those linear search techniques needed in these methods), more appropriate for the problem under consideration, will be covered here. Assuming that the function is continuous and that its first and second derivatives (with respect to the above variables) exist, it may be expanded by means of a general Taylor series

$$
\begin{aligned}
E(\mathbf{X} + \Delta\mathbf{X}) = {} & E(\mathbf{X}) \\
& + (\partial E/\partial X_1)\,\Delta X_1 + (\partial E/\partial X_2)\,\Delta X_2 + \ldots \\
& + (1/2)\,((\partial^2 E/\partial X^2{}_1)\,\Delta X^2{}_1 + (\partial^2 E/\partial X_1\,\partial X_2)\,\Delta X_1\,\Delta X_2 + \ldots) \\
& + \text{higher-order terms}
\end{aligned}
$$

where the (column) vector $\Delta\mathbf{X}$ stands for the set $\Delta X_1, \Delta X_2, \ldots, \Delta X_N$.

This expression may be rewritten as

$$
\begin{aligned}
E(\mathbf{X} + \Delta\mathbf{X}) = {} & E(\mathbf{X}) + \Delta\mathbf{X}^+ \cdot \mathbf{G}(\mathbf{X}) + (1/2)\,(\Delta\mathbf{X}^+ \cdot \mathbf{H}(\mathbf{X}) \cdot \Delta\mathbf{X}) \\
& + \text{higher terms}
\end{aligned}
$$

where the superscript (+) stands for transposition, $\mathbf{G}(\mathbf{X})$ is the (column) gradient vector, with elements

$$G_i = \partial E/\partial X_i$$

and $\mathbf{H}(\mathbf{X})$ denotes the matrix of the second derivatives, with elements

$$H_{ij} = \partial^2 E/\partial X_i\,\partial X_j$$

The minimization techniques, discussed below, use only the first two or the first three terms of the above expansion, implying a linear or quadratic behaviour, respectively, of the function under consideration, such approximations considered 'valid' in the neighbourhood of the minimum.

Steepest-Descent Method

Retaining only the first two terms of the Taylor expansion, one can write

$$E(\mathbf{X} + \Delta\mathbf{X}) \equiv E(\mathbf{X}) + \Delta E(\mathbf{X}) = E(\mathbf{X}) + \Delta\mathbf{X}^+ \cdot \mathbf{G}(\mathbf{X})$$

or

$$\Delta E(\mathbf{X}) = \Delta\mathbf{X}^+ \cdot \mathbf{G}(\mathbf{X}) = \Delta X\, G(X) \cos\omega$$

where ω is the angle between the directions defined by $\mathbf{G}(\mathbf{X})$ and $\Delta\mathbf{X}$.

When searching for a minimum, it must be $\Delta E < 0$ and, therefore, $\cos\omega < 0$. The maximum change, $|\Delta E|$, corresponds to $\cos\omega = -1$, which defines a direction (denoted as direction of steepest descent) opposite to that of the gradient.

A steepest-descent procedure of minimization consists of the following steps:

a) Select a starting point, $\mathbf{X}(0)$.
b) Evaluate the gradient vector, $\mathbf{G}(0)$, at point $\mathbf{X}(0)$.
c) Minimize the functional

$$E(\lambda) = E(\mathbf{X}(0) - \lambda\,\mathbf{G}(0))$$

(where λ is a variable) and determine the value of λ, say $\lambda = \mu$ corresponding to the largest decrease in the value of the function.

d) Stop if the (desired) minimum has been reached. Otherwise, set

$$\mathbf{X}(0) = \mathbf{X}(0) - \mu\,\mathbf{G}(0)$$

and restart at step *b*.

The basic equation for this method, including only the first two terms of the expansion, is only approximately valid near the minimum. The convergence properties, on the other hand, constitute an attractive feature of this method responsible for its wide use in conformational analyses: if the function has a minimum and a uniformly continuous gradient and if $E(\lambda)$ also has a minimum, then the successive function values generated in the iterative steps will be smaller and the successive gradient vectors will converge to zero. The rate of convergence, however, is very slow and, in general, the method does not terminate finitely.

The linear behaviour of the function, implied in this method, represents a rather drastic approximation and, for this reason, it has been replaced by methods based on a quadratic behaviour, such as the Newton-Raphson method.

Newton-Raphson Method

Taking into account that the first derivatives of a function must vanish at a minimum, differentiation of the expansion for $E(\mathbf{X} + \Delta\mathbf{X})$ leads to

$$\mathbf{G}(\mathbf{X} + \Delta\mathbf{X}) = \mathbf{G}(\mathbf{X}) + \mathbf{H}(\mathbf{X}) \cdot \Delta\mathbf{X} = 0$$

if $\mathbf{X} + \Delta\mathbf{X}$ represents the minimum. This equation can be solved for $\Delta\mathbf{X}$ whenever $\mathbf{H}^{-1}(\mathbf{X})$ can be evaluated and the position of the minimum can then be reached in a single step,

$$\mathbf{X}(\min) = \mathbf{X} + \Delta\mathbf{X} = \mathbf{X} - \mathbf{H}^{-1}(\mathbf{X}) \cdot \mathbf{G}(\mathbf{X})$$

or, equivalently,

$$\Delta\mathbf{X} = -\mathbf{H}^{-1}(\mathbf{X}) \cdot \mathbf{G}(\mathbf{X})$$

This equation holds only, however, for functions which are quadratic in the variables. For any other function, one may assume such a behaviour and follow an iterative procedure:

a) Select a starting point, $\mathbf{X}(0)$.

b) Evaluate $\mathbf{G}(0)$ and $\mathbf{H}^{-1}(0)$ at point $\mathbf{X}(0)$.

c) Evaluate the value of $\Delta\mathbf{X}$, as described above.

d) Stop if the (desired) minimum has been reached. Otherwise set

$$\mathbf{X}(0) = \mathbf{X}(0) + \Delta\mathbf{X}$$

and restart at step *b*.

If the initial point lies close to the minimum, where the function has a behaviour closer to quadratic, and if the matrix of the second derivatives, $\mathbf{H}(\mathbf{X})$, is positive definite, the iterations will be quadratically convergent.

The method fails to converge if the initial estimate is poor, in which case an extended Newton-Raphson procedure should be used. Within the framework of this method, as it was the case in the steepest-descent method, the $\Delta\mathbf{X}$, obtained in the above procedure, is used to define the direction of search; the correction in this direction is obtained by minimization of the function

$$E(\lambda) = E(\mathbf{X} + \lambda\,\Delta\mathbf{X}) = E(\mathbf{X} - \lambda\,(\mathbf{H}^{-1}(\mathbf{X}) \cdot \mathbf{G}(\mathbf{X})))$$

with respect to λ, yielding the value $\lambda = \mu$ at the minimum. This method has, in addition to the second-order convergence, a better overall convergence when starting from a poor initial guess. Although it has proven to be very successful in practice, the general conditions of convergence have not been studied very thoroughly; in particular, it could even fail [Powell (1966)] at a point where the gradient exists and the matrix of second derivatives is non-singular.

The reason for its acceptance may be found in its rapid convergence and in the possibility of its improvement through appropriate modifications. One such improved Newton-Raphson procedure (Marquardt-Levenberg method), consists of the following steps [Murray (1969)]:

a) Select a starting point, $\mathbf{X}(0)$.
b) Evaluate $\mathbf{G}(0)$ and $\mathbf{H}(0)$ at point $\mathbf{X}(0)$.
c) If $\mathbf{H}(0)$ is positive definite, go to step *d*. Otherwise set

$$\mathbf{H}(0) = \mathbf{H}(0) + \lambda\,\mathbf{Q}$$

where $\mathbf{Q}$ is a positive definite matrix and λ is a positive scalar.

d) Determine

$$\Delta\mathbf{X} = -\mathbf{H}^{-1}(0) \cdot \mathbf{G}(0)$$

e) Minimize the function

$$E(\lambda) = E(\mathbf{X} + \lambda\,\Delta\mathbf{X})$$

and determine the value $\lambda = \mu$ corresponding to the minimum.

f) Stop if the (desired) minimum has been reached. Otherwise set

$$\mathbf{X}(0) = \mathbf{X}(0) + \mu\,\Delta\mathbf{X}$$

and restart at step *b*.

The most serious objection against any type of Newton-Raphson method, arising from the need of evaluating the second derivatives of the function, has resulted in the development of new techniques that obviate such a need.

Fletcher-Reeves Method

These new algorithms, whereby the curvature of the function is taken into account without the explicit evaluation of the second derivatives, find the exact minimum of a quadratic function through the use of 'conjugate directions'.

Given a general quadratic function

$$E(\mathbf{X}) = A + \mathbf{X}^+ \cdot \mathbf{B} + (1/2)\,(\mathbf{X}^+ \cdot \mathbf{C} \cdot \mathbf{X})$$

where A is a constant, **B** denotes a column vector, **C** is a positive definite symmetric matrix, and **X** represents the column vector of the variables, the directions **D**(i) and **D**(j) are said to be conjugate with respect to the above function if, both being individually non-zero, the condition

$$\mathbf{D}(i)^+ \cdot \mathbf{C} \cdot \mathbf{D}(j) = 0$$

is satisfied. The usefulness of the conjugate directions lies in the simplification of the search procedure: Starting from point **X**, one searches in the direction **D**(i) and finds the minimum of the function at the point **X**(min,i); starting now from this new point, a search in the direction **D**(j) will lead to a minimum **X**(min,j). If, starting from this point, a search were now to be carried out in the direction **D**(i), no minimum would be found, due to the fact that the point **X**(min,i) corresponds to the minimum in the direction **D**(i) if, and only if, the gradient vector at that point is orthogonal to **D**(j).

Therefore, given the N mutually conjugate directions **D**(1), **D**(2), . . . , **D**(N), the (desired) minimum is defined by

$$\mathbf{X}(N) = \mathbf{X}(0) + \Sigma\lambda_i\, \mathbf{D}(i)$$

where **X**(0) is the starting point and the λ_i are obtained by minimization of

$$\begin{aligned} E(\mathbf{X}(N)) &= A + \mathbf{X}^+(N) \cdot \mathbf{B} + (1/2)\,(\mathbf{X}^+(N) \cdot \mathbf{C} \cdot \mathbf{X}(N)) \\ &= E(\mathbf{X}(0)) \\ &\quad + \Sigma(\lambda_i\, \mathbf{D}^+(i) \cdot (\mathbf{B} + \mathbf{C} \cdot \mathbf{X}(0)) + (1/2)\,\lambda^2_i\,(\mathbf{D}^+(i) \cdot \mathbf{C} \cdot \mathbf{D}(i))) \end{aligned}$$

and the search implies the successive steps represented by

$$\mathbf{X}(k) = \mathbf{X}(k-1) + \lambda_k\, \mathbf{D}(k)$$

The gradient at the final point, if it corresponds to a minimum of the function, must be zero and orthogonal to the N search directions. Taking into account that the search directions are mutually conjugate and independent (as they satisfy the condition imposed above), the gradient at the final point is a null vector (in order to be orthogonal to N independent directions), which means that the minimum is reached in those N steps.

The preceding argument applies to a quadratic function and a modified algorithm is required for functions which are not quadratic. Such an algorithm is based on the fact [Zoutendijk (1960)] that when moving from the point **X**(k −

1) to the point $\mathbf{X}(k)$, along the search condition $\mathbf{D}(k)$, the above condition for conjugate directions may be replaced, if the function is quadratic, with

$$(\mathbf{G}(k) - \mathbf{G}(k-1)) \cdot \mathbf{D}(k+1) = 0$$

which can be used to generate the new conjugate search direction without any need of evaluating the second derivatives.

An additional simplification may be achieved [Fletcher and Reeves (1964)] by choosing the starting search direction as that corresponding to the steepest descent and the subsequent search direction, $\mathbf{D}(k)$, as the sum of the gradient $\mathbf{G}(k-1)$ and a linear combination of the search directions $\mathbf{D}(1), \mathbf{D}(2), \ldots, \mathbf{D}(k-1)$, which preserves the property of conjugacy when the function is quadratic. The procedure consists of the following steps:

a) Select an initial point, $\mathbf{X}(0)$.

b) Calculate the gradient, $\mathbf{G}(0)$, at the point $\mathbf{X}(0)$.

c) Define the (first) search direction as

$$\mathbf{D} = -\mathbf{G}(0)$$

d) Minimize the function

$$E(\lambda) = E(\mathbf{X}(0) + \lambda\, \mathbf{D})$$

(with respect to λ) and obtain the value $\lambda = \mu$ corresponding to the minimum. Stop the calculation if the desired minimum has been found.

e) Otherwise, define the point

$$\mathbf{X} = \mathbf{X}(0) + \mu\, \mathbf{D}$$

f) Calculate the gradient, $\mathbf{G}(\mathbf{X})$, at the new point $\mathbf{X}$.

g) Calculate

$$\nu = (\mathbf{G}^{+}(\mathbf{X}) \cdot \mathbf{G}(\mathbf{X})) / (\mathbf{G}^{+}(0) \cdot \mathbf{G}(0))$$

h) Define the next search direction as

$$\mathbf{D}' = -\mathbf{G}(\mathbf{X}) + \nu\, \mathbf{D}$$

i) Set $\mathbf{X}(0) = \mathbf{X}$ and $\mathbf{D} = \mathbf{D}'$ and restart at step *d*.

There are methods similar to the one just described and which, in addition, make use of estimates of the inverse of the matrix of the second derivatives, at each iteration, in order to define the new search directions

$$\mathbf{D}(k) = -\mathbf{H}^{-1}(k) \cdot \mathbf{G}(k-1)$$

One such method [Davidon (1959), Fletcher and Powell (1963)] is discussed below because of its interest in connection with the present work.

Davidon-Fletcher-Powell Method

The search direction at point **X**, corresponding to the steepest descent, can be written as

$$\Delta \mathbf{X} = -\mathbf{G}(\mathbf{X}) = -\mathbf{I} \cdot \mathbf{G}(\mathbf{X})$$

where **I** is the unit matrix. Comparison of this equation with the one defining the search direction in the Newton-Raphson method shows that, in the latter, the unit matrix has been replaced with the inverse of the matrix of the second derivatives; in the Newton-Raphson method, the search direction formed an angle smaller than $\pi/2$ with the gradient direction, while the direction defined by the above condition forms an angle of 2π.

Consequently, it was suggested [Davidon (1959)] that the search directions could be defined by

$$\mathbf{D}(k) = -\mathbf{J}(k-1) \cdot \mathbf{G}(k-1)$$

and the new points by

$$\mathbf{X}(k) = \mathbf{X}(k-1) + \lambda_k \, \mathbf{D}(k)$$

The matrix, $\mathbf{J}(k-1)$, which is positive definite, is to be chosen so that a fast rate of convergence will be achieved. This method only uses the first derivatives of the functions but accounts for the second derivatives by improving the matrix **J** at each iteration.

A procedure [Fletcher and Powell (1963)], based on the preceding definitions, consists of the following steps:

a) Select an initial point, $\mathbf{X}(0)$.

b) Evaluate the gradient, $\mathbf{G}(0)$, at point $\mathbf{X}(0)$.

c) Choose an initial matrix, $\mathbf{J}(0)$, which must be positive definite; a possible choice, a course, is the unit matrix.

d) Define the (first) search direction by

$$\mathbf{D} = \mathbf{J}(0) \cdot \mathbf{G}(0)$$

e) Minimize the function

$$E(\lambda) = E(\mathbf{X}(0) + \lambda \, \mathbf{D})$$

with respect to λ and determine the value $\lambda = \mu$ corresponding to the minimum; this minimization is carried out by cubic interpolation, described below.

f) Stop if the desired minimum has been reached. Otherwise, set

$$\mathbf{X} = \mathbf{X}(0) + \mu \, \mathbf{D}$$

g) Evaluate the gradient, $\mathbf{G}(X)$, at the new point.

h) Update the matrix **J** according to

$$\mathbf{J} = \mathbf{J}(0) + \mu \, (\mathbf{D} \cdot \mathbf{D}^{+})/(\mathbf{D}^{+} \cdot \mathbf{T}) - (\mathbf{J}(0) \cdot \mathbf{T} \cdot \mathbf{T}^{+} \cdot \mathbf{J}(0))/(\mathbf{T}^{+} \cdot \mathbf{J}(0) \cdot \mathbf{T})$$

where

$$\mathbf{T} = \mathbf{G}(\mathbf{X}) - \mathbf{G}(0)$$

i) Define the new search direction as

$$\mathbf{D}' = -\mathbf{J} \cdot \mathbf{G}(\mathbf{X})$$

j) Set $\mathbf{X}(0) = \mathbf{X}$, $\mathbf{D} = \mathbf{D}'$, and $\mathbf{J}(0) = \mathbf{J}$ and restart at step *e*.

The search directions used in this procedure become mutually conjugate when the function is quadratic, in which case only N iterations are needed in order to reach the minimum [Fletcher and Powell (1963)]. It must also be pointed out that, if $\mathbf{J}(0)$ is positive definite, the new $\mathbf{J}$ matrix defined at each iteration is also positive definite.

In an alternate procedure [Broyden (1967), Davidon (1968)], the matrix $\mathbf{J}$ is updated according to

$$\mathbf{J} = \mathbf{J}(0) - \nu\,(\mathbf{D} + \mathbf{J}(0) \cdot \mathbf{T}) \cdot (\mathbf{D} + \mathbf{J}(0) \cdot \mathbf{T})^{+} / (\mathbf{D} + \mathbf{J}(0) \cdot \mathbf{T})^{+} \cdot \mathbf{T}$$

(ν being a scalar factor) instead of the expression given in step *h* of the previous method, and the new point is defined by

$$\mathbf{X}(k) = \mathbf{X}(k-1) - \lambda_k\,(\mathbf{J}(k-1) \cdot \mathbf{G}(k-1))$$

where the value of λ_k is arbitrary, as long as $\mathbf{J}$ is updated as indicated [Broyden (1967)], but it should not cause $\mathbf{J}$ to become singular or ill-conditioned. Consequently, the minimization step for the determination of the value of λ at the minimum is not required; in fact, the value $\lambda = 1$ may be used [Davidon (1968)], but in such a case it may happen that the equation defining $\mathbf{J}$ becomes undefined. An appropriate value of the scalar ν (normally taken as 1) must be chosen whenever $\mathbf{D}$ and $\mathbf{T}$ become such that $\mathbf{J}$ does not remain positive definite.

It is worthwhile to point out that, as a result of the above definition, $\mathbf{J}$ becomes identical to the inverse of the hessian matrix, without need of a linear search; this is a rather important characteristic, not only in terms of computer time, but also taking into account that a linear search may fail whenever no optimal point is to be found along a certain direction. Another important feature arises from the fact that if $E_k < E_{k-1}$, once $\mathbf{J}$ has been updated, one may use $\mathbf{X}(k)$ for the next iteration.

The technique of cubic interpolation, used in the linear search at step *e* of the above procedure, minimizes the (cubic) function

$$E(\mathbf{X}) = E(\mathbf{X}(0) + \lambda\,\mathbf{D}) = a + b\,\lambda + c\,\lambda^2 + d\,\lambda^3$$

with respect to λ in the direction $\mathbf{D}$, starting from the current point $\mathbf{X}(0)$.

The calculation of λ_m, corresponding to the minimum, makes use of the condition(s) $\partial E(\lambda)/\partial\lambda = 0$ (and $\partial^2 E(\lambda)/\partial\lambda^2 > 0$) at that point and requires the values of the expansion coefficients *a*, *b*, *c*, and *d*. Using the notation

$$Y(\lambda) = \partial E(\lambda)/\partial\lambda = b + 2\,c\,\lambda + 3\,d\,\lambda^2$$

for the directional derivative, $\mathbf{G}^+(\mathbf{X}) \cdot \mathbf{D}$, and taking into account the cubic expansion of the function, one can write, for two points $\mathbf{X}(0)$ (with $\lambda_0 = 0$) and $\mathbf{X}(1) = \mathbf{X}(0) + \lambda\,\mathbf{D}$ (with $\lambda_1 = \lambda$),

$$E_0 \equiv E(\mathbf{X}(0)) = a$$
$$Y_0 = b$$
$$E_1 \equiv E(\mathbf{X}(1)) = E_0 + Y_0\,\lambda_1 + c\,\lambda^2_1 + d\,\lambda^3_1$$
$$Y_1 = Y_0 + 2\,c\,\lambda_1 + 3\,d\,\lambda^2_1$$

which can be solved for the expansion coefficients. λ_m is then given by

$$\lambda_m = \lambda_1(1 - (Y_1 + A - B)/(Y_1 - Y_0 + 2A))$$

with

$$A = (B^2 - Y_0\,Y_1)^{1/2}$$
$$B = (3/\lambda_1)\,(E_0 - E_1) + Y_0 + Y_1$$

The steps involved are:

a) Evaluate E_0 and Y_0 at the initial point, $\mathbf{X}(0)$.

b) Estimate a value of E for the minimum (preferably $E < E_{min}$).

c) Determine a value of λ_1 from the condition

$$\lambda_1 = \min(2, -2\,(E_0 - E)/Y_0)$$

d) Evaluate E_1 and Y_1 at the point $\mathbf{X}(1) = \mathbf{X}(0) + \lambda_1\,\mathbf{D}$.

e) With $Y_0 < 0$, a minimum exists between X(0) and X(1) if $Y_1 > 0$ or $E_1 > E_0$. In such a case, evaluate λ_m as indicated above, using the values of λ_1, E_0, E_1, Y_0, and Y_1. Otherwise, proceed to step *f*.

f) Double the value of λ_1 and restart at step *d*.

The methods discussed so far require at least the values of the first partial derivatives of the function, which are used in the selection of the search directions. In those cases where the analytical evaluation of those derivatives is not possible or whenever it involves a considerable computation time, it is desirable to use alternate methods.

There are two possibilities in this connection. On one hand, there are methods that, while still using the technique of linear searches along mutually conjugate directions, do not require a knowledge of the gradient vector. The second option is embodied in those methods that obtain the gradient vector by a finite difference technique; such methods are usually poorer due to truncation and round-off errors which, in addition, affect the search directions. Representative examples are the Smith-Powell-Zangwill and Stewart methods discussed below.

Stewart Method

The usefulness of a method using a numerical evaluation of the gradient vector depends on its handling of the trunction and rounding-off errors. In fact, a careful determination of the intervals in a difference method may yield results equivalent to those obtained in methods using explicitly evaluated derivatives.

Such a technique may be used at step *g* of the Davidon-Fletcher-Powell procedure. The algorithm for the numerical evaluation of **G(X)** makes use [Stewart (1967)] of the value E(**X**) (obtained in step *e*) at the current point **X** and of the gradient vector, **G**′(0), and the matrix, **J**(0), obtained in the previous iteration. The algorithm for the evaluation of each component, G_i, of **G(X)**, consists of the following steps (where η is a threshold value):

a) Evaluate

$$\delta' = 2(|E(\mathbf{X})|/|J_{ii}|)^{1/2}$$

or

$$\delta' = 2(|E(\mathbf{X})|\ |G'_i|/J^2_{ii})^{1/3}$$

depending on whether $(G'_i)^2$ is smaller or greater than $\eta\ |J_{ii}\ E(\mathbf{X})|$.

b) Evaluate

$$\delta = \delta'\ (1 - \delta'\ |J_{ii}|/(3\ \delta'\ |J_{ii}| + 4|G'_i|))$$

or

$$\delta = \delta'\ (1 - 2|G'_i|/(3\ \delta'\ |J_{ii}| + 4|G'_i|))$$

(with the appropriate value of δ') depending again on the same condition of step *a*.

c) Set

$$\delta = \text{sign}\ (J_{ii})\ \text{sign}(G'_i)\ \delta$$

d) If

$$(1/2)\ |\delta\ J_{ii}/G'_i| < 10^{-m}$$

evaluate

$$G_i = (E(+\delta) - E(\mathbf{X}))/\delta$$

and restart at step *a* for the next component. Otherwise go to step *e*.

e) Evaluate δ as the positive root of

$$(1/2)|J_{ii}|\ \delta^2 + |G'_i|\ \delta - 10^m\ |E(\mathbf{X})|\ \eta = 0$$

f) Evaluate

$$G_i = (E(+\delta) - E(-\delta))/2\ \delta$$

using the new value of δ obtained in step *e*, and restart at step *a* for the next component.

[In steps *d* and *f*, $E(+\delta)$ and $E(-\delta)$ stand for the values of the function at the point defined by changing the i-th component of **X** by $+\delta$ or $-\delta$, respectively.]

The elements, J_{ii}, of **J**(0) must be positive and the derivative of the function along the direction considered must be negative; if either condition is violated, **J**(0) is set equal to the identity matrix and the iterations restarted. The linear search along the direction **D**(i) remains as described above, except for the following changes:

a) The initial value λ_1 is chosen as

$$\lambda_1 = \min(1, -2(E_0 - E)/Y_0)$$

b) A second estimate of the minimum is chosen by parabolic interpolation, using the values of E_0, E_1, and Y_0.

Smith-Powell-Zangwill Method

The properties of parallel subspaces for quadratic functions may be used to generate mutually conjugate directions by an algorithm that does not require the evaluation of the gradient vector and is based on the fact that, if **X**(i) and **X**(j) represent the minima of two quadratic functions in parallel subspaces, the minimum in the subspace with a dimension one higher than the two parallel subspaces lies in the direction **X**(i) − **X**(j).

The basic procedure [Smith (1962)] consists of the following steps:

a) Select a starting point, **X**(1), and a set of linearly independent directions **D**′(1), **D**′(2), . . . , **D**′(N).

b) Adopt **D**′(1) as the first conjugate direction, **D**(1).

c) Search along the direction **X**(1) − **D**′(1), obtaining the point **X**(2).

d) Define a new point, **Y**(2) = **X**(2) + λ **D**′(2).

e) Search along the direction **Y**(2) − **D**′(1), parallel to **X**(1) − **D**′(1), obtaining the point **Z**(2). The direction **D**(2) = **Z**(2) − **X**(2) is conjugate to **D**(1).

f) Search along **D**(2), obtaining the point **X**(3).

g) Define a new point **Y**(3) = **X**(3) + λ **D**′(3).

h) Search along the directions **Y**(3) − **D**(1) and **Y**(3) − **D**(2), obtaining the point **Z**(3). The direction **D**(3) = **Z**(3) − **X**(3) is conjugate to both **D**(1) and **D**(2).

i) . . . . .

The minimum of a quadratic function in N variables may thus be obtained in (1/2) N (N + 1) linear minimizations, but this method adolesces of the defect that the first search is used N times while the last one is used only once, a situation that may be crucial if the latter is more important than the former.

Appropriate modifications to remove this deficiency [Powell (1964)] and to maintain the linear independence of the directions [Zangwill (1967)] has resulted in an algorithm consisting of the following steps:

a) Select the starting point, $\mathbf{X}(0)$.

b) Select N initial (normalized) direction vectors, $\mathbf{D}(i)$, corresponding to the coordinate directions.

c) Select the scalar $0 < \epsilon \leqslant 1$ and set $\delta = 1$.

d) Minimize (by quadratic interpolation) the function

$$E(\mathbf{X}(i)) = E(\mathbf{X}(0) + \Sigma'\lambda_j\, \mathbf{D}(j) + \lambda_i\, \mathbf{D}(i))$$

along each direction $\mathbf{D}(i)$, step by step.

e) Define

$$\mathbf{D}(N + 1) = \mathbf{T}/|T|$$

with the total displacement vector, $\mathbf{T}$, given by

$$\mathbf{T} = \Sigma\lambda_i\, \mathbf{D}(i)$$

f) Evaluate λ_{N+1} by minimization of $E(\mathbf{X}(N) + \lambda_{N+1}\, \mathbf{D}(N + 1))$.

g) Set

$$\mathbf{X}(0) = \mathbf{X}(N) + \lambda_{N+1}\, \mathbf{D}(N + 1)$$

h) Set

$$\lambda_{max} = \max(\lambda_1, \lambda_2, \ldots, \lambda_N)$$

and define

$$\delta' = \lambda_{max}\, \delta/|T|$$

i) If $\delta' < \epsilon$, restart at step *d*. Otherwise go to step *j*.

j) Set $\delta = \delta'$, retain all the search directions except the one corresponding to λ_{max}, which is replaced with $\mathbf{D}(N + 1)$, and restart at step *c*.

It must be pointed out that δ represents the determinant of the matrix formed by the (column) direction vectors, so that the replacement of a direction when $\delta' \geqslant \epsilon$ preserves the linear independence of the search directions. This algorithm is well behaved and converges for a strictly convex and continuously differentiable function.

The technique of quadratic interpolation used in the linear search in step *d* minimizes the function

$$E(\mathbf{X}) = E(\mathbf{X}(0) + \lambda\, \mathbf{D}) = a + b\,\lambda + c\,\lambda^2$$

with respect to λ in the direction $\mathbf{D}$, starting from the current point, $\mathbf{X}(0)$. It consists of the following steps:

a) Select a value of λ.

b) Evaluate

$E_1 = E(\mathbf{X}(0)) \equiv E(\mathbf{X}(0) + \lambda_1 \mathbf{D})$ at $\lambda_1 = 0$
$E_2 = E(\mathbf{X}(0) + \lambda \mathbf{D}) \equiv E(\mathbf{X}(0) + \lambda_2 \mathbf{D})$ at $\lambda_2 = \lambda$

c) If $E_1 < E_2$, evaluate

$E_3 = E(\mathbf{X}(0) - \lambda \mathbf{D}) \equiv E(\mathbf{X}(0) + \lambda_3 \mathbf{D})$ at $\lambda_3 = -\lambda$

Otherwise, evaluate

$E_3 = E(\mathbf{X}(0) + 2 \lambda \mathbf{D}) \equiv E(\mathbf{X}(0) + \lambda_3 \mathbf{D})$ at $\lambda_3 = 2 \lambda$

d) Solve the system of equations

$E_1 = E(\mathbf{X}(0) + \lambda_1 \mathbf{D}) = a + b \lambda_1 + c \lambda^2_1$
$E_2 = E(\mathbf{X}(0) + \lambda_2 \mathbf{D}) = a + b \lambda_2 + c \lambda^2_2$
$E_3 = E(\mathbf{X}(0) + \lambda_3 \mathbf{D}) = a + b \lambda_3 + c \lambda^2_3$

for a, b, and c.

e) Find the value

$\lambda_m = -b/2 c$

corresponding to the turning point.

f) Denote by E_s and E_l the smallest and largest, respectively, of E_1, E_2, and E_3, and by λ_s and λ_l the corresponding values of λ. Select the (desired) threshold values ϵ and $\eta > \epsilon$.

g) Go to step *h* or *k* depending on whether $E(\mathbf{X}(0) + \lambda_m \mathbf{D})$ corresponds to a minimum or a maximum.

h) If

$|\lambda_s - \lambda_m| \leqslant \epsilon$

the desired minimum has been found at $\mathbf{X}(0) + \lambda_m \mathbf{D}$ or, equivalently, at $\mathbf{X}(0) + \lambda_s \mathbf{D}$. Otherwise proceed to step *i*.

i) If

$|\lambda_s - \lambda_m| < \eta$

discard E_l and λ_l, unless by doing so the minimum is not bracketed, in which case the appropriate point would be discarded. Adopt $\mathbf{X}(0) + \lambda_m \mathbf{D}$ as the new point, evaluate the function at that point and restart at step *d*. Otherwise proceed to step *j*.

j) If

$|\lambda_s - \lambda_m| > \eta$

discard E_l and λ_l, adopt a new point $\mathbf{X}(0) + (\lambda_s + \eta) \mathbf{D}$, evaluate the function at that point, and return to step *d*.

k) If λ_m corresponds to a maximum, discard E_s and λ_s, define a new point $\mathbf{X}(0) + (\lambda_s + \eta) \mathbf{D}$, evaluate the function at that point and restart at step *d*.

Implementation

The number of iterations constitutes the most important factor in any iterative minimization procedure.

The iterations should be continued, in principle, until all the elements of the gradient vector are zero and the matrix of the second derivatives has become positive definite. Such conditions are too strict, taking into account the rounding and truncation errors, inherent to any calculation in a computer of a given limited word length, and their dependence on the algorithm used. In a compromise between accuracy and computation time, the acceptable accuracy could be decided for each case depending on the characteristics of the problem, the algorithm employed, and the computing facilities available.

In particular, the cut-off point may be defined by the condition that the absolute value of the correction vector, $(\mathbf{X}(k) - \mathbf{X}(k-1))$, be smaller than a given threshold value; alternatively, the condition may be imposed that each component of such a correction vector must be smaller than a certain threshold value. In any case, however, N iterations must at least be performed for a problem with N variables. The importance of the convergence criteria is due to the fact that, depending on how they are chosen, either computing time will be wasted or the final point may be far away from the minimum.

Linear searches are usually performed by cubic interpolation, with extreme caution being advised when in the proximity of the convergence limit. Although these techniques are not unstable near the minimum, it must be realized that the linear minimum estimates are obtained from information affected by rounding errors. The result is that it will not be possible to get closer to the minimum, along a given search direction, whenever the cubic interpolation starts showing inconsistencies.

In this connection, it should also be pointed out that rounding errors may affect the overall performance of a stable, quadratically convergent method, which uses linear searches, even before its effect is observed in the linear search step; they may, for example, destroy the conjugacy of the search direction, with a disturbance of the quadratic convergence.

As mentioned above, the steepest-descent method was used widely in the past. At present, it would seem that the Davidon-Fletcher-Powell or the Stewart methods are the most efficient ones for the problems under consideration; the use of one or the other will depend, essentially, on the time and expense involved in the evaluation of the gradient vector.

The problems implied here are the conformational analysis of large polymers and the study of their interactions. There is, however, an additional problem which requires some additional comments: it is the conformational analysis of the monomeric units.

Usually, such an analysis is carried out by a tabulation procedure involving the determination of a considerable number of values of the molecular energy for different geometries (such as it is done for the determination of the equilibrium internuclear separation in diatomic molecules). This method, however, becomes both expensive and insensitive as the size of the molecule increases. The insensitivity arises as a consequence of the increase of the molecular energy (in absolute value) with the molecular size and the limitations in the accuracy of the energy values, with the result that the energy differences may fall within the range of uncertainty of the calculations for large enough monomeric units.

A possible alternative involves the use of a double iteration procedure [Pancir (1973)]. First, a SCF calculation is carried out for the initial conformation and then an optimization method is used until the coordinates do not change by more than a fixed value. A new SCF calculation is then carried out for the new conformation, with a stricter convergence criterion for the orbital energies. The complete procedure is repeated until the change, between consecutive cycles of the procedure, in the orbital energies becomes smaller than a specified threshold.

PRACTICAL APPLICATIONS

The advances in computer and display hardware technology will have a great impact on the computational aspects of conformation and interaction studies of biopolymers.

These studies could be preceded by a step of computer model-building and visualization in order to determine the optimal, stereochemically compatible configurations. The residues of the biopolymer(s) can be manipulated under visual observation into positions that will avoid steric conflicts and will allow optimum interactions. The resulting structures may then be used as a starting point in the optimization procedure of the conformational analysis or directly in the evaluation of the interaction energy. Display units are now a common fixture at computing centers and programs are readily available [see, e.g., the work of Nir *et al.* (1977) for more details].

These reasons have determined the pictorial emphasis evident throughout this work and, in particular, in the discussion of some examples chosen as illustrative examples of work that may be carried out. The formulations used may be different from the one discussed in this work and, for that reason, the corresponding details are omitted, centering the presentation on the systems that may be treated and the results that may be obtained.

The conformation of pyridoxal amino acid Schiff's bases, the valinomycin-potassium ion complex and the DNA-echinomycin binding, and the catalytic effect of lysozyme have been chosen as examples of work on conformations, interactions and reactions.

Conformations

The pryidoxal amino acid Schiff's bases of alanine, valine, leucine and phenylalanine were studied as dianion species in a three step procedure using semiempirical potential energy functions [Weintraub *et al.* (1976)]. First a

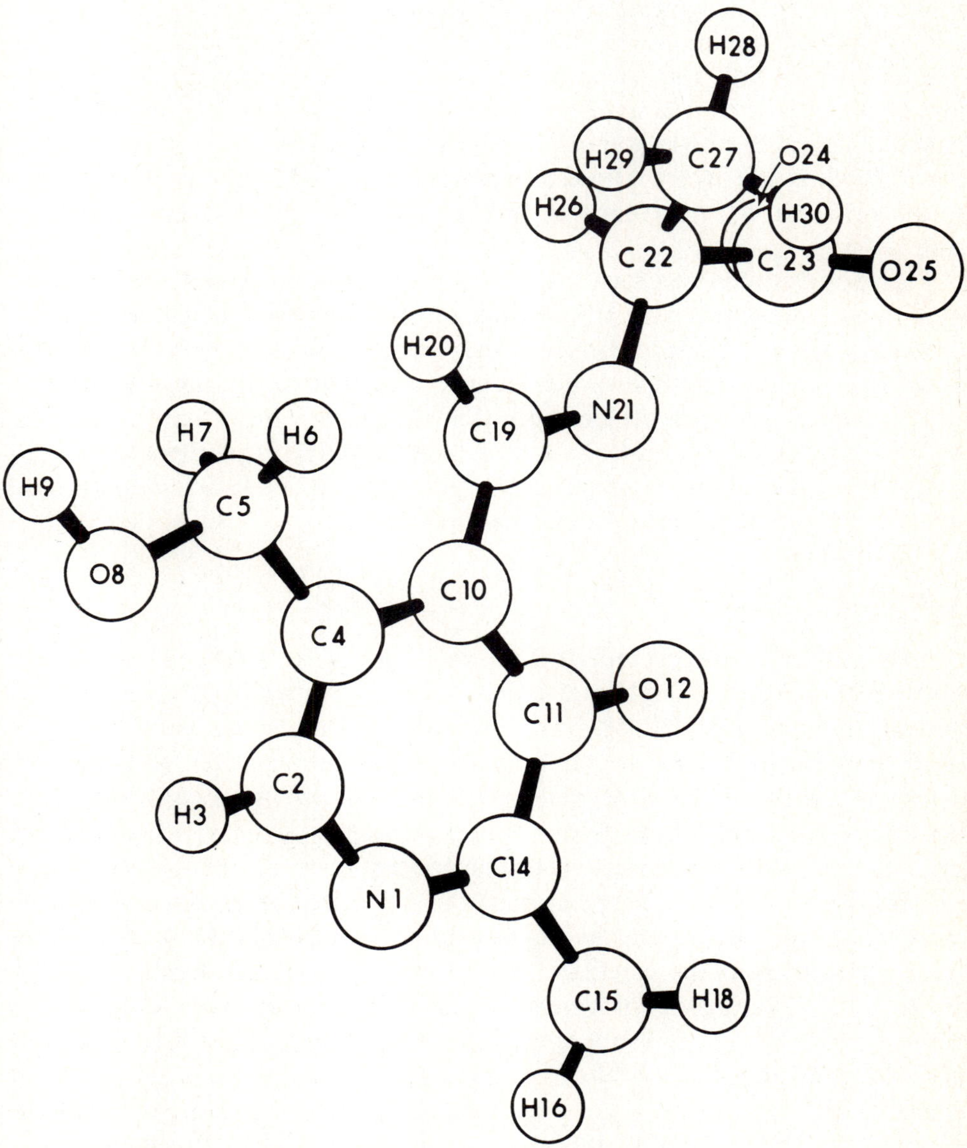

Figure 4-46. Preferred conformation of the pyridoxal alanine Schiff base (dianion), corresponding to the global minimum [Weintraub *et al* (1976)].

sequential scan was performed at 20-30 degree resolution for all the major torsional rotations, providing a mapping of the conformational hyperspace; next, several hundred conformers were studied, allowing all the possible rotations, yielding new conformational maps. Finally, a multi-dimensional minimization (involving all the rotatable bonds) was carried out using a number of starting points chosen by examination of the maps obtained above. The average analysis involved over 5000 conformers.

The global minimum in the four cases corresponded to a 'planar' conformation, with the carboxyl carbon, the α carbon and the nitrogen approximately in the plane of the pyridine ring (Fig. 4-46). In addition, the bases of alanine, phenylalanine and leucine presented three local minima corresponding to conformations in which the H, R, or COO^- group, attached to the α carbon, was approximately perpendicular to the plane of the pyridine ring.

The populations of these perpendicular C(α)-H conformers was in qualitative agreement with the observed order (alanine $>$ phenylalanine $>$ leucine) of the exchange rate; valine does not exchange α protons in the same conditions, in agreement with the absence of the perpendicular conformer.

The semiempirical treatment confirms the suggestion [Dunathan (1966, 1971)] that the conformation about the C(α)-N bond plays an important role in determining the relative susceptibility of a pyridoxal amino acid Schiff's base toward the various reactions that may take place.

Interactions

Valinomycin is a cyclododecadepsipeptide (Fig. 4-47). Some of the conformers have a roughly cylindrical conformation, with the peptides constituting the vertical sides and the esters forming the bases. When a potassium ion is bound by valinomycin, the conformation becomes very symmetric with all the ester group carbonyl oxygens pointing towards the central ion and the vertical peptides supported by six hydrogen-bonds; in the uncomplexed state, three ester groups on one side open up and point outwards.

Taking into account that the primary structure may be written as (*D*-val-*L*-lac-*L*-val-*D*-hy-iv)$_3$ (where the abbreviations stand for valyl, lactyl, and α-hydroxyisovaleryl), the calculations may be simplified [Sundaram (1974)]. First of all, various conformers of the repeat unit were obtained by optimization of the corresponding eight dihedral angles (see Fig. 4-48) and then the full molecule was synthesized using various combinations of the repeat units. The optimization with respect to the side chains was simplified by rotating them simultaneously in equal steps, with a refinement of the results by using smaller steps.

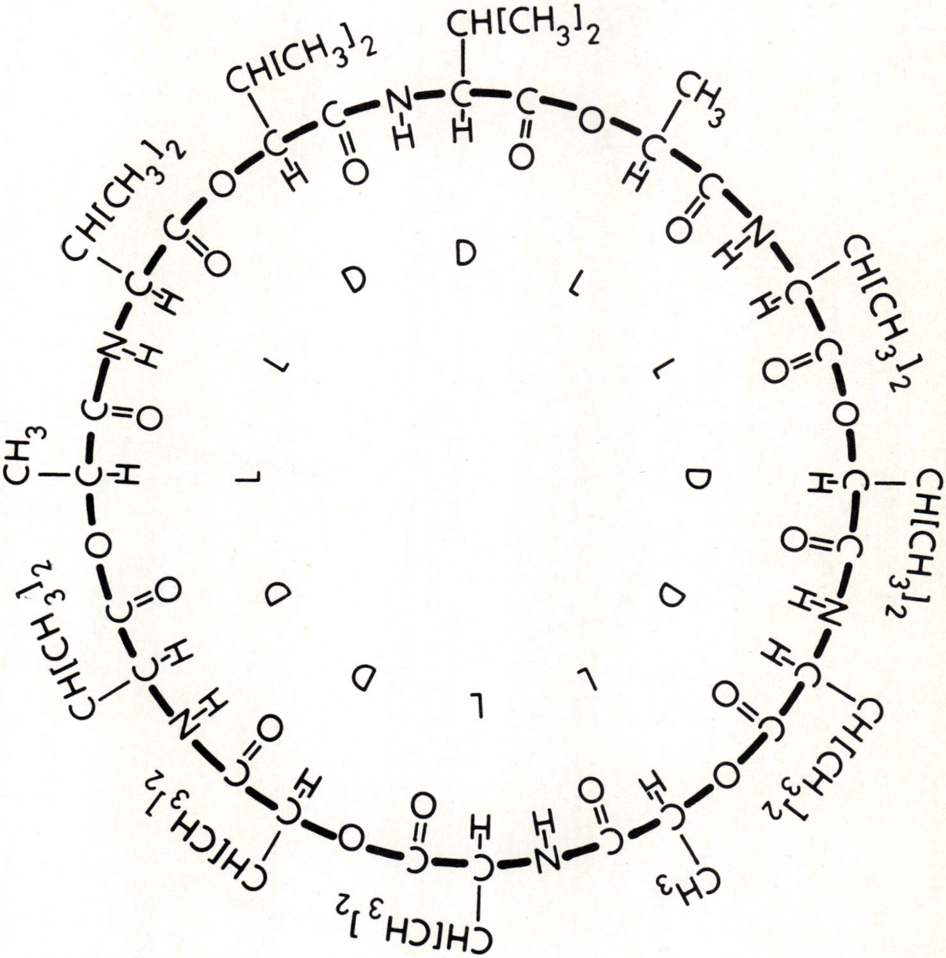

Figure 4-47. Molecular formula of valinomycin [Sundaram (1974)].

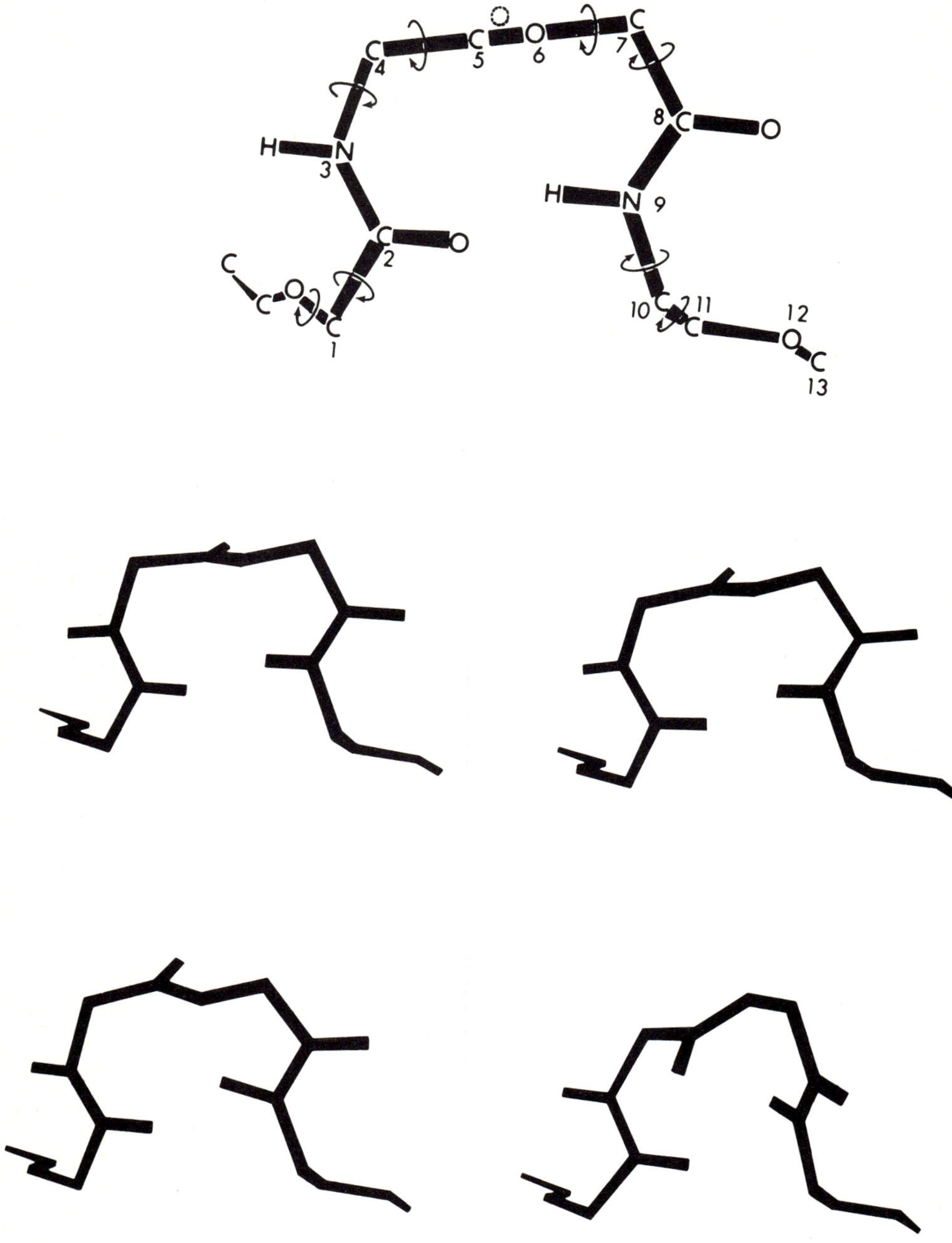

Figure 4-48. Various conformations of a repeat unit of the valinomycin backbone. The top figure represents the closed symmetric conformation and the figure on the right at the bottom, the open conformation. The other three figures represent intermediate conformations [Sundaram (1974)].

Assuming that the binding of the potassium ion does not alter the molecular framework, it is possible then to evaluate the corresponding energy. The results show a strong affinity towards K^+, providing thus a quantitative explanation of the selective cation permeability induced by valinomycin.

Echinomycin is a cyclic-peptide antibiotic (Fig. 4-49), with an antitumor activity assumed to be due to inhibition of the DNA template activity, its binding to DNA involving an intercalation of the quinoxaline residues between the bases.

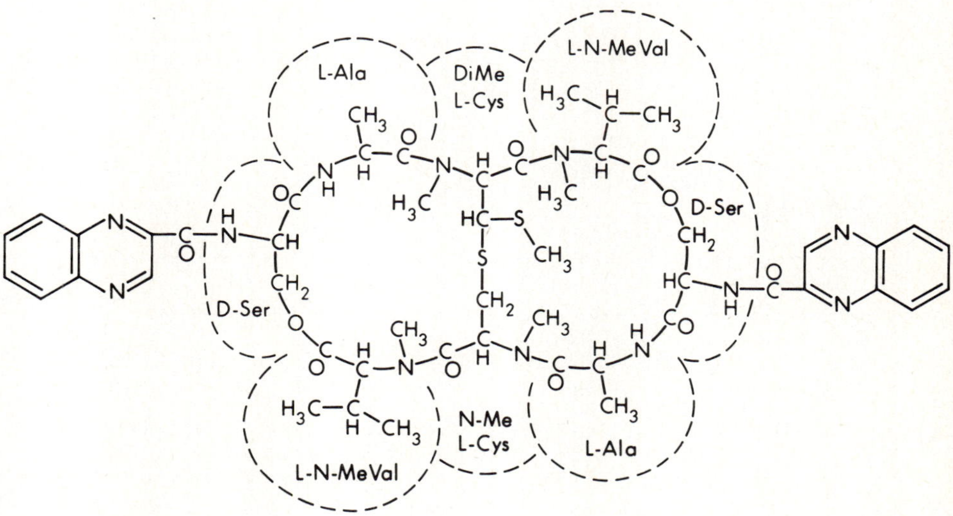

Figure 4-49. Structural formula of echinomycin [Ughetto and Waring (1977)]. (Reprinted with permission. Copyright by Academic Press, Inc.)

The optimization of the potential energy was carried out in steps [Ughetto and Waring (1977)]. First, the conformations of the four amino acid residues were studied. Then, tripeptide fragments were analyzed within the allowed regions, rejecting those conformations unsuitable for ring closure. The molecule was then considered to be composed of two identical halves, except for the S-methyl group, which was not included at this stage, and those conformations which permitted ring formation by the half-molecule were selected and used as a starting point of a refinement process of optimization by variation of dihedral angles. Finally, the conformations for the half-molecule were combined together, the S-methyl group was included, and the conformation of the complete molecule was again optimized.

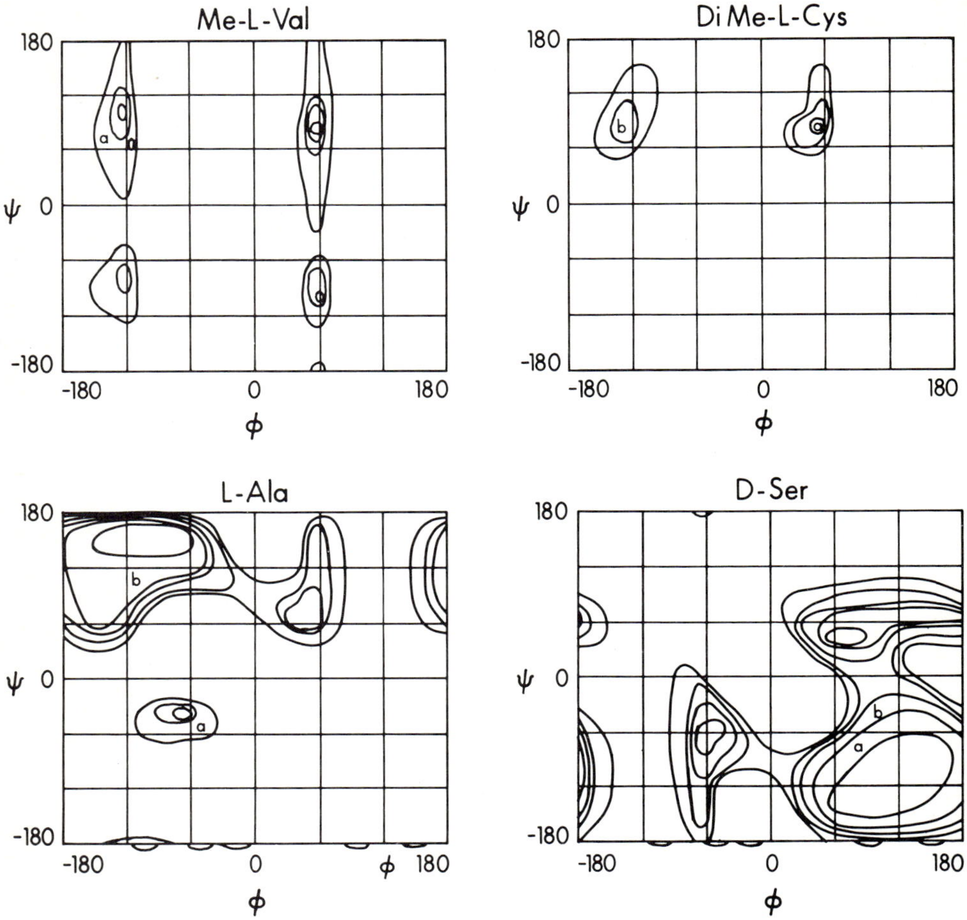

Figure 4-50. Conformational energy maps of N-methyl-*L*-valine peptide lactone unit, N,S-dimethyl-*L*-cysteine dipeptide unit, *L*-alanine dipeptide unit, and *D*-serine dipeptide unit [Ughetto and Waring (1977)]. (Reprinted with permission. Copyright by Academic Press, Inc.)

The conformational maps for the amino acids residues are presented in Fig. 4-50. The external contours correspond to a potential energy of 8 kcal/mole and those inside are for the values 4, 2, 1, and 0.5 kcal/mole. For the rotational angle, about N-C(α), $\phi = 0°$ corresponds to C(α)-C *cis* to N-C, while for rotation about C(α)-C′, $\psi = 0°$ corresponds to C(α)-N *cis* to C′-N; in each case the peptide bond C-N is *trans*. The diagrams are sections of three-dimensional maps corresponding to energy minima along the z-axis, which represents the rotation angle χ about the C(α)-C(β) bond.

The half-molecule ring has only two energetically favourable conformations (Fig. 4-51), with all the peptide bonds *trans* and rotational parameters

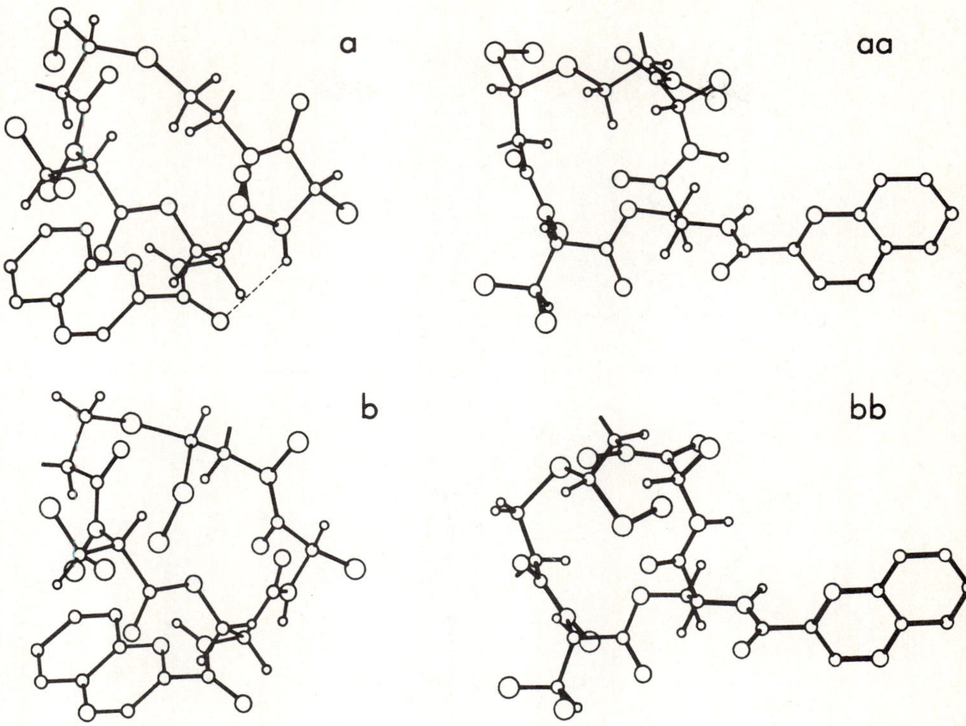

Figure 4-51. Conformations of the half-molecule ring of echinomycin. Broken bonds show the junction with the other half-molecule. The dashed line in conformation *a* represents a possible hydrogen bond. Hydrogens on aromatic rings are omitted [Ughetto and Waring (1977)]. (Reprinted with permission. Copyright by Academic Press, Inc.)

corresponding to the points marked by the corresponding letters in Fig. 4-50. The other two conformations presented in Fig. 4-51 are equivalent except for the rotational angles about the C(α)-C(β) and C(α)-C′ bonds of serine, but possess a higher conformational energy.

The optimum conformation for the complete molecule (Fig. 4-52) was obtained by refinement of a structure having one ring in conformation *a* and the other in conformation *b*. The quinoxaline residues project out of the peptide ring, with their planes not far from parallel and separated by a distance of 10.2 A, so that two stacked DNA base pairs may be accommodated between them; in addition, the molecule presents the same chirality about its quasi-dyad axis (passing through the central sulfur atom of the cross-bridge) as does DNA.

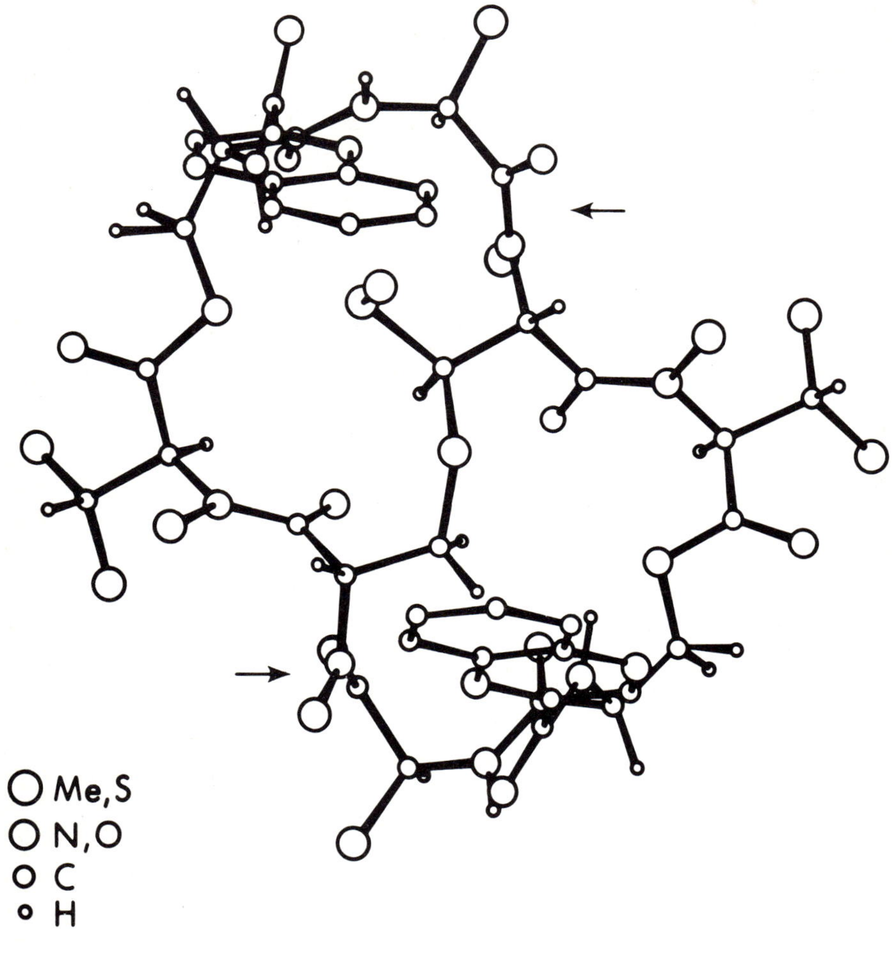

Figure 4-52. Proposed molecular structure of echinomycin [Ughetto and Waring (1977)]. (Reprinted with permission. Copyright by Academic Press, Inc.)

Reactions

Lysozyme lyses certain bacteria by selective hydrolysis of a cell wall mucopolysaccharide, composed of alternating β-(1-4)—linked residues of N-acetyl-glucosamine (NAG) and N-acetyl-*D*-muramic acid (NAM). It is believed [Blake *et al.* (1967)] that: the binding cleft can accommodate up to six sugar residues (rings A through F), the normal substrate being probably $(NAG\text{-}NAM)_3$; the hydrolysis takes place at the bond between C(1) of ring D (of NAM) and the glycosidal oxygen, O(g), between rings D and E (of NAG) (Fig. 4-53); binding of NAM at the site D would be opposed by steric hindrance; the residue 35 (glutamic acid) of lysozyme acts as a general acid catalyst [Blake *et al.* (1965)], involving protonation of the glycosidal oxygen and the rate-determining formation of a carbonium ion intermediate; the residue 52 of lysozyme (aspartic acid) stabilizes the positively charged carbonium ion intermediate at C(1).

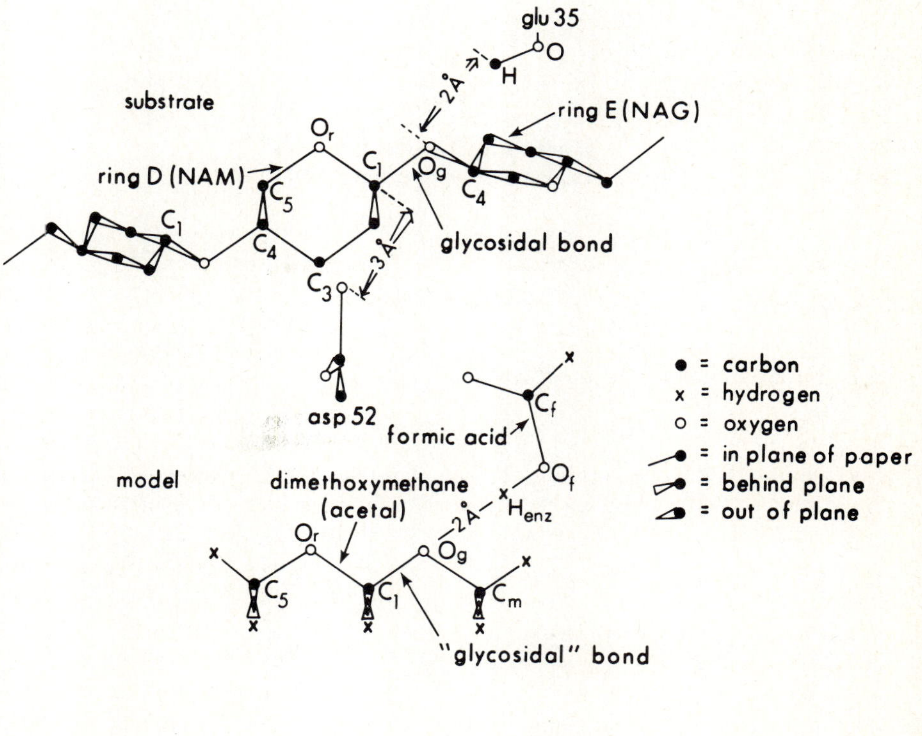

.gure 4-53. Model of substrate polysaccharide around the glycosidal bond and actual model used in the first step of the calculations [Loew and Thomas (1972)]. [Reprinted with permission. Copyright by Academic Press Inc. (London) Ltd.]

Model calculations were carried out in order to determine the role of *glu* 35 in the enzyme-directed general acid catalysis, including all the valence electrons and using formic acid and dimethoxymethane in place of *glu* 35 and the substrate [Loew and Thomas (1972)]. Figure 4-54 shows the two possibilities of accepted mechanisms, involving either a sequence of discrete steps or a concerted reaction, while Fig. 4-55 shows the geometrical arrangements of atoms used as input in the calculations, designed to simulate a sequence of steps along each pathway.

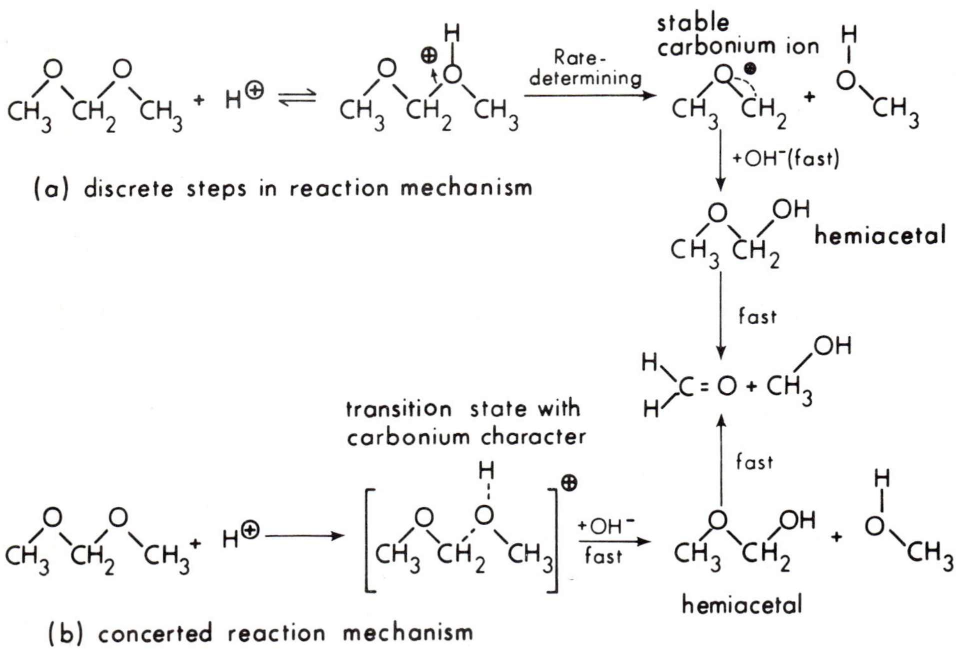

Figure 4-54. Proposed mechanism for the acid-catalyzed hydrolysis of dimethoxymethane [Raftery and Dahlquist (1969), Cordes (1967); as given by Loew and Thomas (1972)]. [Reprinted with permission. Copyright by Academic Press Inc. (London) Ltd.]

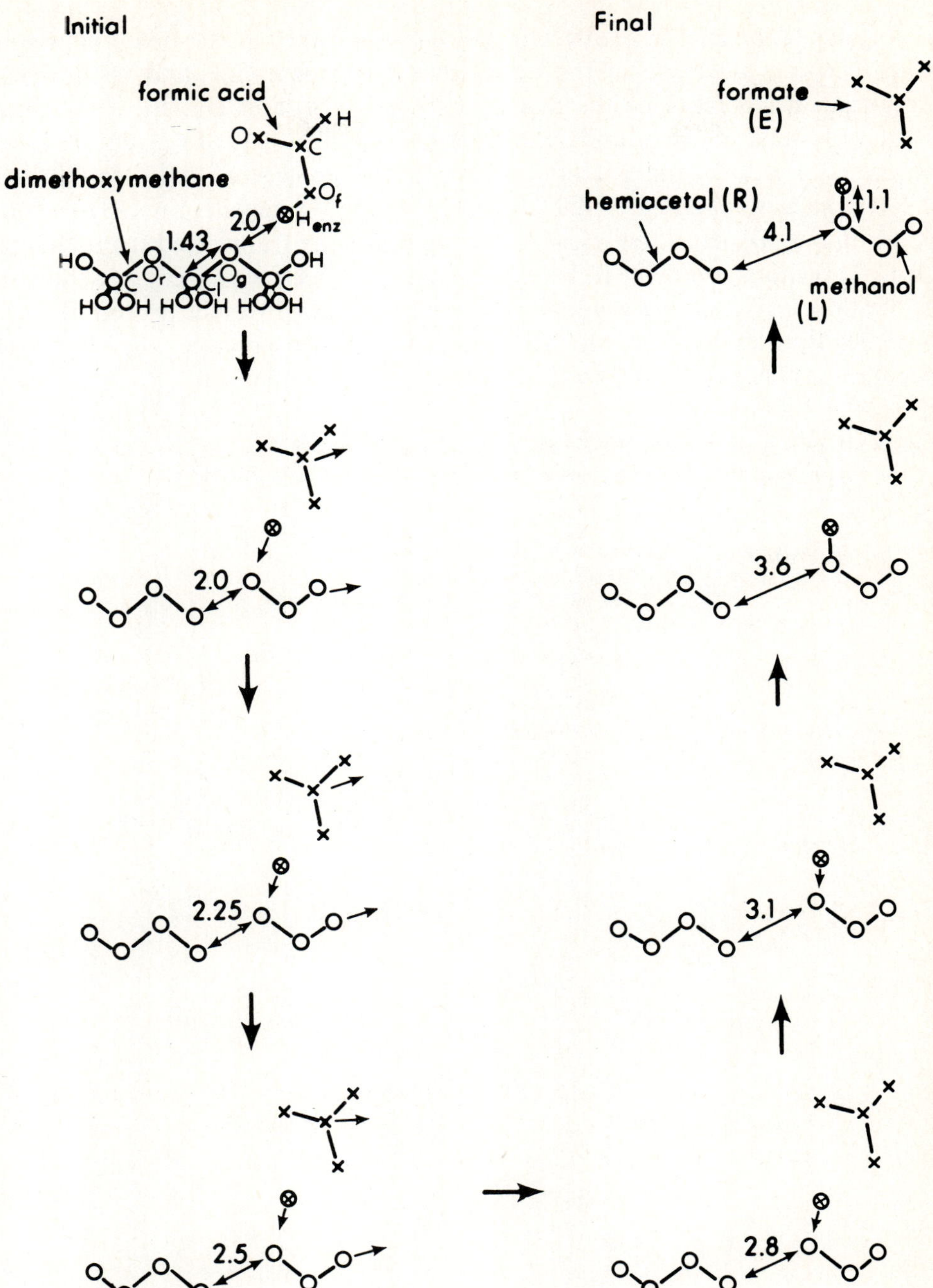

Figure 4-55. Geometry of the reaction pathways, showing the sequence of geometrical configurations used as input for uncatalyzed (not including the formic acid) and the catalyzed reactions; the out-of-plane hydrogen atoms are pictured only for the initial configuration. The bond lengths are given in angstroms [Loew and Thomas (1972)]. [Reprinted with permission. Copyright by Academic Press Inc. (London) Ltd.]

An inspection of the changes in the bond densities as the 'reaction' takes place shows a decrease in the bond density between C(1) and O(g) and a simultaneous increase of the density between O(g) and H(enz). The initial attraction between O(g) and H(enz) destabilizes the C(1)-O(g) bond in the reactants and the subsequent increase in the density of the O(g)-H(enz) facilitates the weakening of the C(1)-O(g) bond. Meanwhile, the O(f)-H(enz) bond also weakens as the proton is being gradually transferred from O(f) to O(g). This transfer may be followed through the effective charges presented in Table 4-42: O(f) becomes gradually more negative as it loses the proton, H(enz) becomes first less positive and then regains its positive charge, and O(g) first becomes less negative and then regains its negative charge.

Table 4-42. Dependence of the effective charges on some atoms as a function of the distance (in A) between C(1) and O(g) during the hydrolysis of dimethoxymethane*

Distance	1.43	2.00	2.25	2.50	2.80	3.10	3.35	3.60	4.10
C(1)	+.10	+.06	+.025	.00	−.004	−.02		−.06	−.14
	+.07	+.07	+.085	+.09	+.08	+.10	+.45		+.55
O(r)	−.44	−.43	−.40	−.41	−.39	−.38		−.42	−.39
	−.43	−.40	−.36	−.32	−.27	−.23	−.25		−.20
O(g)	−.44	−.46	−.45	−.46	−.46	−.46		−.38	−.31
	−.43	−.41	−.385	−.36	−.35	−.39	−.50		−.61
H(enz)	+.42	+.31	+.265	+.23	+.225	+.25	+.33		+.33
O(f)	−.33	−.36	−.42	−.45	−.44	−.50	−.75		−.76

* As given by Loew and Thomas (1972). The first line of values for a given atom corresponds to the uncatalized reaction while the second line corresponds to the catalyzed reaction.

Appendix A1

Coordinates

The position of a point in space may be described by its cartesian coordinates (X, Y, Z) or the position vector **R**, with modulus R and direction angles θ (between the Z-axis and R) and ϕ (between the X-axis and the projection of R on the XY-plane); R, θ, ϕ represent the polar spherical coordinates.

The two sets of coordinates are related by

$$X = R \sin\theta \cos\phi$$
$$Y = R \sin\theta \sin\phi$$
$$Z = R \cos\theta$$
$$R^2 = X^2 + Y^2 + Z^2$$

The position of a point *j*, relative to a point *i*, characterized by the position vectors $\mathbf{R}_j$ and $\mathbf{R}_i$, respectively, is given by

$$\mathbf{R}_{ji} = \mathbf{R}_j - \mathbf{R}_i$$

The corresponding distance between the two points is

$$R_{ji} = |\mathbf{R}_{ji}| = |\mathbf{R}_j - \mathbf{R}_i|$$

Appendix A2

Transformation of Coordinates

Given two molecules, it may be necessary to determine their coordinates in a chosen system of axes for different relative separations and orientations. It is assumed here that their original coordinates are known and, for simplicity, the first molecule M_1 will be kept at its original position. The problem is then reduced to the determination of the new coordinates of M_2 after a given move.

In order to visualize the process, it is convenient to consider a new system of axes, originally coinciding with the basic system but capable of moving with molecule M_2. In such a case, a movement of molecule M_2 may be substituted with a movement of its associated system of axes; the coordinates of the atoms of M_2 remain unchanged relative to this system, and the desired coordinates, relative to the basic system (after the move), may be obtained from the relationships between the two systems of axes.

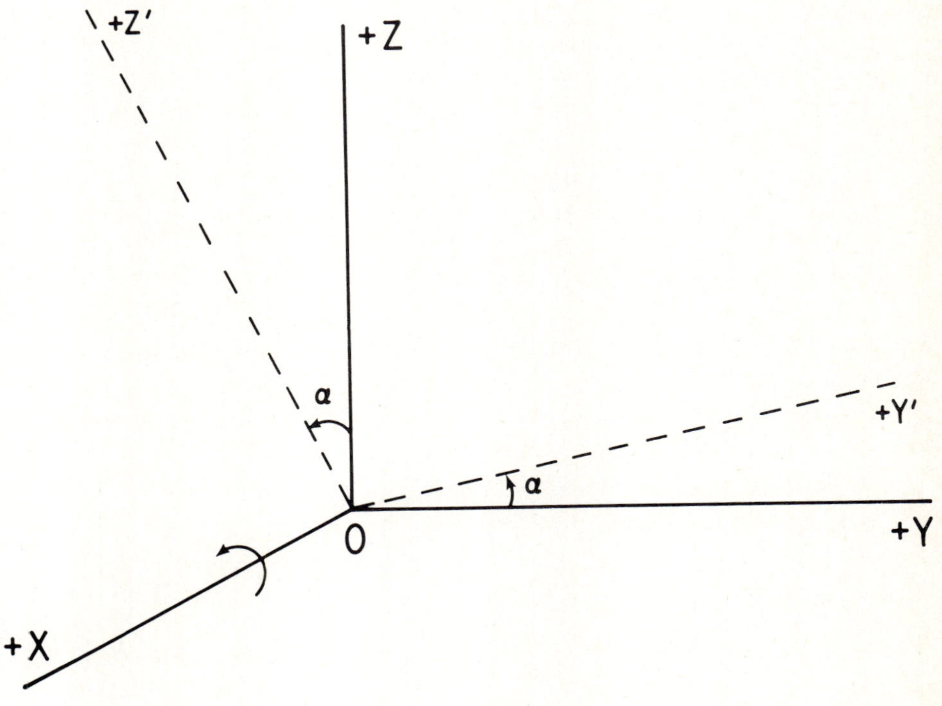

Figure A2-1. Rotation about X axis by an angle α.

The motion may consist of one or more of the following:

a) a rotation by an angle α about the X-axis;

b) a rotation by an angle β about the Y-axis;

c) a rotation by an angle γ about the Z-axis;

d) a translation of the origin of the moving system to a point O′ defined by the polar coordinates R, θ, ϕ with respect to the basic system of coordinates.

Figure A2-1 represents the case of a rotation by an angle α about the X-axis. This rotation is taken as positive when it corresponds to a counterclockwise rotation, as seen when looking from the positive to the negative direction of the X-axis. This rotation transforms the coordinates of the i-th atom into

$$\begin{pmatrix} X'_i \\ Y'_i \\ Z'_i \\ 1 \end{pmatrix} = \begin{pmatrix} 1 & 0 & 0 & 0 \\ 0 & \cos\alpha & -\sin\alpha & 0 \\ 0 & \sin\alpha & \cos\alpha & 0 \\ 0 & 0 & 0 & 1 \end{pmatrix} \begin{pmatrix} X_i \\ Y_i \\ Z_i \\ 1 \end{pmatrix} = \mathbf{A} \begin{pmatrix} X_i \\ Y_i \\ Z_i \\ 1 \end{pmatrix}$$

where it can be observed that only the Y, Z coordinates are affected. In a similar fashion and with analogous conventions for the signs of the angles, one has

$$\begin{pmatrix} X'_i \\ Y'_i \\ Z'_i \\ 1 \end{pmatrix} = \begin{pmatrix} \cos\beta & 0 & \sin\beta & 0 \\ 0 & 1 & 0 & 0 \\ -\sin\beta & 0 & \cos\beta & 0 \\ 0 & 0 & 0 & 1 \end{pmatrix} \begin{pmatrix} X_i \\ Y_i \\ Z_i \\ 1 \end{pmatrix} = \mathbf{B} \begin{pmatrix} X_i \\ Y_i \\ Z_i \\ 1 \end{pmatrix}$$

$$\begin{pmatrix} X'_i \\ Y'_i \\ Z'_i \\ 1 \end{pmatrix} = \begin{pmatrix} \cos\gamma & -\sin\gamma & 0 & 0 \\ \sin\gamma & \cos\gamma & 0 & 0 \\ 0 & 0 & 1 & 0 \\ 0 & 0 & 0 & 1 \end{pmatrix} \begin{pmatrix} X_i \\ Y_i \\ Z_i \\ 1 \end{pmatrix} = \mathbf{C} \begin{pmatrix} X_i \\ Y_i \\ Z_i \\ 1 \end{pmatrix}$$

for a rotation by an angle β about the Y-axis and a rotation by an angle γ about the Z-axis, respectively.

For a translation (Fig. A2-2) with the axes of the moving system parallel to those of the fixed system, one can write

$$\begin{pmatrix} X'_i \\ Y'_i \\ Z'_i \\ 1 \end{pmatrix} = \begin{pmatrix} 1 & 0 & 0 & X_0' \\ 0 & 1 & 0 & Y_0' \\ 0 & 0 & 1 & Z_0' \\ 0 & 0 & 0 & 1 \end{pmatrix} \begin{pmatrix} X_i \\ Y_i \\ Z_i \\ 1 \end{pmatrix} = \mathbf{T} \begin{pmatrix} X_i \\ Y_i \\ Z_i \\ 1 \end{pmatrix}$$

where

$X_0' = R \sin\theta \cos\phi$
$Y_0' = R \sin\theta \sin\phi$
$Z_0' = R \cos\theta$

are the coordinates of point O′ relative to the fixed system; R, θ, and ϕ are the polar spherical coordinates of the point O′ with respect to the basic system of coordinates (see Appendix A1 and Fig. A2-2).

The successive application of various transformations may be expressed in terms of the corresponding transformation matrices, taking care to preserve the order of the operations. Thus, the new coordinates, after a rotation by α about the X-axis, a rotation by β about the Y-axis, a rotation by γ about the Z-axis, and a translation (in that order), are given by

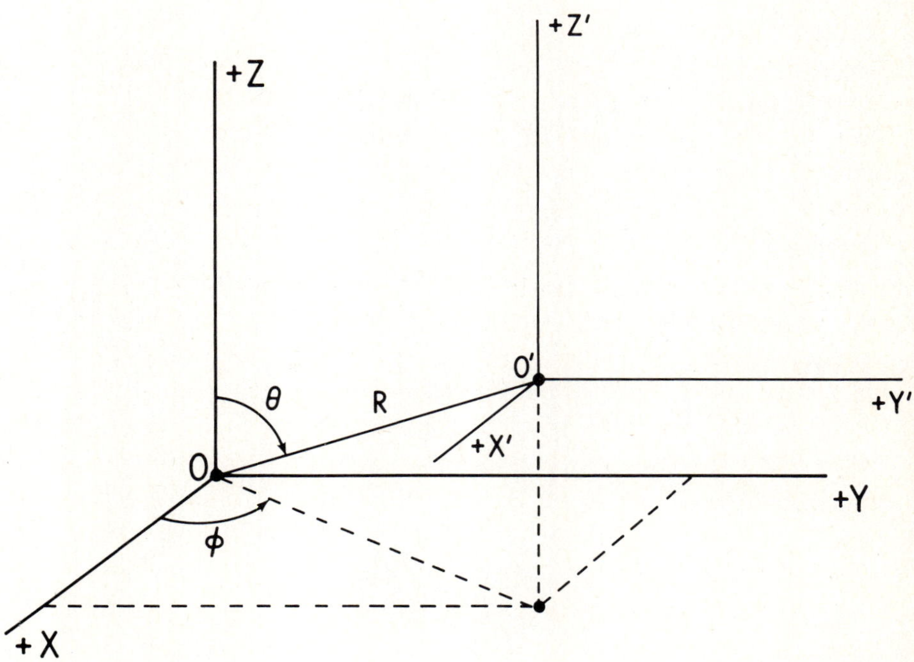

Figure A2-2. Relative position of the coordinate system X′Y′Z′ with respect to the coordinate system XYZ.

$$\begin{pmatrix} X'_i \\ Y'_i \\ Z'_i \\ 1 \end{pmatrix} = \mathbf{T\,C\,B\,A} \begin{pmatrix} X_i \\ Y_i \\ Z_i \\ 1 \end{pmatrix} = \mathbf{V} \begin{pmatrix} X_i \\ Y_i \\ Z_i \\ 1 \end{pmatrix}.$$

The explicit expression of the elements of the matrix are

$V_{11} = \cos\gamma \cos\beta$
$V_{12} = \cos\gamma \sin\beta \sin\alpha - \sin\gamma \cos\alpha$
$V_{13} = \cos\gamma \sin\beta \cos\alpha + \sin\gamma \sin\alpha$
$V_{14} = R \sin\theta \cos\phi$
$V_{21} = \sin\gamma \cos\beta$
$V_{22} = \sin\gamma \sin\beta \sin\alpha + \cos\gamma \cos\alpha$
$V_{23} = \sin\gamma \sin\beta \cos\alpha - \cos\gamma \sin\alpha$
$V_{24} = R \sin\theta \sin\phi$
$V_{31} = -\sin\beta$
$V_{32} = \cos\beta \sin\alpha$
$V_{33} = \cos\beta \cos\alpha$
$V_{34} = R \cos\theta$
$V_{41} = 0$
$V_{42} = 0$
$V_{43} = 0$
$V_{44} = 1$

It is obvious that two transformation matrices, **V** and **W** will yield the same relative position of M_2 with respect to M_1 if

$$\mathbf{VW} = \mathbf{WV} = \mathbf{I}$$

where **I** is the identity matrix.

Appendix A3

Determination of Cartesian Coordinates

The cartesian coordinates of the atoms of the system(s) for which potential energy calculations are to be carried are needed in order to evaluate the corresponding interatomic separations. There are many instances, however, when such coordinates are not available, although the equivalent information may exist in the form of geometrical parameters (bond lengths, bond angles, dihedral angles) or other types of coordinates; as a useful example of this type of information, Fig. A3-1 presents the bond lengths and angles for uridine. The determination of the cartesian coordinates from such data is discussed below.

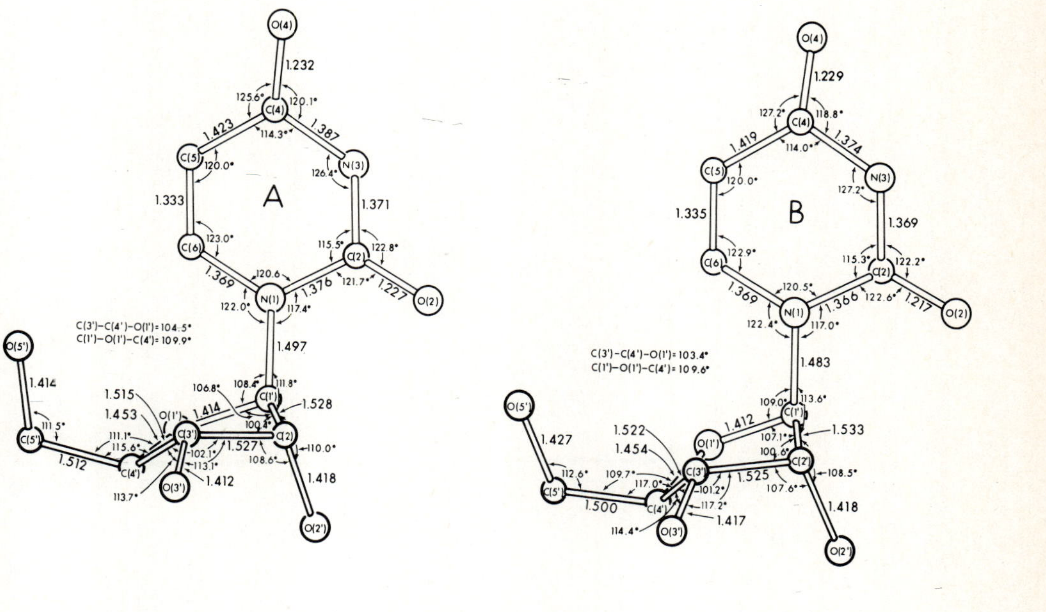

Figure A3-1. Bond distances and bond angles, involving the heavy atoms, of uridine [Green *et al.* (1975)].

CALCULATIONS USING GEOMETRICAL PARAMETERS

The cartesian coordinates of an atom may be determined provided that at least three geometrical parameters, that define its position with respect to three other atoms of the molecule (with known coordinates), are available.

The three atoms of known coordinates relative to the system of axes OXYZ (Fig. A3-2) are labelled 1, 2, and 3, and the atom with unknown coordinates is denoted by N. First of all, it is necessary to define a new system of coordinates O′X′Y′Z′ as follows: the origin O′ is at the position of atom 1; the vector $\mathbf{R}_{2\,1}$ from point 1 to point 2, defines the direction of the positive X′-axis; the direction of the positive Y′-axis is chosen so that atom 3 lies on the plane X′Y′, either in the first or second quadrant; and the direction of the positive Z′-axis is

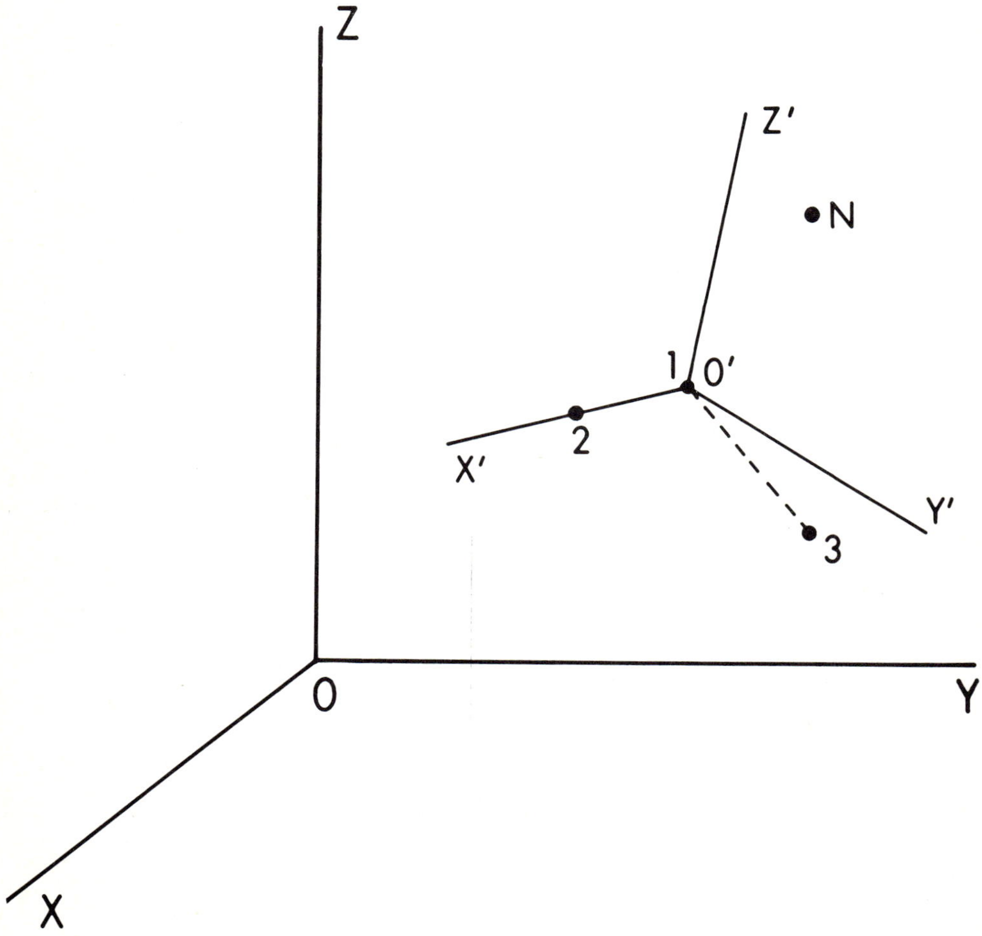

Figure A3-2. Local coordinate system X′Y′Z′ for the determination of the coordinates of atom N from those for atoms 1, 2 and 3.

chosen so that the coordinate system is right-handed. The directions of these three axes relative to the system OXYZ are given by

$$\mathbf{T}_{X'} = \mathbf{R}_{21}/R_{21}$$
$$\mathbf{T}_{Z'} = (\mathbf{R}_{21} \times \mathbf{R}_{32})/(\mathbf{R}_{21} \cdot \mathbf{R}_{32})$$
$$\mathbf{T}_{Y'} = (\mathbf{T}_{Z'} \times \mathbf{T}_{X'})/(\mathbf{T}_{Z'} \cdot \mathbf{T}_{X'})$$

in terms of the coordinates of atoms 1, 2, 3, relative to the system OXYZ. Next, one determines the coordinates of atom N relative to the system O′X′Y′Z′, this step being dependent on the type of structural parameters available for that atom. Many combinations are possible, but here only the four most important cases are described.

a) Data available: one bond length (R), one bond angle (θ), and one dihedral angle (τ).

The bond length corresponds to the distance R_{N1}, the bond angle is the one defined by the points, N, 1, 2, and the dihedral angle is the one formed by the planes N12 and 123, taken in the right-handed screw motion along the path N123 (Fig. A3-3)

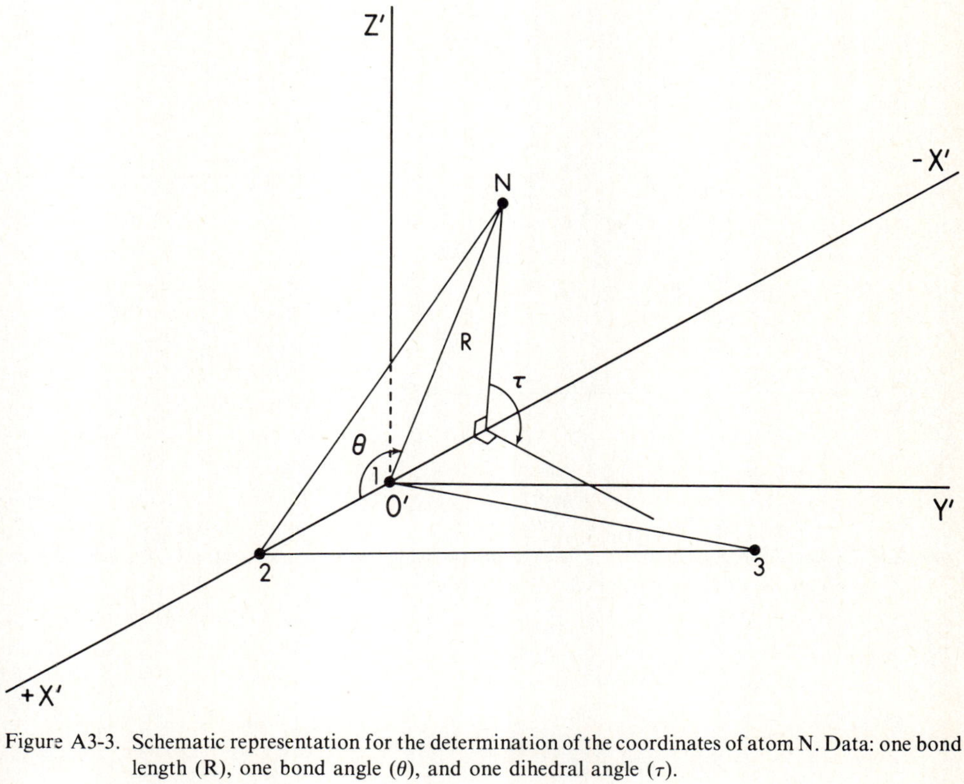

Figure A3-3. Schematic representation for the determination of the coordinates of atom N. Data: one bond length (R), one bond angle (θ), and one dihedral angle (τ).

The coordinates of atom N are then

$$X'_N = R \cos \theta$$
$$Y'_N = R \sin \theta \cos \tau$$
$$Z'_N = -R \sin \theta \sin \tau$$

b) Data available: one bond length (R) and two bond angles (θ, θ')

This case, for which two of the parameters (R, θ) are the same as in the previous case, is represented in Fig. A3-4.

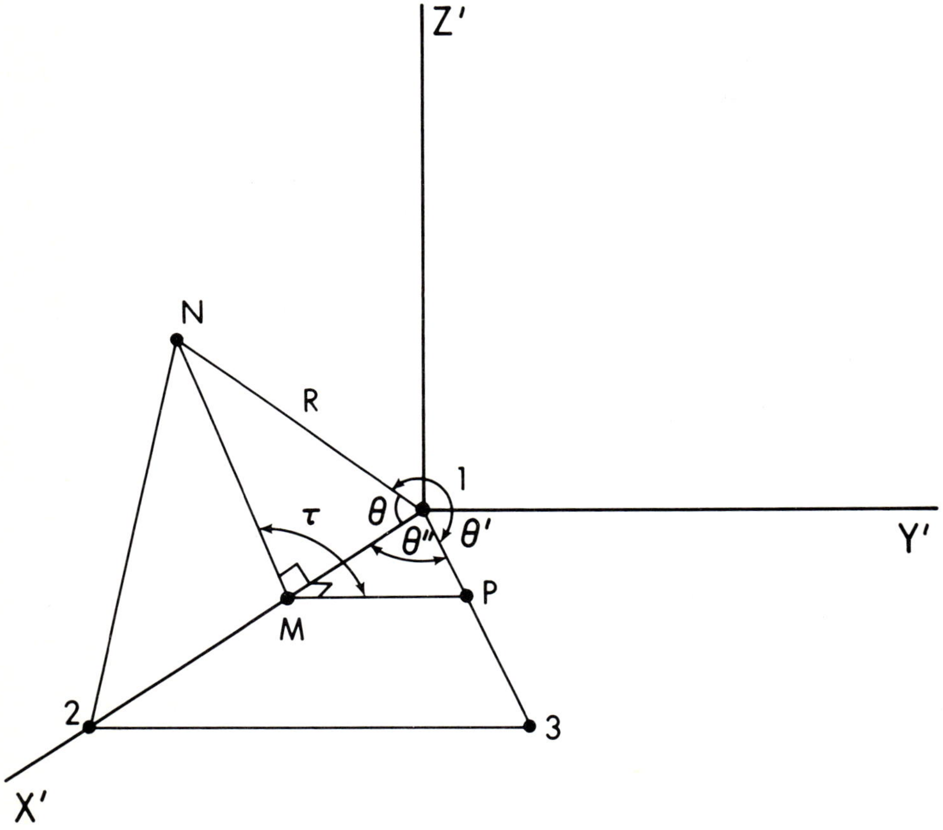

Figure A3-4. Schematic representation for the determination of the coordinates of atom N. Data: one bond length (R) and two bond angles (θ and θ').

The easiest approach involves the determination of the dihedral angle τ, between the planes N12 and 123, and then follow the procedure described in the previous case. One can write

$$O'M = R \cos\theta = R' \cos\theta''$$
$$\cos\theta' = (R^2 + (R')^2 - (R'')^2)/2\,R\,R'$$

where R′ is the distance between O′ and P, R″ denotes the separation between N and P, and θ'' represents the bond angle 213. One can also write

$$\cos\tau = ((NM)^2 + (MP)^2 - (R'')^2)/2\,(NM)\,(MP)$$

with

$$MP = R' \sin\theta''$$
$$NM = R \sin\theta$$

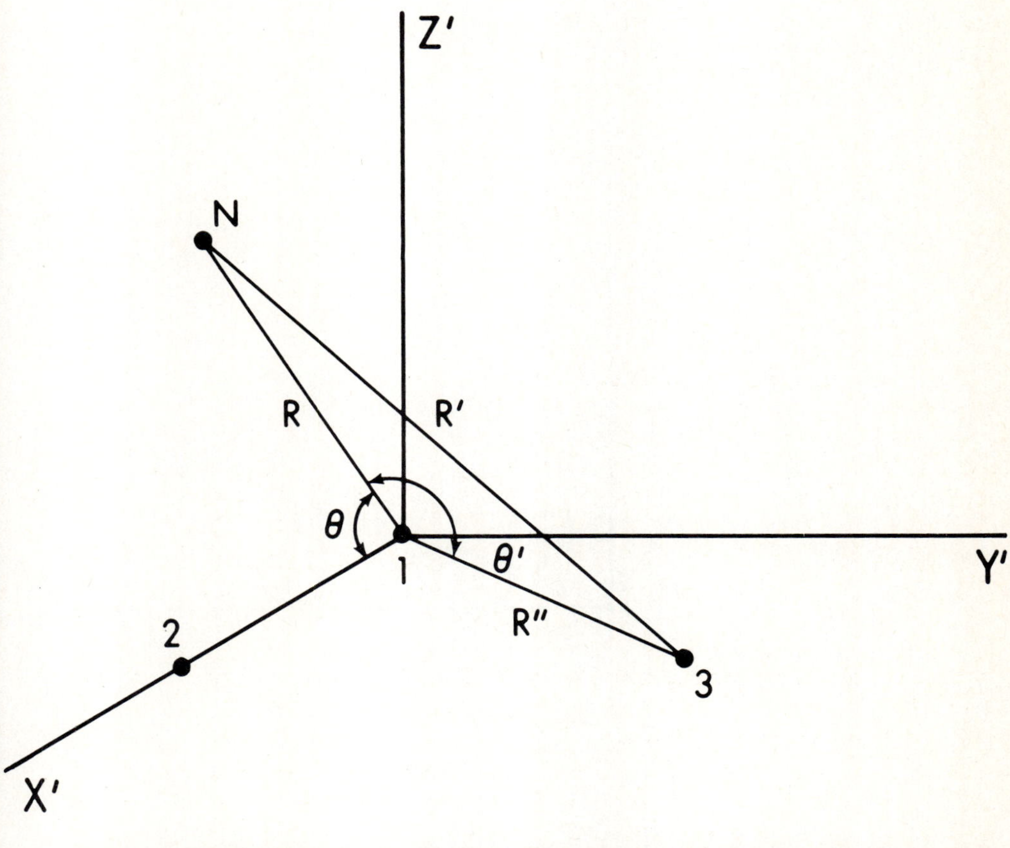

Figure A3-5. Schematic representation of the determination of the coordinates of atom N. Data: two bond lengths (R and R′) and one bond angle (θ).

so that finally one obtains

$$\cos \tau = (\cos \theta' - \cos \theta \cos \theta'')/\sin \theta \sin \theta''$$

It should be noted that θ'' can be easily obtained from the coordinates of the atoms 1, 2, 3.

c) Data available: two bond lengths (R,R′) and one bond angle (θ).

As observed in Fig. A3-5, the angle N13 (= θ') can be easily obtained using the expression

$$\cos \theta' = (R^2 + (R'')^2 - (R')^2)/2\,R\,R''$$

where R″ denotes the distance between atoms 1 and 3, and the problem is then reduced to the previous case.

d) Data available: three bond lengths (R, R′, R″)

This case, represented in Fig. A3-6, may also be transformed to case *b*. The angles N12 (= θ) and N13 (= θ') may be determined from the expressions

$$\cos \theta = (R^2 + (R''')^2 - (R')^2)/2\,R\,R'''$$
$$\cos \theta' = (R^2 + (R'''')^2 - (R'')^2)/2\,R\,R''''$$

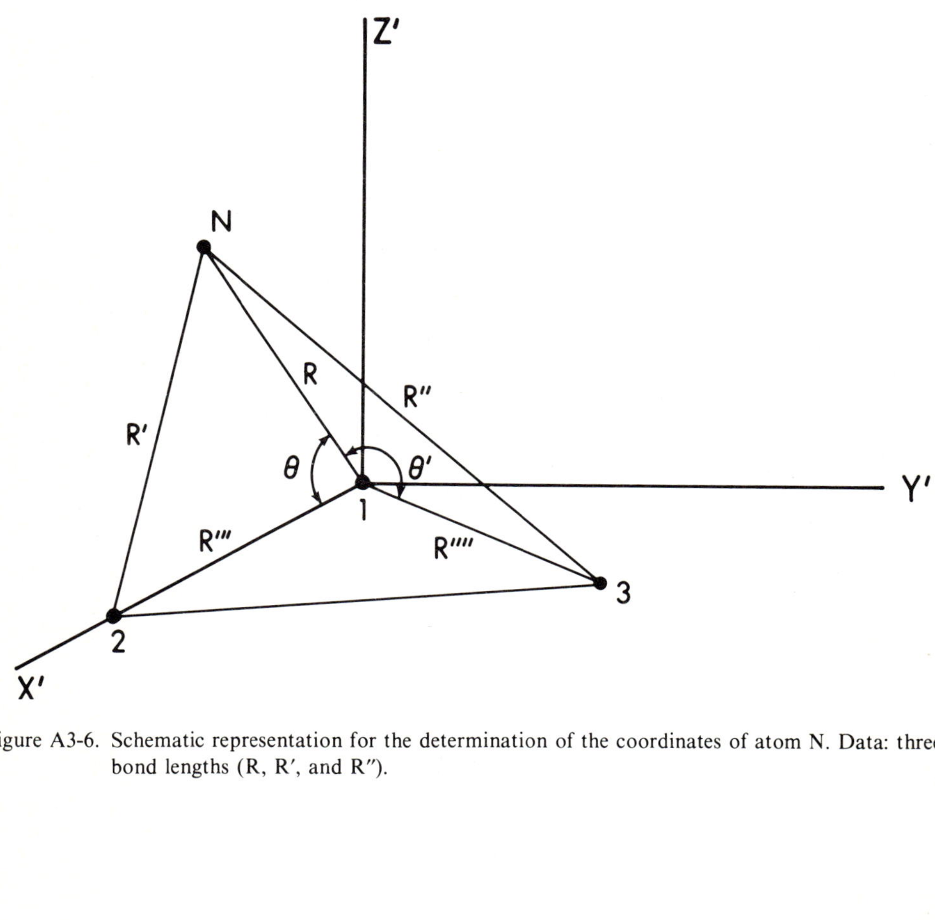

Figure A3-6. Schematic representation for the determination of the coordinates of atom N. Data: three bond lengths (R, R′, and R″).

where R‴ and R⁗ denote the distances between atoms 1 and 2 and 1 and 3, respectively.

Once the coordinates of atom N, relative to the system O′X′Y′Z′ have been obtained, they must be transformed to the main system OXYZ (Fig. A3-7). Adopting a new system O′X″Y″Z″, with origin at O′ (the position of atom 1) and with axes parallel to those of the main system, the coordinates X''_N, Y''_N, and Z''_N are given by

$$\begin{pmatrix} X''_N \\ Y''_N \\ Z''_N \end{pmatrix} = \begin{pmatrix} T(X'_1) & T(Y'_1) & T(Z'_1) \\ T(X'_2) & T(Y'_2) & T(Z'_2) \\ T(X'_3) & T(Y'_3) & T(Z'_3) \end{pmatrix} \begin{pmatrix} X'_N \\ Y'_N \\ Z'_N \end{pmatrix}$$

using the direction cosines described earlier. Finally, it is necessary to translate the origin (O′ or O″) to O so that one obtains

$$X_N = X''_N + X_1$$
$$Y_N = Y''_N + Y_1$$
$$Z_N = Z''_N + Z_1$$

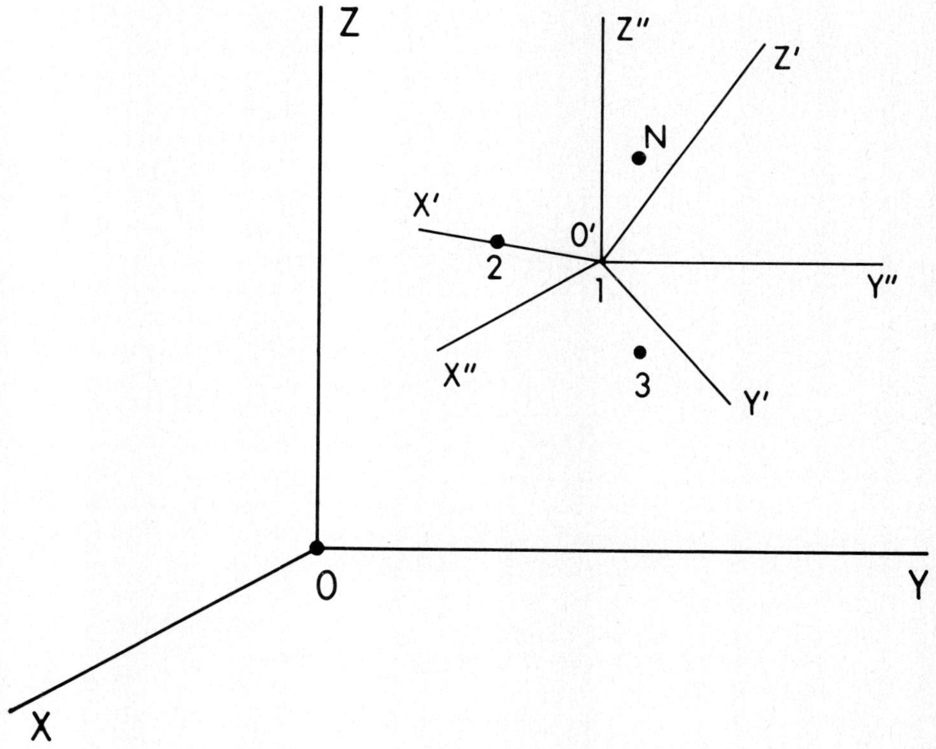

Figure A3-7. Schematic representation of the rotation and translation of the coordinate system X′Y′Z′ with respect to the fixed coordinate system XYZ.

The atomic coordinates are reported in these cases in fractional coordinates, x, y, z, referred to the unit cell of dimensions $a, b, c, \alpha, \beta, \gamma$, where a, b, c are the lengths of the unit cell edges, α is the angle between b and c, β is the angle between a and c, and γ is the angle between a and b.

Choosing the Z-axis in the direction of the c-axis and the Y-axis lying on the bc-plane (Fig. A3-8) the cartesian coordinates, X, Y, Z are then given by

$$\begin{pmatrix} X \\ Y \\ Z \end{pmatrix} = \begin{pmatrix} a \sin\beta \sin\gamma^* & 0 & 0 \\ -a \sin\beta \cos\gamma^* & b \sin\alpha & 0 \\ a \cos\beta & b \cos\alpha & c \end{pmatrix} \begin{pmatrix} x \\ y \\ z \end{pmatrix}$$

where

$\sin\gamma^* = (1 - (\cos\alpha \cos\beta - \cos\gamma)^2/(\sin\alpha \sin\beta)^2)^{1/2}$

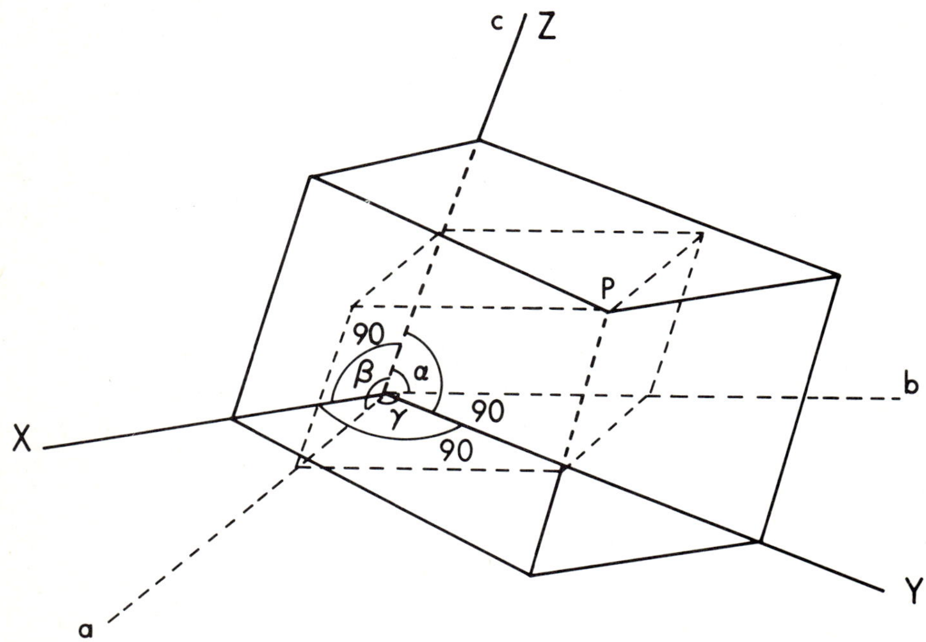

Figure A3-8. X-ray coordinate system (abc, $\alpha\beta\ \gamma$) and the cartesian coordinate system XYZ.

Appendix A4

Spherical Harmonics

The normalized spherical harmonics are defined by

$$Y(l,m;\theta,\phi) = P(l,m;\cos\theta)\ \Phi(m;\phi)$$
$$= (1/2\pi)^{1/2}\ P(l,m;\cos\theta)\ \exp(im\phi)$$

(with $-l \leqslant m \leqslant l$) in terms of normalized associated Legendre polynomials (with m a positive integer)

$$P(l,m;\cos\theta) = (-1)^m\ ((2l+1)\ (l-m)!/2\ (l+m)!)^{1/2}\ (1/(2^l)\ l!)$$
$$((\sin\theta)^m)\ (((d/d(\cos\theta))^{(l+m)})((\cos^2\theta - 1)^l))$$
$$P(l,-m;\cos\theta) = ((2l+1)\ (l-m)!/2\ (l+m)!)^{1/2}\ (1/(2^l)\ l!)$$
$$((\sin\theta)^m)\ (((d/d(\cos\theta))^{(l+m)})((\cos^2\theta - 1)^l))$$

The spherical harmonics to second order, in explicit form, are

$$Y(0,0) = (1/4\pi)^{1/2}$$
$$Y(1,0) = (3/4\pi)^{1/2}\cos\theta = (3/4\pi)^{1/2}\ (Z/R)$$
$$Y(1,\pm 1) = \mp\ (3/8\pi)^{1/2}\sin\theta\ \exp(\pm\ i\phi) = \mp\ (3/8\pi)^{1/2}\ (X \pm iY)/R$$
$$Y(2,0) = (5/16\pi)^{1/2}\ (3\cos^2\theta - 1) = (5/16\pi)^{1/2}\ (3Z^2 - R^2)/R^2$$
$$Y(2,\pm 1) = \mp\ (15/8\pi)^{1/2}\sin\theta\cos\theta\ \exp(\pm\ i\phi) = \mp\ (15/8\pi)^{1/2}\ Z\ (X \pm iY)/R^2$$
$$Y(2,\pm 2) = (15/32\pi)^{1/2}\sin^2\theta\ \exp(\pm\ 2i\phi) = (15/32\pi)^{1/2}\ (X \pm iY)^2/R^2$$

Appendix A5

Basic Statistical — Thermodynamical Formulas

The thermodynamical study of a system consisting of N independent, indistinguishable particles of the same type (e.g., molecules), each of which can exist in any of a number of states characterized by their energies, may be accomplished through the use of a Boltzmann distribution.

The probability associated to a given energy level is given by

$$p_i = (1/f) \exp(-E_i/kT)$$

or

$$p = (1/f) \exp(-E/kT)$$

depending on whether the energy E takes discrete or continuous values; T is the absolute temperature, k denotes the Boltzmann constant, and (1/f) represents a normalization factor, f being the partition function given by

$$f = \Sigma \exp(-E_i/kT)$$

or

$$f = \int \exp(-E/kT)\, dv$$

(with integration over all the variables), respectively.

The internal energy of the system is defined by

$$U = \Sigma E_i\, p_i = (1/f)\Sigma E_i \exp(-E_i/kT)$$

or

$$U = (1/f)\int E \exp(-E/kT)\, dv$$

Taking into account that

$$\delta f/\delta T = (f/kT^2)\, U$$

one can write

$$U = (kT^2/f)\,(\delta f/\delta T) = kT^2\,(\delta \ln f/\delta T)$$

The information entropy is

$$S' = -\,N\Sigma p_i \ln p_i$$

or

$$S' = -\,N\int p \ln p\, dv$$

the final result being in both cases

$S' = N \ln f + NU/kT$

The entropy is then given (see below) by

$S = kN \ln(f/N) + NU/T + kN$

and for a mole of the system one finally has

$$U = RT^2 (\delta \ln f/\delta T)$$
$$S = R \ln(f/N) + U/T + R$$

where U and S are now the molar quantities, N stands for Avogadro number, and R is the universal gas constant. [The definition of the entropy, given above, has been corrected for the indistinguishability of the particles; this correction will be carried on hereafter for completeness, although in the end it must be dropped.]

The derivation of the expressions for the remaining thermodynamic (molar) functions, which are given in terms of the internal energy and the entropy, is now straightforward. Taking into account that

$F = U - TS = -RT \ln(f/N) - RT$

and that the pressure may be defined by

$P = -\delta F/\delta V = RT (\delta \ln(f/N)/\delta V)$ (at constant T)

one obtains

$$H = U + PV = RT^2 (\delta \ln f/\delta T) + RTV (\delta \ln(f/N)/\delta V)$$
$$G = H - TS = -RT \ln(f/N) + RTV (\delta \ln(f/N)/\delta V) - RT$$

where V is the volume; the partial derivatives are taken at constant V or T, respectively.

The change in the standard Gibbs free energy is related

$\Delta G^\circ = -RT \ln K$

to the corresponding equilibrium constant, so that the evaluation of the latter is immediate whenever the partition functions of all the systems involved in the reaction are known.

The formulation may be simplified for ideal gases. In such a case one can write

$G = F + PV = F + RT = -RT \ln(f/N) = -RT \ln f + RT \ln N$

and

$\Delta G = -RT \Delta(n \ln f) + RT (\ln N) \Delta n$

where

$$\Delta[n \ln f] = \Sigma n_p \ln f_p - \Sigma n_r \ln f_r$$
$$\Delta n = \Sigma n_p - \Sigma n_r$$

with the summations running over the products (p) and reactants (r) in the reaction, n_p and n_r being the corresponding stoichiometric coefficients. Taking into account the accessibility of the states, it can be shown that the second term in ΔG must be dropped, so that one has

$$\Delta G = -RT\ \Delta(n \ln f)$$

so that comparison of the two expressions for ΔG finally yields

$$\ln K = \Delta[n \ln f]$$

For the particular case of a reaction of the form

$$A + B \leftrightarrow AB$$

(used within the main text for the discussion of molecular associations) one has then

$$K = (\Sigma \exp(-E_{AB\,i}/kT))\ ((\Sigma \exp(-E_{A\,j}/kT))\ (\Sigma \exp(-E_{B\,m}/kT)))^{-1}$$

Referring the energy levels of each system to the ground state of AB one can write

$$\ln K = -\Delta E/kT + (\ln f_{AB} - \ln f_A - \ln f_B)$$

where

$$\Delta E = E_{AB\,0} - E_{A\,0} - E_{B\,0}$$

represents the difference between the ground state molecular energies of the product and the reactants.

Because of the difficulties in the evaluation of the partition functions, recourse is some times made to the very drastic approximation of assuming all the partition functions equal to unity. In such a case one has

$$\Delta G = N\ \Delta E$$
$$\ln K = -\Delta E/kT$$

Appendix A6

Electric Fields and Moments

There are two different definitions for electric fields and moments arising from the method (Neumann or Taylor expansion) used for the multipole expansion (see below). Although the conventions arising from the Neumann expansion are far more elegant, for the purpose of this work the definitions corresponding to the Taylor expansion are of interest because of existing formulations that make use of them. For this reason, and also taking into account that both appear in the literature, they are both discussed here.

ELECTRIC FIELDS

The (scalar) potential, V, at a point (x,y,z), due to a charge Q at position (X,Y,Z), is defined by

$$V = Q/((x - X)^2 + (y - Y)^2 + (z - Z)^2)^{1/2} = Q/R\ .$$

The (vector) electric field, **F**, at the same point, is characterized by its components

$$F_x = -\delta V/\delta x = Q\ (x - X)/R^3$$
$$F_y = -\delta V/\delta y = Q\ (y - Y)/R^3$$
$$F_z = -\delta V/\delta z = Q\ (z - Z)/R^3$$

and a total magnitude

$$F = Q/R^2$$

The (tensor) electric field gradient, **F′**, at the given point has components

$$F_{xx} = \delta^2 V/\delta x^2 = Q\ (3(x - X)^2 - R^2)/R^5$$
$$F_{xy} = F_{yx} = \delta^2 V/\delta x \delta y = 3Q\ (x - X)\ (y - Y)/R^5$$

...etc. The trace of the field gradient, with a value $2Q/R^3$, is invariant under rotation of the system of coordinates.

In the particular case when the charge Q is at position (0,0,R), all the components of the field and field gradient at the origin of coordinates vanish except for

$$F_z = -Q/R^2$$
$$F_{zz} = -2F_{yy} = -2F_{xx} = 2Q/R^3$$

The general definition in terms of spherical harmonics for the field, field gradients, etc., at the origin, due to the charge Q at a point characterized by the position vector **R**, is

$$F(l,m;R) = (4\pi/(2l+1))^{1/2}\, Y(l,m;\theta,\phi)\, (Q/R^{l+1})$$

The relations between the cartesian components derived from this definition and those given above may be obtained using the expressions given in Appendix A4 for the spherical harmonics.

ELECTRIC MOMENTS

A charge distribution [with charges Q_i at positions x_i, y_i, z_i] is characterized by its total charge

$$Q' = \Sigma Q_i$$

and its moments.

The (vector) dipole moment, μ, has components

$$\mu_x = \Sigma Q_i\, x_i$$
$$\mu_y = \Sigma Q_i\, y_i$$
$$\mu_z = \Sigma Q_i\, z_i$$

and a total magnitude

$$\mu = \Sigma Q_i\, R_i$$

The (tensor) quadrupole moment, **Q**, is characterized by its components

$$Q_{xx} = \Sigma Q_i\, x^2_i$$
$$Q_{xy} = Q_{yx} = \Sigma Q_i\, x_i\, y_i$$

...etc. The trace of the quadrupole moment with a value

$$\Sigma Q_i\, R^2_i$$

is invariant under a rotation of the axes.

A generalized definition in terms of spherical harmonics is

$$M(l,m) = (4\pi/(2l+1))^{1/2} \Sigma Q_i\, R^l_i\, Y(l,m;\theta_i,\phi_i)$$

The cartesian components of the moments may be obtained from this definition using the expressions given in Appendix A4 for the spherical harmonics.

MULTIPOLE EXPANSION

The electrostatic potential interaction between a charge Q, with position vector **R**, and a charge distribution [charges Q_i with position vectors $\mathbf{R}_i$] may be written (Neumann expansion) as

$$\Delta E = Q\Sigma Q_i/|\mathbf{R}_i - \mathbf{R}| = Q\Sigma Q_i \Sigma(r^k(<)/r^{k+1}(>))\, P(k;\cos\omega_i)$$

where r(<) and r(>) are the lesser and the greater of R and R_i and ω_i is the angle between **R** and $\mathbf{R}_i$; $P(k;\cos\omega_i)$ is a Legendre polynomial. The second summation extends to $0 \leqslant k \leqslant \infty$.

If R is larger than all R_i, the preceding equation may be rewritten as

$$\Delta E = Q\Sigma Q_i \Sigma(R^k{}_i / R^{k+1})\ P(k;\cos\omega_i)$$

Taking into account that

$$P(k;\cos\omega_i) = (4\pi/(2k+1))\Sigma Y(k,m;\theta,\phi)\ Y(k,m;\theta_i,\phi_i)$$

(with $-k \leqslant m \leqslant k$) one obtains

$$\begin{aligned}\Delta E &= \Sigma\Sigma(4\pi/(2k+1))^{1/2}\ Y(k,m;\theta,\phi)\ (Q/R^{k+1})\\ &\quad (4\pi/(2k+1))^{1/2}\ \Sigma Q_i\ R^k{}_i\ Y(k,m;\theta_i,\phi_i)\\ &= \Sigma\Sigma\ F(k,m;R)\ M(k,m)\end{aligned}$$

in terms of products of the moments associated with the charge distribution and the potential, field, field gradients, etc., at the origin due to the charge Q. When R coincides with the z-axis, the above expression yields (for $k \leqslant 2$)

$$\Delta E = (Q/R)\Sigma Q_i + (Q/R^2)\Sigma Q_i\ z_i + (1/2)(Q/R^3)\Sigma Q_i\ (3z^2{}_i - R^2{}_i)$$

A Taylor series expansion, involving the cartesian coordinates given explicitly above, leads to the same result. In practice, it has been customary to use the conventions

$$\begin{aligned}F_z &= -Q/R^2\\ F_{zz} &= -2F_{yy} = -2F_{xx} = 2Q/R^3\\ \mu_z &= \Sigma Q_i\ z_i\\ Q_{zz} &= -(1/2)\Sigma Q_i\ (3z^2{}_i - R^2{}_i)\\ Q_{xy} &= Q_{yx} = -(3/2)\Sigma Q_i\ x_i\ y_i\end{aligned}$$

...etc., in which case the above expression may be cast in the form

$$\Delta E = Q'\ V - \mu_z\ F_z - (1/3)(Q_{xx}\ F_{xx} + Q_{yy}\ F_{yy} + Q_{zz}\ F_{zz})$$

The above convention for the diagonal components of the quadrupole moment has the advantage of showing more clearly the isotropy/anisotropy of the charge distribution.

Appendix A7

Density Distributions: Population Analysis and Properties

The electron density, due to the i-th electron, for a molecule in a given electronic state, is defined by

$$\rho_i = (1/N)\Sigma N_m (|m\rangle\langle m|)$$

N being the total number of electrons and where the summation extends to all the occupied orbitals $|m\rangle$, occupied by N_m electrons; $\langle m|$ denotes the complex conjugate of $|m\rangle$. Integration over all space yields, of course,

$$\int \rho_i \, dv_i = (1/N)\Sigma N_m \langle m|m\rangle = (1/N)\Sigma N_m = 1$$

Within the expansion method, the orbitals are expressed in terms of appropriate basis functions, with coefficients determined in the self-consistent field (SCF) calculations. Using the notation $|u\rangle$ for the basis functions, one can write

$$|m\rangle = \Sigma|u\rangle c_{um} = uc_m$$

for the orbital $|m\rangle$; u represents a row vector, whose elements are the basis functions, and c_m is a column vector, with elements c_{um}. The basis functions normally used are either Slater-type atomic orbitals (STO) or gaussian functions (GTO), usually centered on the various nuclei of the molecule; when the latter information will be needed, the notation will be changed to $|u_a\rangle$, indicating that the basis function is centered on nucleus a.

POPULATION ANALYSIS

The electronic population in the m-th orbital may be written as

$$N_m = N_m\Sigma c^2_{u(a)m} + 2N_m\Sigma\Sigma' S_{u(a)v(b)} \, c_{u(a)m} \, c_{v(b)m}$$

with the summations extending to all the basis functions, assumed to be normalized; $c_{u(a)m}$) is the coefficient in the m-th orbital of the basis function $|u_a\rangle$, centered on atom a; $S_{u(a)v(b)}$ represents the overlap between the two basis functions, $|u_a\rangle$ and $|v_b\rangle$. The first term in the above expression can be considered to represent an atomic population and the second one corresponds to an overlap population.

The atomic populations (which are the ones of interest for the purposes of this work) can be formulated in terms of partial contributions [Mulliken

(1955*a*, *b*, *c*, *d*)]. The total gross population on atom *a* is given by

$$N_a = \Sigma N_{ma}$$

where the summation extends to all the occupied molecular orbitals. N_{ma} denotes the subtotal gross population on atom *a* due to the m-th occupied orbital, defined by

$$N_{ma} = \Sigma N_{mu(a)}$$

with the summation running over all the basis functions. The partial gross population, $N_{mu(a)}$ is finally given by

$$N_{mu(a)} = N_m c_{u(a)m} (c_{u(a)m} + \Sigma' c_{v(b)} S_{u(a)v(b)})$$

Hereafter, the total gross population on atom *a*, N_a, which gives its total electronic charge, will be denoted by Q'_a.

PROPERTIES

The electronic density distribution may be characterized by its moments and its potential, electric field and field gradient at a given point of space (Appendix A6). The total charge and the components of the dipole and quadrupole moments are given by

$$Q = \Sigma \int \rho_i \, dv_i = N$$
$$\mu_p = \Sigma \int p_i \, \rho_i \, dv_i = \Sigma N_m < m|p|m >$$
$$Q_{pp'} = -(1/2)\Sigma \int (3\, p_i\, p'_i - R^2_i\, \delta_{pp'})\, \rho_i \, dv_i$$
$$= \Sigma N_m < m|-(1/2)(3pp' - R^2\, \delta_{pp'})|m >$$

where the summations extend to all the electrons in the molecule or the occupied orbitals, respectively, and *p*,*p'* stand for any cartesian coordinate. In a similar way one obtains

$$V_0 = \Sigma \int (1/R_i)\, \rho_i \, dv_i = \Sigma N_m < m|1/R|m >$$
$$F_p = -\Sigma \int (1/R^3_i)\, \rho_i \, dv_i = \Sigma N_m < m|1/R^3|m >$$
$$F_{pp'} = 3\Sigma \int ((p_i\, p'_i - R^2_i)/R^5_i)\, \rho_i \, dv_i = \Sigma N_m < m|3(pp' - R^2)/R^5|m >$$

for the potential at the origin and for the components of the electric field and field gradient at the origin, with the same notation as for the moments; R_i, abbreviated as R in the bra-ket notation of the integrals, denotes the distance of the i-th electron from the origin.

The expressions for the quantities considered above in terms of integrals over basis functions and expansion coefficients may be obtained in a straightforward manner and for that reason they are omitted here, and only the formulation for the electrostatic potential, because of its use in the evaluation of electrostatic interactions will be considered.

The potential at position p (with position vector **R**, referred to the origin of coordinates) may be written as

$$V_R = \Sigma \int (1/|\mathbf{R}_i - \mathbf{R}|)\, \rho_i \, dv_i = \Sigma N_m \Sigma \Sigma c_{rm} < r|1/R_p|u > c_{um}$$

where R_p indicates the distance between the electron and point *p*, and the summations extend over the occupied orbitals and the basis functions. In the latter no indication is given explicitly of the nuclei on which they are centered; when such a distinction is made the above equation may be given as

$$V_r = \Sigma N_m \Sigma \Sigma \Sigma \Sigma c_{r(b)m} < r_b|1/R_p|u_a > c_{u(a)m}$$

where the summations extend now over the occupied orbitals, the centers (*a* and *b*), and the basis functions centered on them (u_a and r_b).

The practical problem in connection with the evaluation of the electrostatic potential at a given point arises, as for the evaluation of other quantities, from the large number of integrals to be evaluated. Several alternatives have been proposed in order to simplify the calculation of the integrals. In practice, it has been found that the best results are obtained by means of an expansion of the basis functions centered on one atom in terms of a set of basis functions centered on another.

At a low level of sophistication, the electrostatic molecular potential may be evaluated by

$$V_R = \Sigma Q_a / |\mathbf{R}_a - \mathbf{R}|$$

where the summation extends to all the atoms in the molecule, with effective (net) charges, Q_a, defined as

$$Q_a = Z_a - Q'_a$$

where Z_a is the effective nuclear charge and Q'_a is the gross electronic charge, defined above. At this level of approximation, the theoretical description of a molecule is embodied in the set of its effective charges which, one might say, represent the ‘bytes’ of information stored in the carrier.

Appendix A8

Bond-Energy Analysis

The energy quantities (molecular Hartree-Fock energy, orbital energy, and one-electron energy) may be written [Clementi (1967)] as

$$E = \Sigma E_a + \Sigma\Sigma E_{ab} + \Sigma\Sigma\Sigma E_{abc} + \Sigma\Sigma\Sigma\Sigma E_{abcd}$$
$$\epsilon_i = \Sigma\epsilon_{i\,a} + \Sigma\Sigma\epsilon_{i\,ab} + \Sigma\Sigma\Sigma\epsilon_{i\,abc} + \Sigma\Sigma\Sigma\Sigma\epsilon_{i\,abcd}$$
$$h_i = \Sigma h_{i\,a} + \Sigma\Sigma h_{i\,ab} + \Sigma\Sigma\Sigma h_{i\,abc} + \Sigma\Sigma\Sigma\Sigma h_{i\,abcd}$$

where a,b,c,d denote the atoms of the molecule, i indicates the i-th orbital, and in the multiple summations it is $a \neq b$, $a \neq b \neq c$, $a \neq b \neq c \neq d$, respectively. The four terms in each expression represent the one-, two-, three- and four-centre contribution to the corresponding energy.

The terms E_{abc} and E_{abcd} are smaller than E_a and E_{ab}, the latter including the nuclear repulsion. The one-centre term, E_a, represents the energy of atom a when forming part of the molecule under consideration and has been designated as the molecular orbital valency state (MOVS) and may be evaluated from

$$E_a = \Sigma\eta_{i\,a} = \Sigma(\epsilon_{i\,a} + h_{i\,a})$$

where $\eta_{i\,a}$ denotes the contribution from atom a to the so-called electron energy.

The MOVS have been used as a criterion for the characterization of atoms in molecules, under the assumption that similar values for a given atom in different molecules indicate a similarity of molecular environments. Considering the possibility of the adoption in future calculations of the MOVS criterion, Table A8-1 presents a summary of upper and lower values of MOVS, as found in amino acids and purine and pyrimidine bases.

Table A8-1. Limiting and most frequent values of MOVS*

Atom	*Class*	*Lower Limit*	*Upper Limit*	*Most Frequent Value*
H	1	0.20	0.32	0.28
	2	0.26	0.31	0.28
	3	0.26	0.30	0.29
	4	0.26	0.33	0.32
C	5	1.35	1.55	1.41
	6	0.92	1.06	
	7	0.88	1.09	1.06
	8	1.05	1.20	1.08
O	9	0.42	0.57	0.52
	10	0.36	0.43	0.38
N	11	0.83	1.06	0.91
	12	0.82	0.94	0.83
	13	1.02		
C	14	1.07	1.15	1.15
N	15	1.13	1.34	1.14
H	16	0.28	0.35	0.30
C	17	1.13	1.29	
	18	1.10	1.15	1.11
	19	1.09	1.16	
	20	1.12	1.30	1.30
	21	0.94	0.97	0.97
S	22	0.19	0.22	0.19
H	23	0.22		
C	24	1.50	1.50	
	25	1.20	1.43	
	26	1.60	1.67	1.66
O	27	0.33	0.35	0.33
C	28	1.22		

* From the values given by Clementi and co-workers for the amino acids and puric and pyrimidinic bases; see Chapter 4 for references and Table 4-29 for details on the classes of atoms.

References

The references mentioned within the text are grouped by chapter and listed in alphabetical order within each chapter. In those cases where several references have the same author(s) they are given in chronological order, distinguished by the notation a, b, . . . , if the year is the same. The titles are given in order to show their relevance more clearly.

General

E. Fraga and S. Fraga.
Biomoleculas: Estudios Teoricos
Editorial Alhambra, S.A., Madrid (1976)

Chapter 1

P.H. Abelson. Ann. N.Y. Acad. Sci. *69*, 276 (1957)
Some aspects of paleobiochemistry.

P.H. Abelson. Proc. Natl. Acad. Sci. (U.S.) *55*, 1365 (1966)
Chemical events on the primitive earth.

E. Anders, R. Hayatsu, and M.H. Studier. Science *182*, 781 (1973).
Organic compounds in meteorites.

C.B. Anfinsen and H.A. Scheraga. Adv. Protein Chem. *29*, 205 (1975)
Experimental and theoretical aspects of protein folding.

A. Babloyantz. Biopolymers *11*, 2349 (1972)
Far from equilibrium synthesis of 'prebiotic' polymers.

A. Babloyantz and G. Nicolis. J. Theor. Biol. *34*, 185 (1972)
Chemical instabilities and multiple steady state transitions in Monod-Jacob type models.

A.F. Bartholomay. Bull. Math. Biophys. *20*, 97 (1958a)
On the linear birth and death processes of biology as Markoff chains

A.F. Bartholomay. Bull. Math. Biophys. *20*, 175 (1958b)
Stochastic models for chemical reactions: I. Theory of the unimolecular reaction process.

A.F. Bartholomay. Bull. Math. Biophys. *21*, 363 (1959)
Stochastic models for chemical reactions. II. The unimolecular rate constant.

R. Blumenthal, J.P. Changeux, and R. Lefever. J. Membrane Biol. *2*, 351 (1970)
Membrane excitability and dissipative instabilities.

M.H. Briggs. Nature *191*, 1137 (1961)
Organic constituents of meteorites.

H.G. Bungenberg de Jong
Crystallisation — Coacervation — Flocculation, in 'Colloid Science', edited by H.R. Kruyt.
Elsevier Scientific Publishing Co., Amsterdam (1949).

D. Burger. Can. J. Soil Science *49*, 11 (1968)
Calcium release from 11 minerals of fine-sand size by dilute sulfuric acid.

M. Calvin. Chem. Eng. News *39*, part 2, 96 (1961)
The chemistry of life. 3. How life originated on earth and in the world beyond.

R. Carbo, J.M. Fornos, and M. Martin. Unpublished results
The origin of the genetic code: a theoretical discussion of (L)Ala-(L)Ala, (L)Ala-(D)Ala and (L)Ala-Gly interactions.

E. Clementi. Proc. Natl. Acad. Sci. (U.S.) *69*, 2942 (1972)
Computation of large molecules with the Hartree-Fock model.

E. Clementi, J.M. Andre, M. Cl. Andre, D. Klint, and D. Hahn. Acta Phys. Acad. Sci. Hung. *27*, 493 (1969)
Study of the electronic structure of molecules. X.

E. Clementi, J. Mehl, and W. von Niessen. J. Chem. Phys. *54*, 508 (1971)
Study of the electronic structure of molecules. XII. Hydrogen bridges in the guanine-cytosine pair and the dimeric form of formic acid.

R.E. Christoffersen. Adv. Quantum Chem. *6*, 333 (1972)
Ab initio calculations on large molecules.

H. Craig. Geochim. Cosmochim. Acta *3*, 53 (1953)
The geochemistry of the stable carbon isotopes.

F.H.C. Crick. Sci. Am. *215*, 55 (1966)
The genetic code.

F.H.C. Crick and J.D. Watson. Proc. Roy. Soc. (London) *A223*, 80 (1954)
The complementary structure of deoxyribonucleic acid.

H. de Voe and I. Tinoco. J. Mol. Biol. *4*, 518 (1962).
The hypochromism of helical polynucleotides.

M. Eigen. Naturwiss. *58*, 465 (1971)
Self-organization of matter and the evolution of biological macromolecules.

J.P. Ferris. Science *161*, 53 (1968)
Cyanovinyl phosphate: a prebiological phosphorylating agent?

J.P. Ferris, J.E. Kuder, and A.W. Catalano. Science *166*, 765 (1969)
Photochemical reactions and the chemical evolution of purines and nicotinamide derivatives.

R.L. Flurry, Jr. and D. Breen. Nature *225*, 729 (1970)
Huckel-level calculations on the band structure of Poly-AT.

S.W. Fox. Naturwiss. *56*, 1 (1969)
Self-ordered polymers and propagative cell-like systems.

S.W. Fox. Chem. Eng. News *49*, 46 (1971)
Chemical origins of cells.

S.W. Fox. Pure Appl. Chem. *34*, 641 (1973)
Molecular evolution to the first cells.

S.W. Fox. BioSystems *7*, 22 (1975)
Stereomolecular interactions in microsystems in experimental protobiogenesis.

S.W. Fox. J. Mol. Evol. *8*, 301 (1976a)
Thermal polypeptides. Comments.

S.W. Fox. Origins of Life *7*, 49 (1976b)
The evolutionary significance of phase-separated microsystems.

S.W. Fox and K. Harada. Science *128*, 1214 (1958)
Thermal copolymerization of amino acids to a product resembling protein.

S.W. Fox and K. Harada. J. Am. Chem. Soc. *82*, 3745 (1960)
Thermal copolymerization of amino acids common to protein.

S.W. Fox and K. Dose.
Molecular Evolution and the Origin of Life
Freeman and Co., San Francisco (1972).

S.W. Fox and C.R. Windsor. Science *170*, 984 (1970)
Synthesis of amino acids by the heating of formaldehyde and ammonia.

S.W. Fox, K. Harada, P.E. Hare, G. Hinsch, and G. Mueller. Science *167*, 767 (1970)
Bio-organic compounds and glassy microparticles in lunar fines and other materials.

S.W. Fox, K. Harada, and P.E. Hare. Geochim. Cosmochim. Acta *40*, 1069 (1976)
Amino acid precursors in lunar fines: limits to the contribution of jet exhaust.

S. Fraga and C. Valdemoro. Structure and Bonding *4*, 1 (1968)
Quantum chemical studies on the submolecular structure of the nucleic acids.

H. Frohlich. Int. J. Quantum Chem. *2*, 641 (1968)
Long-range coherence and energy storage in biological systems.

N.W. Gabel and C. Ponnamperuma. Nature *216*, 453 (1967)
Model for the origin of monosaccharides.

E. Gelpi and J. Oro. Geochim. Cosmochim. Acta *34*, 981 (1970)
Organic compounds in meteorites. IV. Gas chromatographic-mass spectromic studies on the isoprenoids and other isomeric alkanes in carbonaceous chondrites.

P. Glansdorff and I. Prigogine
Thermodynamic Theory of Structure, Stability and Fluctuations
John Wiley & Sons, Inc., New York (1971).

R.J. Goldacre
Surface films, their collapse on compression, the shapes and sizes of cells and the origin of life, in 'Surface Phenomena in Chemistry and Biology', edited by J.F. Danielli, K.G.A. Pankhurst and A.C. Riddiford
Pergamon Press, London (1958).

A. Goldbeter and G. Nicolis. Biophysik *8*, 212 (1972)
Far from equilibrium synthesis of small polymer chains and chemical evolution.

R.H. Hall. Progr. Nucl. Ac. Res. Molec. Biol. *10*, 57 (1970)
N^6-(Δ^2-isopentenyl)adenosine: chemical reactions, biosynthesis, metabolism and significance to the structure and function of t-RNA.

P.B. Hamilton. Nature *205*, 284 (1965)
Amino acids on hands.

P.B. Hamilton and B. Nagy. Space Life Sci. *3*, 432 (1972)
Problems in the search for amino acids in lunar fines.

K. Harada. Nature *214*, 479 (1967)
Formation of amino acids by thermal decomposition of formamide and olygomerization of hydrogen cyanide.

K. Harada and S. W. Fox. Nature *201*, 335 (1964)
Thermal synthesis of natural amino acids from postulated primitive terrestrial atmosphere.

P.E. Hare, K.H. Harada, and S.W. Fox. Proc. Apollo II Lunar Sci. Conf., Geochim. Cosmochim. Acta (Suppl. 1) *2*, 1799 (1970)
Analysis for amino acids in lunar fines.

R. Hayatsu. Science *146*, 1291 (1964)
Orgueil meteorite: organic nitrogen contents.

R. Hayatsu, M.H. Studier, A. Oda, K. Fuse, and E. Anders. Geochim. Cosmochim. Acta *32*, 175 (1968)
Origin of organic matter in early solar system. II. Nitrogen compounds.

T.A. Hoffman and J. Ladik. Adv. Chem. Phys. *7*, 84 (1964)
Quantum mechanical considerations on some properties of DNA.

N.H. Horowitz and S.L. Miller. Fortschr. Chem. Org. Naturstoffe *20*, 423 (1962)
Current theories on the origin of life.

K.A. Kvenvolden. Space Life Science *3*, 330 (1972)
Review of methods used in lunar organic analysis: extraction and hydrolysis techniques.

J. Ladik, S. Suhai, P. Otto, and T.C. Collins. Int. J. Quantum Chem., Quantum Biology Symp. *4*, 55 (1977)
Estimation of the band structure of DNA on the basis of an *ab initio* SCF LCAO band structure of polycytosine.

R. Langridge, H.R. Wilson, C.W. Hooper, M.H.F. Wilkins, and L.D. Hamilton. J. Mol. Biol. *2*, 19 (1960)
The molecular configuration of deoxyribonucleic acid. I. X-ray diffraction study of a crystalline form of the lithium salt.

C. Lifschitz, E.D. Bergman, and B. Pullman. Tetrahedron Lett. 4583 (1967)
The ionization potentials of biological purines and pyrimidines.

B. Maigret and B. Pullman. Theoret. Chim. Acta *35*, 113 (1974)
Molecular orbital calculations on the conformation of polypeptides and proteins. XI. Conformational studies on 'tripeptide' models.

B. Maigret, D. Perahia, and B. Pullman. Biopolymers *10*, 491 (1971)
Molecular orbital calculations on the conformation of polypeptides and proteins. V. Conformational energy maps and stereochemical rotational states of aliphatic residues.

O. Matsuoka, C. Tosi, and E. Clementi. Biopolymers (1978)
Conformational studies on polynucleotide chains. I. Description of the Hartree-Fock energies for non-bonded interactions with a Lennard-Jones potential.

C.N. Matthews and R.E. Moser. Proc. Natl. Acad. Sci. (U.S.) *56*, 1087 (1966)
Prebiological protein synthesis.

C.N. Matthews and R.E. Moser. Nature *215*, 1230 (1967)
Peptide synthesis from hydrogen cyanide and water.

R. Mayer and L. Jaschke. Angew. Chem. *72*, 635 (1960)
Zur Umwandlung von Formaldehyd in Kohlehydrate.

B. Mely and A. Pullman. Theoret. Chim. Acta *13*, 278 (1969)
Ab initio calculations on cytosine, thymine, and adenine.

S.L. Miller. Science *117*, 528 (1953)
A production of amino acids under possible primitive Earth conditions.

K. Morokuma. Chem. Phys. Lett. *9*, 129 (1971)
Electronic structure of linear polymers: CNDO/2 calculation for polyglycine.

A. Munster.
Statistische Thermodynamik kondensierter Phasen, in 'Handbuch der Physik', edited by S. Flugge
Springer-Verlag, Berlin (1959).

M.E. Murphy, V.E. Modzeleski, B. Nagy, W.M. Scott, M. Young, C.M. Drew, P.B. Hamilton, and H.C. Urey. Proc. Apollo II Lunar Sci. Conf., Geochim. Cosmochim. Acta (Suppl. 1) *2*, 1879 (1970)
Analysis of Apollo II lunar samples by chromatography and mass spectrometry: pyrolysis products, hydrocarbons, sulfur, amino acids.

B. Nagy and C.M. Bitz. Arch. Biochem. Biophys. *101*, 240 (1963)
Long chain fatty acids from the Orgueil meteorite.

B. Nagy, C.M. Drew, P.B. Hamilton, V.E. Modzeleski, M.E. Murphy, W.M. Scott, H.C. Urey, and M. Young. Science *167*, 770 (1970)
Organic compounds in lunar samples: pyrolysis products, hydrocarbons, amino acids.

M. Nirenberg.
On the translation of the genetic code, in 'Biology and the Physical Sciences', edited by S. Devons
Columbia University Press, New York (1969).

A.I. Oparin
Proiskhozdeni Zhizny
Izd. Moskovshii Rabochii, Moscow (1924).

A.I. Oparin. Adv. Enzymol. *27*, 347 (1965)
The origin of life and the origin of enzymes.

A.I. Oparin.
The Origin and Initial Development of Life
Meditsina Publishing House, Moscow (1966).

A.I. Oparin
Life: Its Nature, Origin and Development
Academic Press, Inc., New York (1969).

A.I. Oparin, K.L. Gladilini, D.B. Kirpotin, G.V. Chertikhin, and A.F. Orlovskii. Dokl. Akad. Nauk. SSSR *232*, 485 (1977)
Coacervate drops containing polyphosphates: a model of protobiont division.

A.F. Orlovskii, K.L. Gladilin, V. Ya. Voronstova, D.B. Kirpotin, and A.I. Oparin. Dokl. Akad. Nauk. SSSR *232*, 236 (1977)
Stabilization of drops by orthophosphates and nucleotides.

J. Oro. Nature *191*, 1193 (1961)
Mechanism of synthesis of adenine from hydrogen cyanide under possible primitive earth conditions.

J. Oro. Nature *197*, 862 (1963)
Synthesis of organic compounds by electric discharges.

J. Oro. Space Life Sci. *3*, 507 (1972)
Extraterrestrial organic analysis.

J. Oro and A.P. Kimball. Arch. Biochem. Biophys. *94*, 217 (1961)
Synthesis of purines under possible primitive Earth conditions. I. Adenine from hydrogen cyanide.

J. Oro, W.S. Updegrove, J. Gibert, J. McReynolds, E. Gil-Av, J. Jbanez, A. Zlatkis, D.A. Flory, R.L. Levy, and C. Wolf. Science *167*, 765 (1970)
Organogenic elements and compounds in surface samples from the Sea of Tranquillity.

J. Oro, S. Nokaparskin, H. Lichtenstein, and E. Gil-Av. Nature *230*, 107 (1971)
Configurations of amino acids in carbonaceus chondrites and a precambrian chart.

C. Palm and M. Calvin. J. Am. Chem. Soc. *84*, 2115 (1962)
Primordial organic chemistry. I. Compounds resulting from electron irradiation of $^{14}CH_4$.

D. Perahia, A. Pullman, and H. Berthod. Theor. Chim. Acta *40*, 47 (1975)
An *ab initio* study of the effect of 3d orbitals of phosphorus on the properties of the phosphate group.

C. Ponnamperuma.
A biological synthesis of some nucleic acid constituents, in 'The Origins of Prebiological Systems and of Their Molecular Matrices', edited by S.W. Fox.
Academic Press Inc., New York (1965).

C. Ponnamperuma and S. Chang
The role of phosphates in chemical evolution, in 'Chemical Evolution and the Origin of Life', edited by R. Buvet and C. Ponnamperuma
North Holland Publishing Co., Amsterdam (1971).

C. Ponnamperuma and P. Kirk. Nature *203*, 400 (1963)
Synthesis of deoxyadenosine under simulated primitive Earth conditions.

C. Ponnamperuma, R.M. Lemmon, R. Mariner, and M. Calvin. Proc. Natl. Acad. Sc. (U.S.) *49*, 737 (1963)
Formation of adenine by electron irradiation of methane, ammonia and water.

C. Ponnamperuma, K. Kvenvolden, S. Chang, R. Johnson, G. Pollock, D. Philpott, I. Kaplan, J. Smith, J.W. Sehopf, C. Gehrke, G. Hodgson, J.A. Breger, B. Halpern, A. Duffield, K. Krauskopf, E. Barghoorn, H. Holland, and K. Keil. Science *167*, 760 (1970)
Search for organic compounds in the lunar dust from the Sea of Tranquillity.

P.K. Ponnuswamy, P.K. Warme, and H.A. Scheraga. Proc. Natl. Acad. Sci. (U.S.) *70*, 830 (1973)
Role of medium-range interactions in proteins.

I. Prigogine
Etude thermodynamique des phenomenes irreversibles
Dosoer, Liege (1947).

A. Pullman and H. Berthod. Int. J. Quantum Chem., Quantum Biology Symp. *4*, 327 (1977)
An *ab initio* study of a nucleoside: uridine.

B. Pullman, B. Maigret, and D. Perahia. Theoret. Chim. Acta *18*, 44 (1970)
Molecular orbital calculations on the conformation of polypeptides and proteins. III. Conformational maps of amino acid residues: A preliminary comparison of empirical, quantum-mechanical and experimental results.

J. Rabinowitz, S. Chang, and C. Ponnamperuma. Nature *218*, 442 (1968)
Phosphorylation by way of inorganic phosphate as potential prebiotic process.

C. Reid and L.E. Orgel. Nature *216*, 455 (1967)
Synthesis of sugar in potentially prebiotic conditions.

C. Reid, L.E. Orgel, and C. Ponnamperuma. Nature *216*, 936 (1967)
Nucleoside synthesis under potentially prebiotic conditions.

V. Renugopalakrishnan and D.W. Urry. Int. J. Quantum Chem., Quantum Biology Symp. *3*, 13 (1976)
Conformational energy calculations demonstrating the β turn and other secondary structural features in the repeat pentapeptide of elastin.

W.W. Rubey. Geol. Soc. Ann. Bull. *62*, 111 (1951)
Geological history of sea water.

R.A. Sanchez, J.P. Ferris, and L.E. Orgel. Science *153*, 72 (1966)
Conditions for purine synthesis: did prebiotic synthesis occur at low temperatures?

R.A. Sanchez, J.P. Ferris, and L.E. Orgel. J. Mol. Biol. *30*, 223 (1967)
Studies in prebiotic synthesis. II. Synthesis of purine precursors and amino acids from aqueous hydrogen cyanide.

R.A. Sanchez, J.P. Ferris, and L.E. Orgel. J. Mol. Biol. *38*, 11 (1968)
Studies in prebiotic synthesis. IV. Conversion of 4-aminoimidazole-5-carbonitrile derivatives to purines.

A.W. Schwartz
Phosphate: solubilization and activation on the primitive Earth, in 'Chemical Evolution and the Origin of Life', edited by R. Buvet and C. Ponnamperuma
North Holland Publishing Co., Amsterdam (1971).

A. Schwartz and C. Ponnamperuma. Nature *218*, 443 (1968)
Phosphorylation of adenosine with linear polyphosphate salts in aqueous solution.

M.P. Schweizer, G.B. Chheda, L. Baczynskyj, and R.H. Hall. Biochem. *8*, 3283 (1969)
Aminoacyl nucleosides. VII. N-(purin-6-ylcarbamoyl) threonine. A new component of transfer ribonucleic acid.

L. Seprodi, G. Biczo, and J. Ladik. Int. J. Quantum Chem. *3*, 621 (1969)
The effect of electric field on the electronic structure of DNA.

P.G. Seybold. Int. J. Quantum Chem., Quantum Biology Symp. *3*, 39 (1976)
Why are there four bases in DNA?

D.O. Shah
The origin of membranes and related surface phenomena, in 'Exobiology', edited by C. Ponnamperuma
North Holland Publishing Co., Amsterdam (1972).

M.A. Sweeney, A.P. Toste, and C. Ponnamperuma. Origins of Life *7*, 187 (1976)
Formation of amino acids by cobalt-60 irradiation of hydrogen cyanide solutions.

C. Tosi, E. Clementi, and O. Matsuoka. Biopolymers (1978a)
Conformational studies on polynucleotide chains. I. Analysis of steric interactions and derivation of potential functions for internal rotations.

C. Tosi, E. Clementi, and O. Matsuoka. Biopolymers (1978b)
Conformational studies on polynucleotide chains. III. Intramolecular energy maps and comparison with experiments.

H.C. Urey. Proc. Natl. Acad. Sci. (U.S.) *38*, 351 (1952)
On the early chemical history of the Earth and the origin of life.

D. Vasilescu, H. Broch, H. Peronneau, D. Perahia, and B. Pullman. Folia Chim. Theoret. Latina *5*, 69 (1977)
Analyse conformationelle des acides nucleiques par etude PCILO du disucre-phosphate.

C.C. Yang and J. Oro
Synthesis of nitrogen organic compounds, in 'Chemical Evolution and the Origin of Life', edited by R. Buvet and C. Ponnamperuma
North Holland Publishing Co., Amsterdam (1971).

Chapter 2

G. Alagona, R. Cimiraglia, E. Scrocco and J. Tomasi. Theoret. Chim. Acta *25*, 103 (1972)
Theoretical investigations on the solution process. II. The monohydrated associates of some three-membered ring molecules.

D.F. Bradley. J. Macromol., Sc.-Chem. *A4*, 741 (1970)
Recognition polymers.

R.H. Davies. Int. J. Quantum Chem., Quantum Biology Symp. *4*, 413 (1977)
A steady-state model of drug-receptor interaction *in vivo* applied to antagonists of the β-adrenergic receptor.

I.V. Krivoshey and L.A. Sleta. Theoret. Chim. Acta *43*, 165 (1976)
A topological stochastic approach to the study of multidimensional potential energy surfaces of chemical reactions.

Z. Simon. Rev. Roum. Biochim. *5*, 3 (1968)
Specific interactions between molecules and the concept of complementarity relations.

Z. Simon. Rev. Roum. Biochim. *9*, 337 (1972)
Molecular size and specificity of interactions.

Chapter 3

F.F. Abraham. J. Chem. Phys. *61*, 1221 (1974)
Monte Carlo simulation of physical clusters of water molecules.

F.F. Abraham, M.R. Mruzik, and G. Marshall Pound. Faraday Discussions *61*, 34 (1976)
The thermodynamics and structure of hydrated halide and alkali ions.

J.A. Barker and R.O. Watts. Chem. Phys. Letters *3*, 144 (1969)
Structure of water; a Monte Carlo calculation.

R. Bonaccorsi, C. Petrongolo, E. Scrocco, and J. Tomasi. Theoret. Chim. Acta *20*, 331 (1971)
Theoretical investigations on the solvation process. I. A simple model for the dimeric water associate.

R. Bonaccorsi, E. Scrocco, and J. Tomasi. Theoret. Chim. Acta *43*, 63 (1976)
A representation of the polarization term in the interaction energy between a molecule and a point-like charge.

A.D. Buckingham
Permanent and induced molecular moments and long-range intermolecular forces, in 'Intermolecular Forces', edited by J.O. Hirschfelder (Advances in Chemical Physics, vol. 12)
Interscience Publishers, New York (1967).

J. Caillet, P. Claverie, and B. Pullman. Theoret. Chim. Acta *47*, 17 (1978)
Effect of the crystalline environment upon the rotational conformation about the N-C and C-C′ bonds (ϕ and ψ) in amides and peptides.

R. Carbo and J.M. Riera
A General SCF Theory. (Lecture Notes in Chemistry)
Spring-Verlag, Berlin (1978).

I.G. Csizmadia
Theory and Practice of MO Calculations on Organic Molecules
Elsevier Scientific Publishing Co., Amsterdam (1976).

L.B. Kier. J. Medicinal Chem. *11*, 915 (1968)
The conformation of 20-oxopregnane hormones from molecular orbital calculations and a consideration of the cortisol receptor.

H. Kistenmacher, H. Popkie, and E. Clementi. J. Chem. Phys. *58*, 1689 (1973a)
Study of the structure of molecular complexes. II. Energy surfaces for a water molecule in the field of a sodium or potassium cation.

H. Kistenmacher, H. Popkie, and E. Clementi. J. Chem. Phys. *58*, 5627 (1973b)
Study of the structure of molecular complexes. III. Energy surface of a water molecule in the field of a fluorine or chlorine anion.

H. Kistenmacher, H. Popkie, and E. Clementi. J. Chem. Phys. *59*, 5842 (1973c)
Study of the structure of molecular complexes. V. Heat of formation for the Li^+, Na^+, K^+, F^-, and Cl^- ion complexes with a single water molecule.

H. Kistenmacher, H. Popkie, and E. Clementi. J. Chem. Phys. *60*, 4455 (1974)
Study of the structure of molecular complexes. VII. Effect of correlation energy corrections to the Hartree-Fock water-water potential on Monte Carlo simulations of liquid water.

A.I. Kitaigorodsky. Tetrahedron *14*, 230 (1961)
The interaction curve for non-bonded carbon and hydrogen atoms and its application.

E.R. Lippincott and R. Schroeder. J. Chem. Phys. *23*, 1099 (1955)
One-dimensional model of the hydrogen bond.

R.F. McGuire, F.A. Momany, and H.A. Scheraga. J. Phys. Chem. *76*, 375 (1971)
Energy parameters in polypeptides. V. An empirical hydrogen bond potential function based on molecular orbital calculations.

N. Metropolis, A.W. Rosenbluth, M.N. Rosenbluth, A.H. Teller, and E. Teller. J. Chem. Phys. *21*, 1087 (1953)
Equation of state calculations by fast computing machines.

W.G. Moulton and R.A. Kromhout. J. Chem. Phys. *25*, 34 (1956)
Nuclear magnetic resonance: structure of the amino group. II.

T. Ooi, R.A. Scott, G. Vanderkooi, and H.A. Scheraga. J. Chem. Phys. *46*, 4410 (1967)
Conformational analysis of macromolecules. IV. Helical structures of poly-*L*-alanine, poly-*L*-valine, poly-β-methyl-*L*-aspartate, poly-γ-methyl-*L*-glutamate, and poly-*L*-tyrosine.

H. Popkie, H. Kistenmacher, and E. Clementi. J. Chem. Phys. *59*, 1325 (1973)
Study of the structure of molecular complexes. IV. The Hartree-Fock potential for the water dimer and its application to the liquid state.

H.A. Scheraga. Ann. N.Y. Acad. Sci. *303*, 2 (1977)
Intermolecular potential for water and the hydration of proteins.

R. Schroeder and E.R. Lippincott. J. Phys. Chem. *61*, 921 (1957)
Potential function model of hydrogen bonds. II.

R.A. Scott and H.A. Scheraga. J. Chem. Phys. *45*, 2091 (1966)
Conformational analysis of macromolecules. III. Helical structures of polyglycine and poly-*L*-alanine.

K. Sundaram. Int. J. Quantum Chem. *8*, 565 (1974)
Non-bonded interactions in membrane active biopolymers. I. Valinomycin-potassium ion complex.

Chapter 4

C. Altona and M. Sundaralingam. J. Am. Chem. Soc. *94*, 8205 (1972)
Conformational analysis of the sugar ring in nucleosides and nucleotides. A new description using the concept of pseudorotation.

C.C. Blake, D.F. Koenig, G.A. Mair, A.C.T. North, D.C. Phillips, and V.R. Sarma. Nature *206*, 757 (1965)
Structure of hen egg-white lysozyme.

C.C. Blake, L.N. Johnson, G.A. Mair, A.C.T. North, D.C. Phillips, and V.R. Sarma. Proc. Roy. Soc. (London) *B167*, 378 (1967)
Crystallographic studies of the activity of hen egg-white lysozyme.

G. Bolis and E. Clementi. J. Amer. Chem. Soc. *99*, 5550 (1977)
Analytical potentials from *ab initio* computations for the interaction between biomolecules. 3. Reliability and transferability of the pair potentials.

C.G. Broyden. Math. Comp. *21*, 368 (1967)
Quasi-Newton methods and their application to function minimization.

E. Clementi
Determination of Liquid Water Structure. Co-ordination Numbers for Ions and Solvation for Biological Molecules. (Lecture Notes in Chemistry, No. 2)
Springer-Verlag, Berlin (1976).

E. Clementi, F. Cavallone, and R. Scordamaglia. J. Amer. Chem. Soc. *99*, 5531 (1977)
Analytical potentials from *ab initio* computations for the interaction between biomolecules. 1. Water with amino acids.

E.H. Cordes. Progr. Phys. Org. Chem. *4*, 1 (1967)
Mechanism and catalysis for the hydrolysis of acetals, ketals and ortho esters.

A. Dalgarno. Adv. Phys. *11*, 281 (1962)
Atomic polarizabilities and shielding factors.

W.C. Davidon. A.E.C. Res. Develop. Report. ANL-5990 (1959)
Variable Metric Method for Minimization.

W.C. Davidon. Comput. J. *10*, 406 (1968)
Variance algorithm for minimization.

L.C.W. Dixon and G.P. Szego (editors)
Towards Global Optimization. (Proceedings of a workshop held at the University of Cagliari, Italy, 1974)
North Holland and Elsevier Scientific Publ. Co., Amsterdam (1975).

H.C. Dunathan. Proc. Natl. Acad. Sci. (U.S.) *55*, 712 (1966)
Conformation and reaction specificity in pyridoxal phosphate enzymes.

H.C. Dunathan. Adv. Enzymology *35*, 79 (1971)
Stereochemical aspects of pyridoxal phosphate catalysis.

P. Eddington and M.H. Harding. Acta Crystallogr. *B30*, 204 (1974)
The crystal structure of *DL*-histidine.

R. Fletcher and M.J.D. Powell. Comput. J. *6*, 163 (1963)
A rapidly convergent descent method for minimization.

R. Fletcher and C.M. Reeves. Comput. J. *7*, 149 (1964)
Function minimization by conjugate gradients.

S. Fraga, J. Karwowski, and K.M.S. Saxena
Handbook of Atomic Data
Elsevier Scientific Publishing Co., Amsterdam (1976).

M.N. Frey, T.F. Koetzle, M.S. Lehmann, and W.C. Hamilton. J. Chem. Phys. *58*, 2547 (1973a)
Precision neutron diffraction structure determination of protein and nucleic acid components. X.

M.N. Frey, M.S. Lehmann, T.F. Koetzle, and W.C. Hamilton. Acta Crystallogr. *B29*, 876 (1973b)
Precision neutron diffraction structure determination of protein and nucleic acid components. XI.

L. Golic and W.C. Hamilton. Acta Crystallogr. *B28*, 1265 (1972)
The structure of Di-*L*-leucine hydrochloride: an x-ray diffraction study.

G.V. Gurskaya. Sov. Phys.-Crystallogr. *9*, 709 (1965)
Crystal structure of *L*-phenylalanine hydrochloride. Refinement and discussion of structure.

P.-G. Jonsson and A. Kvick. Acta Crystallogr. *B28*, 1827 (1972)
Precision neutron diffraction structure determination of protein and nucleic acid components. III.

D.D. Jones, L. Bernal, M.N. Frey, and T.F. Koetzle. Acta. Crystallogr. *B30*, 1220 (1974)
Precision neutron diffraction structure determination of protein and nucleic acid components. XIV.

T.F. Koetzle, M.S. Lehmann, J.J. Verbist, and W.C. Hamilton. Acta Crystallogr. *B28*, 3207 (1972)
Precision neutron diffraction structure determination of protein and nucleic acid components. VII.

T.F. Koetzle, M.S. Lehmann, and W.C. Hamilton. Acta Crystallogr. *B29*, 231 (1973a)
Precision neutron diffraction structure determination of protein and nucleic acid components. IX.

T.F. Koetzle, M.N. Frey, M.S. Lehmann, and W.C. Hamilton. Acta. Crystallogr. *B29*, 2571 (1973b)
Precision neutron diffraction structure determination of protein and nucleic acid components. XIII.

M.S. Lehmann, T.F. Koetzle, and W.C. Hamilton. J. Am. Chem. Soc. *94*, 2657 (1972)
Precision neutron diffraction structure determination of protein and nucleic acid components. I.

M.S. Lehmann, J.J. Verbist, W.C. Hamilton, and T.F. Koetzle. J. Chem. Soc., Perkin Trans. *2*, 133 (1973)
Precision neutron diffraction structure determination of protein and nucleic acid components. V.

G.H. Loew and D.D. Thomas. J. Theor. Biol. *36*, 89 (1972)
Molecular orbital calculations of the catalytic effect of lysozyme. 1. Glu 35 as general acid catalyst.

B. Maigret, B. Pullman, and J. Caillet. Biochem. Biophys. Res. Commun. *40*, 808 (1970a)
The conformational energy map of an alanyl residue preceding proline. A quantum mechanical approach.

B. Maigret, B. Pullman, and M. Dreyfus. J. Theor. Biol. *26*, 321 (1970b)
Molecular orbital calculations on the conformation of polypeptides and proteins. I. Preliminary investigations and simple dipeptides.

B. Maigret, D. Perahia, and B. Pullman. J. Theor. Biol. *29*, 275 (1970c)
Molecular orbital calculations on the conformation of polypeptides and proteins. IV. The conformation of prolyl and hydroxyprolyl residues.

B. Maigret, B. Pullman, and D. Perahia. Biopolymers *10*, 107 (1971a)
Molecular orbital calculations on the conformation of polypeptides and proteins. II. Conformational energy maps and stereochemical rotational states of aromatic residues.

B. Maigret, D. Perahia, and B. Pullman. Biopolymers *10*, 491 (1971b)
Molecular orbital calculations on the conformation of polypeptides and proteins. V. Conformational energy maps and stereochemical rotational states of aliphatic residues.

B. Maigret, B. Pullman, and D. Perahia. J. Theor. Biol. *31*, 269 (1971c)
Molecular orbital calculation on the conformations of polypeptides and proteins. VII. Refined calculations on the alanyl residue.

B. Maigret, D. Perahia, and B. Pullman. Biopolymers *10*, 1649 (1971d)
Molecular orbital calculations on the conformation of polypeptides and proteins. VIII. The conformational energy maps and stereochemical rotational states of the asparaginyl, glutaminyl, aspartyl and glutamyl residues.

A. McL. Mathieson. Acta Crystallogr. *5*, 332 (1952)
The crystal structures of the dimorphs of *DL*-methionine.

Y. Mitsui, M. Tsuboi, and Y. Iitaka. Acta Crystallogr. *B25*, 2182 (1969)
The crystal structure of *DL*-proline hydrochloride.

F.A. Momany, L.M. Carruthers, R.F. McGuire, and H.A. Scheraga. J. Phys. Chem. *78*, 1595 (1974)
Intermolecular potentials from crystal data. III. Determination of empirical potentials and application to the packing configurations and lattice energies in crystals of hydrocarbons, carboxylic acids, amines, and amides.

F.A. Momany, R.F. McGuire, A.W. Burgess, and H.A. Scheraga. J. Phys. Chem. *79*, 2361 (1975)
Energy Parameters in polypeptides. VII. Geometric parameters, partial atomic charges, nonbonded interactions, hydrogen bond interactions and intrinsic torsional potentials for the naturally occurring amino acids.

W. Murray
Numerical Methods for Unconstrained Optimization
Academic Press, Inc., New York (1969).

S. Nir, R. Garduno, R. Rein, Y. Coeckelenbergh, R.D. MacElroy, and J.T. Egan. Int. J. Quantum Chem., Quantum Biol. Symp. *4*, 135 (1977)
Display, manipulation and simulation of molecules.

J. Pancir. Theoret. Chim. Acta *29*, 21 (1973)
Study of reaction mechanisms by semiempirical methods. I. Optimization of the geometry of the molecule in the framework of a single calculation of the energy function.

R.A. Pasternak. Acta Crystallogr. *9*. 341 (1950)
The crystal structure of glycyl-*L*-tryptophan dihydrate.

D. Perahia, B. Maigret, and B. Pullman. Theoret. Chim. Acta *19*, 121 (1970)
Molecular orbital calculations on the conformation of polypeptides and proteins. VI. The conformational energy maps and stereochemical rotational states of the seryl and threonyl residues.

D.C. Phillips, C.C. Blake, G.A. Mair, A.C.T. North, and V.R. Sarma. Nature *206*, 757 (1965)
Structure of hen egg-white lysozyme.

K.S. Pitzer. Adv. Chem. Phys. *2*, 59 (1959)
Inter and intramolecular forces and molecular polarizability.

M.J.D. Powell. Comput. J. *7*, 155 (1964)
An efficient method for finding minimum of a function of several variables without calculating derivatives.

M.J.D. Powell
Minimization of Functions of Several Variables, in 'Numerical Analysis. An Introduction', edited by J. Walsh
Academic Press, Inc., New York (1966).

B. Pullman, B. Maigret, and D. Perahia. Theoret. Chim. Acta *18*, 44 (1970)
Molecular orbital calculations on the conformation of polypeptides and proteins. III. Conformational maps of amino acid residues: A preliminary comparison of empirical, quantum-mechanical and experimental results.

B. Pullman, J.L. Coubeils, P. Courriere, and D. Perahia. Theoret. Chim. Acta *22*, 11 (1971)
Molecular orbital calculations on the conformation of polypeptides and proteins. IX. The conformational energy maps of the lysyl and arginyl residues.

M.A. Raftery and F.W. Dahlquist. Fortschr. Chem. Org. Naturstoffe *27*, 340 (1969)
The chemistry of lyzozyme.

M. Ramanadham, S.K. Sikka, and R. Chidambaram. Acta Crystallogr. *B28*, 3000 (1972)
Structure of *L*-asparagine monohydrate by neutron diffraction.

S.T. Rao. Acta Crystallogr. *B29*, 1718 (1973)
Refinement of *DL*-aspartic acid.

A. Saran, D. Perahia, and B. Pullman. Theoret. Chim. Acta *30*, 31 (1973)
Molecular orbital calculations on the conformation of nucleic acids and their constituents. VII. Conformation of the sugar ring in β-nucleosides: the pseudorotational representation.

R. Scordamaglia, F. Cavallone, and E. Clementi. J. Am. Chem. Soc. *99*, 5545 (1977)
Analytical potentials from 'ab initio' computations for the interaction between biomolecules. 2. Water with the four bases of DNA.

R.A. Scott and H.A. Scheraga. J. Chem. Phys. *45*, 2091 (1966)
Conformational analysis of macromolecules. III. Helical structures of polyglycine and poly-*L*-alanine.

A. Sequeira, H. Rajagopal, and R. Chidambaram. Acta Crystallogr. *B28*, 2514 (1972)
A neutron diffraction study of the structure of *L*-glutamic acid HCl.

D.P. Shoemaker, J. Donohue, V. Shomaker, and R.B. Coray. J. Am. Chem. Soc. *72*, 2328 (1950)
The crystal structure of L_s-threonine.

C.S. Smith
The Automatic Computation of Maximum Likelihood Estimates
N.C.B.Sc. Department Report SC846/MR/40 (1962).

M. Spencer. Acta. Crystallogr. *12*, 59 (1959)
The stereochemistry of deoxyribonucleic acid. I. Covalent bond lengths and angles.

G.W. Stewart III. J. Assoc. Comput. Math. *14*, 72 (1967)
A modification of Davidon's minimization method to accept difference approximation of derivatives.

K. Sundaram. Int. J. Quantum Chem. *8*, 565 (1974)
Non-bonded interactions in membrane active cyclic biopolymers. I. Valinomycin-potassium ion complex.

K. Torii and Y. Iitaka. Acta Crystallogr. *B26*, 1317 (1970)
The crystal structure of *L*-valine.

K. Torii and Y. Iitaka. Acta Crystallogr. *B27*, 2237 (1971)
The crystal structure of *L*-isoleucine.

G. Ughetto and M.J. Waring. Molec. Pharmacol. *13*, 579 (1977)
Conformation of the DNA-binding peptide antibiotic echinomycin based on energy calculations.

H.J.R. Weintraub, M.D. Tsai, S.R. Byrn, C.J. Chang, and H.G. Floss. Int. J. Quantum Chem., Quantum Biology Symp. *3*, 99 (1976)
Conformational analysis of pyridoxal amino acid Schiff's bases.

W.I. Zangwill. Comput. J. *10*, 293 (1967)
Minimizing a function without calculating derivatives.

G. Zoutendijk
Methods of Feasible Directions
Elsevier Scientific Publishing Co., Amsterdam (1960).

Appendix A3

E.A. Green, R.D. Rosenstein, R. Shiono, D.J. Abraham, B.L. Trus, and R.E. Marsh. Acta Cryst. *B31*, 102 (1975)
The crystal structure of uridine.

Appendix A7

R.S. Mulliken. J. Chem. Phys. *23*, 1833 (1955a)
Electronic population analysis on LCAO-MO molecular wave functions. I.

R.S. Mulliken. J. Chem. Phys. *23*, 1841 (1955b)
Electronic population analysis on LCAO-MO molecular wave functions. II. Overlap population, bond orders and covalent bond energies.

R.S. Mulliken. J. Chem. Phys. *23*, 2338 (1955c)
Electronic population analysis on LCAO-MO molecular wave functions. III. Effects of hybridization on overlap and gross AO populations.

R.S. Mulliken. J. Chem. Phys. *23*, 2343 (1955d)
Electronic population analysis on LCAO-MO molecular wave functions. IV. Bonding and antibonding in LCAO and valence bond theories.

Appendix A8

E. Clementi. J. Chem. Phys. *46*, 3842 (1967)
Study of the electronic structure of molecules. I. Molecular wave functions and their analysis.

Author Index

Subject Index